Radio-Frequency Capacitive Discharges

Yuri P. Raizer
Mikhail N. Shneider
Nikolai A. Yatsenko

CRC Press
Taylor & Francis Group
Boca Raton London New York

CRC Press is an imprint of the
Taylor & Francis Group, an **informa** business

CRC Press
Taylor & Francis Group
6000 Broken Sound Parkway NW, Suite 300
Boca Raton, FL 33487-2742

First issued in paperback 2019

ISBN-13: 978-0-8493-8644-2 (hbk)
ISBN-13: 978-0-367-40186-3 (pbk)

This book contains information obtained from authentic and highly regarded sources. Reasonable efforts have been made to publish reliable data and information, but the author and publisher cannot assume responsibility for the validity of all materials or the consequences of their use. The authors and publishers have attempted to trace the copyright holders of all material reproduced in this publication and apologize to copyright holders if permission to publish in this form has not been obtained. If any copyright material has not been acknowledged please write and let us know so we may rectify in any future reprint.

Library of Congress Cataloging-in-Publication Data

Catalog record is available from the Library of Congress.

Visit the Taylor & Francis Web site at
http://www.taylorandfrancis.com

and the CRC Press Web site at
http://www.crcpress.com

Contents

Preface

At present, there is a growing interest in radio-frequency (RF) capacitive gas discharges of low and moderate pressures. This is one of the most exciting areas in fundamental and applied physics of gas discharge and gas electronics. An RF discharge is produced by applying alternating voltage of the MHz range, typically 13.56 MHz, to metallic or dielectric-coated electrodes. Two major applications have stimulated interest in the study of physical phenomena occurring during an RF capacitive discharge: (i) the use of moderate-pressure discharges ($p \approx 10$–100 Torr) to create an active medium in high-efficiency, reliable and small-size CO_2 lasers, and (ii) the use of low-pressure discharges ($p \approx 10^{-3}$–1 Torr) in plasma and etching technologies for various treatments of semiconductor materials, for thin film deposition, etc. Nowadays, about a quarter of the world's commercial CO_2 lasers operate on RF discharges, which are in many respects superior to direct current discharges. RF reactors for ion bombardment of semiconductor materials are widely employed in the newest industrial technologies.

Publications on RF discharges in current periodicals are quite numerous. Nearly each issue of the world's leading physics journals contains a paper or two on this or related subjects, because the processes involved in RF discharges are very complex and exhibit many specific features that are not easy to analyze and interpret. However, there is still no monograph that would systematize and discuss the accumulated data in terms of a modern physical theory and would also describe experimental techniques and RF plasma diagnostics. To our knowledge, only two recent books contain short chapters on the fundamentals of RF capacitive discharges. One of them (B. Chapman 1980) largely describes low-pressure discharges, and the other (Yu. Raizer 1987, 1991) describes moderate-pressure discharges. Since many research physicists and engineers today deal with these discharges and many others are making their first steps, the authors have made an attempt to present the available knowledge and experience in a monograph that is intended to serve as a textbook and a reference book for those interested in this rapidly developing area of physics.

The list of references is fairly large but does not claim to be exhaustive. We could not include or comment on the numerous original data available in the literature. Our presentation is primarily based on material, appropriately transformed in the authors' minds, that can provide the reader with a clear understanding of essential phenomena and the underlying mechanisms involved in RF capacitive discharges.

This book employs the units of measure conventionally used in gas discharge physics: pressure is expressed in Torr, electrical quantities are taken in volts, amperes, ohms, etc.; particle energy and electron temperature are expressed in electron volts, and gas temperature in Kelvin degrees. The electrodynamic equations and their implications are written in the CGS system generally accepted in theoretical physics.

The authors are very grateful to L. N. Smirnova for the translation of the book and to the staff of the Editorial Office at the Ioffe Physico-Technical Institute for the preparation of a camera-ready copy.

Authors

Yuri P. Raizer, Doctor of Sciences (Phys. and Math.) is Head of the Physical Gasdynamics Department of the Institute for Problems in Mechanics, Russian Academy of Sciences, Moscow (since 1965), as well as a Professor of Physics at the Moscow Institute of Physics and Technology (since 1968). He is an academician of the Russian Academy of Natural Science. Born in 1927, he graduated from Leningrad Polytechnical Institute in 1949, received his Ph.D. in 1953, and received his DSc in 1959 from the Chemical Physics Institute, USSR Academy of Sciences. Dr. Raizer has worked in various fields of gasdynamics, explosion physics, gas discharge physics, and interaction of laser radiation with ionized gases. He is an author of approximately 150 papers including about 20 on RF discharges, 3 inventions and 4 books, 3 of which were published in English. His 'Physics of Shock Waves and High Temperature Hydrodynamic Phenomena' (in co-authorship with Ya. B. Zel'dovich), Academic Press, 1968, is a well-known handbook for researchers and students in the world. Its citation index is extremely high. His books 'Laser Induced Discharges,' Consultants Bureau, 1978, and 'Gas Discharge Physics,' Springer-Verlag, 1991, are also very popular. Dr. Raizer is a Laureate of the Lenin Prize (this was the highest scientific award in the former USSR). He is a Penning Prize winner for 1993. One person is honored once in two years with this Award by the International Scientific Committee. The Penning Award is presented during the International Conferences on Phenomena in Ionized Gases. Dr. Raizer is a member of the Editorial Board of 'Plasma Sources Science and Technology.'

Mikhail N. Shneider, Ph.D. is a researcher at the High Voltage Research Center of the Electrotechnical Institute, Istra, Moscow region. Dr. Shneider was born in Chernowtsy, USSR, in 1958. He received a master's degree in theoretical physics from the Kazan State University in 1980, and a Ph.D. degree in plasma physics from All-Union Electrotechnical Institute in 1990, Moscow. His present research interests are in the theoretical study (including computer simulation) on gas discharge physics, mainly, RF and glow discharges and gasdynamic processes of after-spark channel cooling. He has more than 30 publications.

Nicolai A. Yatsenko, Ph.D., Doctor of Sciences (Phys. and Math.) is a senior researcher at the Institute for Problems in Mechanics, Russian Academy of Sciences, and a Professor of Physics at the Russian Institute of Textile and Light Industry, Moscow. Dr. Yatsenko was born in 1948, graduated from the Moscow Institute of Physics and Technology (MIPT) in 1973, received a Ph.D. from MIPT in 1978 and a Doctor of Science degree from the Institute for High Temperatures, Russian Academy of Sciences, in 1992. Since 1975 he investigated radio-frequency capacitive and combined discharges for laser and plasma technology at the Institute for Problems in Mechanics. 'The fact that two forms of RF capacitive discharge exist, the properties of these forms, and the characteristics of transition of one form into the other have been thoroughly studied since 1978 by N. A. Yatsenko, who succeeded in greatly clarifying the nature of these phenomena.' (Yuri P. Raizer, 'Gas Discharge Physics,' Springer-Verlag, p. 391, 1991.) He discovered the effect of normal current density in radio-frequency discharge and limits on the existence of different modes and gave their interpretation. He suggested (1981) slab CO-lasers with diffusion cooling and low-current mode RF discharge excitation. Dr. Yatsenko is an author of approximately 60 papers, 10 patents, and is a co-author of the book 'Low Temperature Plasma Diagnostic Techniques,' Science, Novosibirsk, 1994.

1

Basic Principles of the RF Capacitive Discharge

This chapter is an introduction to the physics of radio-frequency (RF) capacitive discharges. It describes various techniques for the excitation of RF field in a gas and the behavior of electrons (the majority charge carriers in fast oscillating electric fields), the electrodynamic characteristics of discharge plasma and their influence on an oscillating field. The production and loss of electrons and the plasma maintenance are discussed briefly. Basic data on the structure and behavior of RF discharges are given, and the formation of space charge sheaths at the electrodes and of constant potential in RF plasma is explained. A simplified RF discharge model is analyzed in order to provide a basis for further discussion of experimental data. Evidently, one cannot do an experiment and understand the results obtained without a simplified initial model of the phenomenon under study. The model will also serve as a starting point for consideration of the details of more complicated theories.

1.1 Excitation of an RF discharge

The RF range commonly used in discharge practice is $f = \omega/2\pi \simeq 1$–100 MHz. RF discharges can be subdivided into inductive and capacitive discharges differing in the way an RF field is induced in the discharge space. Inductive methods are based on electromagnetic induction so that the created electric field is a vortex field with closed lines of force. In capacitive methods, the voltage from an RF generator is applied to the electrodes, the lines of force strike them and the resultant field is essentially a potential field.

A simple and commonly used schematic representation of the inductive discharge is shown in Figure 1.1(a). RF current from an external source is passed through a coil which in practice may have even one or a few turns. The magnetic field in the coil is alternating and directed along the coil axis. A circular electric

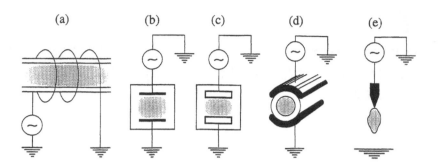

FIGURE 1.1
Basic ways of exciting RF discharges: (a) inductive discharge; (b)–(e) capacitive
discharges with (b) plane naked electrodes, (c) coated electrodes, (d) external electrodes,
and (e) with the earth as the other 'electrode' (torch single-electrode discharge).

field is induced in the coil and its lines of force are concentric with the primary
RF current. This electric field can initiate and maintain a gas discharge. For this
purpose, a dielectric tube or a vessel filled with a gas under study at a required
pressure is placed inside the coil, or a gas is pumped through the tube to produce
a plasma jet. The inductive discharge is in principle electrodeless.

A schematic diagram of a conventional capacitive discharge is shown in Figure 1.1(b,c). Two parallel disc electrodes are placed in a vessel filled with a gas
at a certain pressure, and RF voltage is applied to the electrodes. These may be
identical or vary in diameter; the latter will be shown below to be advantageous for
some applications. Since the ac circuit should not necessarily be closed and may
contain some nonconducting sections, the electrodes can be insulated from the
conducting discharge plasma with dielectric material. This provides a possibility,
often used in practice, to mount the electrodes outside the dielectric discharge
vessel, as is shown in Figure 1.1(d).

The discharge that is excited between the insulated electrodes [Figure 1.1(c,d)]
may also be called electrodeless in the sense that the plasma does not come into
contact with the electrodes. If, however, we will look into the physical phenomena occurring in both types of discharge, we will find no essential difference
between discharges produced with electrodes [Figure 1.1(b)] and without them
[Figure 1.1(c,d)], because both the metal and the dielectric, when in contact with
an RF discharge gas, behave similarly in basic processes.

For some reasons, the world's practice is such that the inductive methods have
been used to maintain high-pressure discharges, which are primarily applied to
produce a pure (free from contamination with electrode material) low-temperature
dense equilibrium plasma of the arc type with a pressure $p \approx 1$ atm and a
temperature $T \approx 10,000$ K. The principal application of inductive discharges is
the production of high-purity refractory materials, abrasive powders, etc. The
capacitive methods are generally used to maintain RF discharges at moderate

pressures $p \simeq 1$–100 Torr and at low pressures $p \simeq 10^{-3}$–1 Torr. The plasma in them is weakly ionized and nonequilibrium, like that of a glow discharge. Moderate-pressure discharges have found application in laser technology to excite CO_2 lasers, while low-pressure discharges are used for ion treatment of materials and in other plasma technologies.

The torch discharge is a capacitive discharge with a single electrode to which RF voltage is applied [Figure 1.1(e)] to excite a plasma torch. In fact, there is the other electrode—the earth, or the grounded reactor walls, to which the plasma torch is connected with a capacitive (reactive) current. The electric field in the electrode-earth system is strongly nonuniform, and the discharge is formed as a plasma torch at the electrode where the field strength is the greatest, so it is similar to a corona discharge visible only at the tip.

Here we will describe the physics and applications of capacitive RF discharges only. Inductive RF discharges, which are simpler and poorer in physical effects, will be left outside the scope of this book. For general features of these discharges the reader is referred to a monograph [1.1].

1.2 Electron motion in an oscillating electric field

Electrons are the major charge carriers in gas discharges. In an oscillating electric field, electrons move randomly and simultaneously perform periodic oscillations, which determine much of the RF discharge behavior. Consider briefly the motion of electrons in an oscillating field and the electrodynamic characteristics of an ionized medium. Since the equations for electron motion and the electrodynamic equations are linear, the superposition principle is quite valid for the interaction of an electron gas with an oscillating field. Therefore, it is sufficient to limit our consideration to the case of monochromatic field normally applied in practice. Because the electron velocities in discharges are nonrelativistic, the Lorentz force is much smaller than the electric force in the absence of a strong additional magnetic field and can be neglected. Keeping in mind that the field amplitude changes but little at a distance equal to the oscillatory displacement of the electron, we will assume the field to be uniform in the space $E = E_a \sin \omega t$ with $E_a = $ const.

1.2.1 Velocity and displacement

The equation of motion for the electron velocity v ensemble-averaged over collisions with the gas atoms has the form [1.1]

$$m\dot{v} = -eE_a \sin \omega t - mv\nu_m \tag{1.1}$$

The second term in the right-hand side of equation (1.1) allows for the loss of electron momentum per second due to the scattering by atoms or molecules. Here,

$\nu_{\mathrm{m}} = N\bar{v}\sigma_{\mathrm{m}}$ is the effective collision frequency of an electron or the collision frequency for the momentum transfer. It corresponds to the transport cross section $\sigma_{\mathrm{m}} = \sigma_{\mathrm{c}}(1 - \overline{\cos\theta})$, where σ_{c} is the elastic collision cross section, $\overline{\cos\theta}$ is the average scattering angle cosine, N is the density of atoms, and $\bar{v}$ is the mean random velocity of an electron which is generally taken to be large compared to the velocity of oriented, oscillatory motion $|v|$.

By integrating equation (1.1) and the equation for oscillatory displacement $\dot{r} = v$, we find

$$v = \frac{eE_{\mathrm{a}}}{m\sqrt{\omega^2 + \nu_{\mathrm{m}}^2}}\cos(\omega t + \phi)$$

$$\phi = \arctan\frac{\nu_{\mathrm{m}}}{\omega} \qquad r = \frac{eE_{\mathrm{a}}}{m\omega\sqrt{\omega^2 + \nu_{\mathrm{m}}^2}}\sin(\omega t + \phi) \qquad (1.2)$$

The oscillatory motion with the velocity amplitude $u_{\mathrm{a}} = eE_{\mathrm{a}}/m\sqrt{\omega^2 + \nu_{\mathrm{m}}^2}$ and the displacement amplitude $A = eE_{\mathrm{a}}/m\omega\sqrt{\omega^2 + \nu^2}$ is superimposed with the random motion. Since v is a collision-averaged velocity and the mean vector of the random velocity v_T is zero, the integration constant in the expression for v is also zero.

In the absence of collisions ($\nu_{\mathrm{m}} = 0$), electrons perform free oscillations with the amplitudes of velocity u_{a} and of displacement a

$$u_{\mathrm{a}} = \frac{eE_{\mathrm{a}}}{m\omega} \qquad a = \frac{eE_{\mathrm{a}}}{m\omega^2} \qquad (1.3)$$

the velocity being $\pi/2$ out of phase with the field. Collisions disturb the strictly harmonic mode of free oscillations, 'knocking' them out of phase. A sharp change in the direction of motion due to the scattering prevents the electron from making a maximum displacement imposed by the applied force, because after each impact it starts oscillating with a new phase and at a different angle with respect to the instantaneous velocity vector. This reduces the mean amplitudes of velocity and displacement by a factor of $\sqrt{1 + \nu_{\mathrm{m}}^2/\omega^2}$ as compared to free oscillations. The reduction becomes greater with increasing collision frequency, and an additional phase shift also occurs.

In the limiting case of very frequent collisions $\nu_{\mathrm{m}}^2 \gg \omega^2$, the oscillation velocity of the electron

$$v \approx -\frac{eE_{\mathrm{a}}}{m\nu_{\mathrm{m}}^2}\sin\omega t = -\frac{eE(t)}{m\nu_{\mathrm{m}}} = -\mu_{\mathrm{e}}E(t) = v_{\mathrm{d}} \qquad (1.4)$$

coincides, at each moment of time, with the drift velocity in a dc field equivalent to the instantaneous field ($\mu_{\mathrm{e}} = e/m\nu_{\mathrm{m}}$ is electron mobility); in other words, the electron, which behaves as if it were in a dc field, follows a relatively slow evolution of E. It seems natural to term electron oscillations in the mobility mode as drift oscillations. The displacement amplitude of drift oscillations

$$A = \frac{eE_a}{m\nu_m\omega} = \frac{\mu_e E_a}{\omega} \qquad r = A\cos\omega t \qquad (1.5)$$

is by a factor of $\nu_m/\omega \gg 1$ less than the free oscillation amplitude a in the same field. For example, at the frequency 13.56 MHz the collision frequency $\nu_m \approx 3 \times 10^9 p\,[\mathrm{Torr}]\,\mathrm{s}^{-1}$ exceeds $\omega = 0.86 \times 10^8\,\mathrm{s}^{-1}$ starting with $p \approx 3 \times 10^{-1}$ Torr. This means that even at $p > 10^{-1}$ Torr at this frequency, the oscillations have a drift nature, and only at $p < 10^{-2}$ Torr do they become free. The respective electron mobility is $\mu_e \approx 6 \times 10^5/p\,\mathrm{cm}^2\,\mathrm{V}^{-1}\,\mathrm{s}^{-1}$. Generally, weakly ionized moderate-pressure plasmas are maintained by reduced fields $E/p \approx 1$–10 V cm^{-1} Torr^{-1} with the lower values for inert gases and the larger values for molecular gases. The drift oscillation amplitudes in this E/p range $A \approx 7 \times 10^{-2}$–$7 \times 10^{-3}$ cm are small as compared to a typical electrode spacing $L \approx 1$–5 cm. The oscillation amplitudes for ions are two orders of magnitude smaller than for electrons, because the ion mobility is $\mu_+ \approx \mu_e/300$.

We would like to make three general comments.

(i) The frequency $f = 13.56$ MHz often used in RF discharges and the respective wavelength $\lambda = 22$ m belong to the shortwave radio-frequency range. Power RF generators produce noises interfering with radio broadcasting and communications. For this reason, an international convention has specified for them a few short frequency ranges including 13.56, 27.12, 40.68, 81.36 MHz.

(ii) When a heavy particle gas may be considered as a cold gas, we will use pressure in Torr instead of density, as is generally done in gas discharge physics. At $p = 1$ Torr and $T = 20\,^\circ$C, $N = 3.295 \times 10^{16}$ cm^{-3}.

(iii) In all numerical formulas in this book, pressure p will be expressed in Torr.

1.2.2 Electron energy

The electron oscillation velocity may be represented as a sum of two components, one of which is proportional to the field itself and the other to its variation rate $\dot{E} = \omega E_a \cos\omega t$

$$v = -\frac{\nu_m e E_a \sin\omega t}{m\left(\omega^2 + \nu_m^2\right)} + \frac{\omega e E_a \cos\omega t}{m\left(\omega^2 + \nu_m^2\right)} \qquad (1.6)$$

The time average work done by the field on an electron is $\langle -eEv \rangle$ per second, where $\langle\,\rangle$ designate averaging over a cycle. In the absence of collisions, the work is zero, because the velocity is $\pi/2$ phase-shifted relative to the field. At the moment of the field switch-on, the electric force swings the electron so that its energy starts pulsating, on the average remaining constant. The ensemble-averaged electron energy is a combination of the random energy and the oscillation energy, which, for free oscillations, is

$$\varepsilon_{\text{free osc}} = \left\langle \frac{mv^2}{2} \right\rangle = \frac{e^2 E_a^2}{4m\omega^2} = \frac{mu_a^2}{4} \tag{1.7}$$

In the absence of collisions, the field energy does not dissipate.

Regular energy transfer to the electrons, accompanied by the field energy dissipation and Joule heat release, occurs only through electron-scattering events.[1] According to equation (1.6), the work

$$\langle -e\boldsymbol{E}\boldsymbol{v} \rangle = \frac{e^2 E_a^2}{2m(\omega^2 + \nu_m^2)}\nu_m = \Delta\varepsilon_E \nu_m \tag{1.8}$$

is defined by the velocity component that oscillates being proportional to the field. In a single effective collision, the electron acquires from the field the energy $\Delta\varepsilon_E$ that is twice as large as the mean kinetic energy of oscillations $\langle mv^2/2 \rangle$.

This result may be interpreted as follows. Between collisions, the electron acquires a certain oscillation energy, though smaller than the energy of free oscillations. On collision, the electron sharply and arbitrarily changes the direction of motion without practically changing its absolute velocity and energy. The energy it possesses at that moment should be referred to as random energy, since the electron starts oscillating anew. So, in each collision event, the energy acquired between collisions, which is about the mean oscillation energy, is transformed to random energy, or heat.

In a single effective collision, the electron transfers to an atom of mass M a portion of its energy ε equal to $\delta = 2m/M$ [1.1]. Therefore, if we ignore inelastic collisions, which in monatomic gases are feasible only at energies exceeding the potentials of excitation E^* and of ionization I of the atoms, the electron energy ε will change as

$$\frac{d\varepsilon}{dt} = (\Delta\varepsilon_E - \delta\varepsilon)\nu_m \tag{1.9}$$

Under steady-state conditions, after the electrons have transferred to atoms all the energy they gained from the field, they possess the mean energy

$$\bar{\varepsilon} = \frac{\Delta\varepsilon_E}{\delta} = \frac{2\varepsilon_{\text{osc}}}{\delta} = \frac{e^2 E_a^2}{2m\left(\omega^2 + \nu_m^2\right)\delta} \tag{1.10}$$

Since $\delta \ll 1$, the mean electron energy, or the energy of random motion, is much larger than the oscillation energy. The same is true for the ratio between the mean random $\bar{v}$ and the oscillation v_{osc} velocities. Ignoring the difference between the real and Maxwellian energy spectra of electrons, one can conveniently use electron temperature $T_e = 2\bar{\varepsilon}/3k$. In the RF range at moderate pressures, $\omega^2 \ll \nu_m^2$ and

---

[1]We will show in Section 1.7 that the field energy dissipation and electron heating may be caused not only by the scattering due to collisions with atoms and ions but also by the recoil from the moving boundary of a strong repulsive field.

the similarity principle $\bar{\varepsilon} = f(E_a/p)$ is valid. At low pressures, when $\omega^2 \gg \nu_m^2$, $\bar{\varepsilon} = f(E_a/\omega)$ and is independent of p.

If $\omega^2 \ll \nu_m^2$, the mean electron energy in an RF field $\bar{\varepsilon} = e^2 E_a^2/2m\nu_m^2\delta$ is the same as in a dc field of strength E equal to an rms value of the RF strength, $E = E_a/\sqrt{2}$, $\bar{\varepsilon} = e^2 E^2/m\nu_m^2\delta$. When calculating $\bar{\varepsilon}$, it is then convenient to use a constant electron path approximation $l = \bar{v}/\nu_m = $ const instead of a constant collision frequency approximation. It is more reasonable to employ the latter when the formulas contain the term $(\omega^2 + \nu_m^2)$. From (1.10) in the $l = $ const approximation, $\bar{\varepsilon} = e^2 E^2 l^2/\delta m\bar{v}^2$, and taking, for certainty, the ratio between $\bar{v}^2$ and $\overline{v^2}$ to be the same as for the Maxwellian spectrum, we can write instead of (1.10)

$$\bar{\varepsilon} = \frac{\sqrt{3\pi}}{4} \frac{eEl}{\sqrt{\delta}} \approx 0.8 \frac{eEl}{\delta} \tag{1.10a}$$

This approximate expression is more convenient than (1.10) owing to the linear relation between the electron temperature and the field.

1.3 Electrodynamic plasma characteristics and interaction with oscillating fields

1.3.1 Current components

We will start with the canonical form of Maxwell's equations on the only assumption that the medium is nonmagnetic, which plasma indeed is (the magnetic vectors B and H then coincide)

$$\text{rot} H = \frac{4\pi}{c} j + \frac{1}{c} \frac{\partial D}{\partial t} \tag{1.11}$$

$$\text{div} D = 4\pi\rho \tag{1.12}$$

$$\text{rot} E = -\frac{1}{c} \frac{\partial H}{\partial t} \tag{1.13}$$

$$\text{div} H = 0 \tag{1.14}$$

Here, j is the current density; $D = E + 4\pi P$ is the electric induction vector; P is the electric polarization vector, or the dipole moment per unit volume; ρ is the free charge density. The displacement current $(1/4\pi)\partial D/\partial t$, added to the current density j, provides the fulfillment of the charge conservation law

$$\frac{\partial \rho}{\partial t} + \text{div} j = 0 \tag{1.15}$$

In steady or slowly varying fields, the current density and the dipole moment are proportional to the electric field strength, $j = \sigma E$, $P = \chi E = [(\varepsilon - 1)/4\pi] E$, where the conductivity σ, the polarizability χ and the dielectric permittivity $\varepsilon = 1 + 4\pi\chi$ are electrodynamic characteristics of the medium.

The first of the two components of the displacement current, $(1/4\pi)\partial E/\partial t$, which may be called 'vacuum' displacement current, is by no means associated with the existence or motion of local charges, so it is not current in the strict sense of the word. The other component is real current, because the electrical dipole moment changes due to the motion of charges in the medium. As for plasma, which exhibits both conductor and dielectric properties, it is natural to unite the term j, corresponding to the conduction current and denoted further as j_{cond}, and the portion of polarization current associated with free charge motion, because together these quantities comprise the total current of free charges: $j_t = j_{\text{cond}} + \partial P/\partial t$. The only difference between the two components of the charge current is that the conduction current is proportional to the field itself and obeys Ohm's law, while the polarization current is proportional to the time variation of the field, $\partial E/\partial t \equiv \dot{E}$, like the vacuum displacement current. Generally, the vector P contains a small component due to the polarization of atoms and ions, i.e., it is associated with displacement of electrons bound to an atom. Since this component corresponds to polarizability of unionized gases at moderate pressures and is, therefore, negligible, $\varepsilon_{\text{bound}} - 1 \lesssim 10^{-5}$, we will ignore it completely.

1.3.2   High-frequency plasma conductivity and dielectric permittivity

In rapidly oscillating fields, the simple relations $j = \sigma E$ and $P = \chi E$ are generally invalid because of delay effects. For example, if a field suddenly reverses its direction, the electrons still continue their motion in the initial direction for some time. Therefore, the current and the dipole moment depend not only on the E value at a given moment but also on the history of E. The problem is, however, much simplified because the equations for charge motion and the electrodynamic equations are linear. In this case, an oscillating field can be represented as a Fourier integral; and since the superposition principle is still valid, one can deal only with the harmonic components. This is convenient because monochromatic fields are normally used in practice. For instance, the current $j = j_a \sin(\omega t + \varphi)$ in the field $E = E_a \sin \omega t$ is also harmonic but may be out of phase with the field. It is, therefore, represented as a sum of terms proportional to $\sin \omega t$ and $\cos \omega t$, or E and $\dot{E}$. Denoting all harmonic components with the subscript ω and retaining the former designations for the proportionality factors with the assigned subscript ω, one can write a relation for the total charge current in the same form as before:

$$j_{t\,\omega} = j_{\text{cond}\,\omega} + \frac{\partial P_\omega}{\partial t} = \sigma_\omega E_\omega + \frac{\varepsilon_\omega - 1}{4\pi}\dot{E}_\omega \qquad E_\omega = E_{a\,\omega} \sin \omega t$$

The quantities σ_ω and ε_ω represent the conductivity and dielectric permittivity of a medium for a field of frequency ω, on which they generally depend. We can rewrite this equation using a complex representation of the harmonic quantities and omitting further all subscripts ω

$$j_t = \left(\sigma + i\frac{\omega(\varepsilon - 1)}{4\pi}\right) E \qquad E = E_a e^{i\omega t} \qquad (1.16)$$

Since equation (1.16) has the form of Ohm's law, we can apply the concept of complex conductivity of a medium σ_t for the case of free charge current. It is also convenient to introduce complex conductivity σ' with respect to the total current j including the vacuum displacement current

$$j_t = \sigma_t E \qquad j = \sigma' E \qquad \sigma_t = \sigma + i\omega\frac{\varepsilon - 1}{4\pi} \qquad \sigma' = \sigma + i\frac{\omega\varepsilon}{4\pi} \qquad (1.17)$$

A comparison of phenomenological relation (1.16) and a microscopic expression for the total electron current density $j_t = -en_e v$, where n_e is the density of electrons and v is their velocity from equation (1.6), can yield the plasma conductivity and dielectric permittivity in an oscillating field

$$\sigma = \frac{e^2 n_e \nu_m}{m\left(\omega^2 + \nu_m^2\right)} \qquad (1.18)$$

$$\varepsilon = 1 - \frac{4\pi e^2 n_e}{m\left(\omega^2 + \nu_m^2\right)} \qquad (1.19)$$

In the limit of relatively low frequencies or relatively high pressures, when $\omega^2 \ll \nu_m^2$, equation (1.18) reduces to the plasma conductivity in a steady field $\sigma = e^2 n_e / m\nu_m$ proportional to the electron mobility $\mu_e = e/m\nu_m$ ($\sigma = e\mu_e n_e$).

1.3.3 Plasma frequency

In the limit of relatively high frequencies or relatively low pressures, when $\omega^2 \gg \nu_m^2$ (the case of 'collisionless' plasma), the conductivity $\sigma = e^2 n_e \nu_m / m\omega^2$ is proportional to a small but finite frequency of electron collisions, and the dielectric permittivity takes the value

$$\varepsilon = 1 - \frac{\omega_p^2}{\omega^2 + \nu_m^2} \rightarrow 1 - \frac{\omega_p^2}{\omega^2} \qquad (1.20)$$

$$\omega_p = \left(\frac{4\pi e^2 n_e}{m}\right)^{1/2} = 5.65 \times 10^4 \sqrt{n_e[\text{cm}^{-3}]}\, \text{s}^{-1} \qquad (1.21)$$

Here ω_p is the plasma frequency for an electron gas. At this frequency, the electron gas will oscillate relative to the ionic gas if the electrons are displaced relative to the ions at the initial moment, thus polarizing the plasma. This is the frequency of intrinsic oscillations of a free (unlimited by the vessel walls) plasma. Indeed, if the electron gas in a homogeneous plasma with 'immobile' ions is displaced as a whole in a direction x through a distance Δx, space charges of different signs and of surface density $en_e\Delta x$ will appear at the boundary, inducing an electric field $E = 4\pi en_e\Delta x$. The electron will be acted upon by the restoring force $-eE$, producing oscillations in the x direction at frequency ω_p; so in accordance with the equation of motion

$$m\overset{..}{\Delta}x + 4\pi en_e\Delta x = 0 \qquad \overset{..}{\Delta}x + \omega_p^2\Delta x = 0$$

The velocity of oriented electrons becomes random due to collisions with atoms; in other words, the induced oscillations are damped at a frequency ν_m if no force is applied. Therefore, we can speak of plasma oscillations only if the frequency of these oscillations is substantially higher than that of collisions, $\omega_p \gg \nu_m$. This criterion is similar to the requirement that the electron path length $l = \bar{v}/\nu_m$ should be much larger than the Debye radius

$$\lambda_D = \left(\frac{kT_e}{4\pi e^2 n_e}\right)^{1/2} \tag{1.22}$$

or that there should be no collisions in the Debye sphere. This will become clear from equations (1.21) and (1.22):

$$\omega_p\lambda_D = \left(\frac{kT_e}{m}\right)^{1/2} = \left(\frac{\pi}{8}\right)^{1/2}\bar{v} \qquad \frac{\nu_m}{\omega_p} = \frac{\bar{v}\lambda_D}{l\omega_p\lambda_D} = \left(\frac{8}{\pi}\right)^{1/2}\frac{\lambda_D}{l}$$

The electron density in RF discharge plasmas normally lies within the range $n_e \sim 10^8 - 10^{11}$ cm^{-3}, to which the plasma frequency range $\omega_p \sim (1-20)\times10^9$ s^{-1} corresponds. Field frequencies generally used in RF discharges are usually lower than plasma frequencies; for instance, at $f = 13.56$ MHz, $\omega = 0.86 \times 10^8$ s^{-1} is one or two orders lower. This means that at low pressures, when $\nu_m^2 \ll \omega^2$, ε has a large negative value. In this case, the polarization current has a much larger amplitude than the vacuum displacement current.

1.3.4 The Joule heat of current

The work done by the field on a single electron per second is given by equation (1.8). All electrons within 1 cm^3 gain per second the energy

$$n_e\langle -eEv\rangle = \langle j_tE\rangle = \sigma\langle E^2\rangle \tag{1.23}$$

where σ has the form of (1.18). This quantity represents the Joule heat of current (per 1 cm^3 per second), and its value is determined only by the conduction current. In order to find the power P dissipated in a discharge, this value should be integrated over the whole discharge space. If $d\bar{l}$ is the length element along the current line and dS is the area of an elementary current tube, then

$$P = \left\langle \int j_tEdldS\right\rangle = \left\langle \int jdS\int Edl\right\rangle = \langle iV\rangle \tag{1.24}$$

where i is the discharge current and V is the discharge voltage. Here we have added the vacuum displacement current, making no contribution to the power integral, to j_t in order to get a common phenomenological expression for the power. We have also used the conservation law for the total current along the tube $j \, \mathrm{d}S = \mathrm{const}$, which follows from the equation

$$\mathrm{div} \left(j_{\mathrm{cond}} + \frac{1}{c} \frac{\partial D}{\partial t} \right) = \mathrm{div} j = 0$$

derived from equations (1.15) and (1.12). It is clear that j is the sum of the conduction and displacement currents, while in equation (1.15) j represents only the conduction current j_{cond}.

1.3.5 The ratio of conduction and displacement currents

Displacement current, or polarization current as its part, does not cause field energy dissipation, because it is $\pi/2$ phase-shifted relative to the field. Only a portion of the discharge current is involved in the Joule heat release. This portion is characterized by the amplitude ratio of the conduction current $j_{a\,\mathrm{cond}}$ and the displacement current $j_{a\,\mathrm{dis}}$, or of the total current j_a. However, because of the phase shift we sum up the squared current components instead of their amplitudes: $j_a = (j_{a\,\mathrm{cond}} + j_{a\,\mathrm{dis}})^{1/2}$. From equations (1.17)–(1.19) the following amplitude ratios are valid:

$$\frac{j_{a\,\mathrm{cond}}}{j_{a\,\mathrm{pol}}} = \frac{4\pi\sigma}{|\varepsilon - 1|\omega} = \frac{\nu_m}{\omega} \tag{1.25}$$

$$\frac{j_{a\,\mathrm{cond}}}{j_{a\,\mathrm{dis}}} = \frac{4\pi\sigma}{\omega|\varepsilon|} = \frac{\nu_m}{\omega} \frac{\omega_p^2}{|\omega^2 + \nu_m^2 - \omega_p^2|} \tag{1.26}$$

The relative contributions of the conduction and polarization currents are defined merely by the ratio of the frequencies ν_m and ω, while the contributions of the conduction and total displacement currents depend on the ratio between the three frequency scales inherent to the RF plasma: ω, ν_m, and ω_p. Below, we give approximate ranges for these frequencies commonly used in the RF discharge practice. For convenience, the wide range of pressures that determine the collision frequency of electrons is subdivided into low and moderate pressures:

$$
\begin{array}{llll}
n_e & \sim & 10^8 - 10^{11} \, \mathrm{cm}^{-3} \quad & \omega_p & \sim & 10^9 - 2 \times 10^{10} \, \mathrm{s}^{-1} \\
f & \sim & 1 - 100 \, \mathrm{MHz} & \omega & \sim & 6 \times 10^6 - 6 \times 10^8 \, \mathrm{s}^{-1} \\
p & \sim & 10^{-3} - 1 \, \mathrm{Torr} & \nu_m & \sim & 3 \times 10^6 - 3 \times 10^9 \, \mathrm{s}^{-1} \\
p & \sim & 1 - 100 \, \mathrm{Torr} & \nu_m & \sim & 3 \times 10^9 - 3 \times 10^{11} \, \mathrm{s}^{-1}
\end{array}
$$

One can see from this list of parameters that $\omega \ll \omega_p$ in the majority of cases. Besides, in moderate-pressure discharges, $\omega \ll \nu_m$, but ν_m may be higher or lower than ω_p. If $\omega, \nu_m \ll \omega_p$, $j_{a\,\mathrm{cond}}/j_{a\,\mathrm{dis}} \approx \nu_m/\omega \gg 1$; i.e., in a highly ionized plasma the conduction current dominates. If $\nu_m > \omega_p$ but the frequency is not high so that $\omega \ll \omega_p$ and $\omega \nu_m < \omega_p^2$, as is usually the case, the conduction current still dominates, though the displacement current may be fairly large.

In low-pressure discharges, $\omega \ll \omega_p$ and quite often $\nu_m < \omega_p$. Situations are, however, possible with $\omega > \nu_m$, then the displacement current takes over; when $\omega < \nu_m$, the conduction current is larger.

1.3.6 Criteria for electric field quasipotentiality

In an inductive RF discharge, the electric field naturally has a vortex, nonpotential character. In capacitive discharges, induction effects are usually not very pronounced in spite of the fairly high frequencies, and the electric field is close to a potential and a quasistationary field. This means that the field is induced only by the voltage applied to the electrodes and by space charges. It is defined, in terms of electrostatic laws, by the instantaneous values of these quantities as if they were steady-state. Speaking of 'voltage' as a source of field, we should keep in mind that a voltage source redistributes the charges in the external circuit, producing surface charges at the electrodes. The field in the discharge space is induced by these and space charges. The validity of the assumption of field quasipotentiality, important for both theory and experiment, can be evaluated by establishing an adequate criterion.

Consider as an illustration a cylindrical plasma column of radius R between two plane electrodes of a height close to the electrode spacing L (Figure 1.2). The magnetic lines of force of the field represent concentric rings around the cylinder axis. By integrating equation (1.11) over the column cross section and using Stocks' theorem, we can find the order of magnitude of the magnetic field $H \sim (2\pi/c)|\sigma'|ER$, where $|\sigma'|$ is the total complex conductivity modulus of equation (1.17). The vortex electric field E_{vort} induced by the alternating magnetic field obeys equation (1.13). The lines of force of the vortex field lie in the planes intercepting the current cylinder and crossing its axis. In each plane, the lines form a closed contour on each side of the cylinder axis. By integrating equation (1.13) over the area RL of a large contour enveloping a cylinder of perimeter $(2R+2L)$ as far as its edges and using again Stocks' theorem, we obtain:

$$E_{\text{vort}}\, 2(R+L) \sim \frac{\omega}{c}HRL \sim \frac{2\pi\omega|\sigma'|}{c^2}R^2LE$$

$$\frac{E_{\text{vort}}}{E} \sim \left(\frac{\Lambda}{\sqrt{2}\delta'}\right)^2 \qquad \Lambda = R\sqrt{\frac{L}{R+L}} \qquad \delta' = \frac{c}{\sqrt{2\pi|\sigma'|\omega}} \tag{1.27}$$

where Λ equal to $\sqrt{RL}$ at $R \gg L$ and to R at $R \ll L$ is the characteristic size of the discharge cylinder.

Let us analyze relation (1.27) for the vortex-potential field ratio in two limiting cases when the quantity δ' has a definite physical meaning. Suppose the conduction current dominates over the displacement current. According to the results obtained from equation (1.26) and described in the previous section, this usually happens in

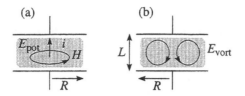

FIGURE 1.2
Sketches for estimation of a vortex electric field in the discharge column: (a) magnetic
field geometry, (b) induced electric field geometry.

moderate-pressure RF discharges. Then $\sigma' \approx \sigma$ and δ' coincides with the known
skin-layer thickness

$$\delta = \frac{c}{\sqrt{2\pi\sigma\omega}} = \frac{5.03}{\left(\sigma[\Omega^{-1}\mathrm{cm}^{-1}]f[\mathrm{MHz}]\right)^{1/2}} \mathrm{cm} \tag{1.28}$$

This quantity describes the penetration depth of an oscillating field in a good
conductor. Therefore, the potentiality criterion here is the small characteristic size
of the discharge as compared to the skin-layer thickness.

The physical meaning of this criterion will be discussed in the next section,
but here we will illustrate these results numerically. If $p = 30$ Torr and $n_e =
10^{10}$ cm^{-3}, then $\nu_m \approx 3 \times 10^9 p = 10^{11}$ s^{-1}, $\omega_p \approx 5.6 \times 10^9$ s^{-1}. At the frequency
$f = 13.56$ MHz, $\omega = 0.86 \times 10^8$ s$^{-1} \ll \nu_m$. Although $\nu_m \gg \omega_p$, $\sqrt{\omega\nu_m} =
2.9 \times 10^9$ s$^{-1} \lesssim \omega_p$. At these parameters $\varepsilon \approx 1 - 3 \times 10^{-3} \approx 1$, $\sigma \approx 10^{-13}$
$n_e/p\,\Omega^{-1}$ cm$^{-1} \approx 3 \times 10^{-5}\,\Omega^{-1}cm^{-1} \approx 3 \times 10^7$ s^{-1} and the conduction current
has an amplitude $4\pi\sigma/\omega|\varepsilon| \approx 4.3$ times larger than that of the displacement
current. The field potentiality criterion is satisfied with a good accuracy, because
$\delta \approx 2.5$ m is much larger than the commonly used dimensions of R and L.

Consider now the case of dominant displacement current typical of low-pressure
discharges. Here $|\sigma'| \approx \omega|\varepsilon|/4\pi$. Taking into account that $\nu_m \ll \omega \ll \omega_p$ at low
pressures and that $\varepsilon \approx -(\omega_p/\omega)^2$ from equation (1.20), we find $\delta' \approx \sqrt{2}c/\omega_p \approx
\sqrt{2}\lambda_{cr}/2\pi$, where λ_{cr} is the critical vacuum wavelength which corresponds to the
field frequency ω coinciding with the plasma frequency ω_p and turning ε to zero:

$$\lambdabar_{cr} = \frac{\lambda_{cr}}{2\pi} = \frac{c}{\omega_p} = \left(\frac{mc^2}{4\pi e^2 n_e}\right)^{1/2} = 3.32 \times 10^6 n_e^{-1/2} \mathrm{cm} \tag{1.29}$$

The criterion for the field potentiality in this case is the small characteristic size of
the plasma as compared to $2\lambda_{cr}$; the criterion validity is independent of the field
frequency if the inequality $\nu_m \ll \omega \ll \omega_p$ is fulfilled.

Leaving the physical interpretation of this criterion for the next section, we
offer a numerical illustration. If $p = 3 \times 10^{-3}$ Torr and $n_e = 3 \times 10^9$ cm^{-3},
then $\nu_m \approx 10^7$ s^{-1} and $\omega_p = 3.2 \times 10^9$ s^{-1}. For $f = 13.56$ MHz the necessary
inequalities are fulfilled. At these parameters $\varepsilon \approx -1400$, the conduction current

is $(4\pi\sigma/\omega|\varepsilon|)^{-1} = 8.5$ times smaller than the displacement current and $2\lambda_{cr} = 1.2$ m, which is also much larger than the common discharge size.

The criteria for field potentiality are more rigid than the trivial condition of unimportance of wave effects, which means that a system size is to be small as compared to the vacuum wavelength. In both situations, the last $\lambda_0 = 2\pi c/\omega = 22$ m is an order of magnitude larger than δ'. Induction (wave) effects may prove to be essential at frequencies as high as 81 MHz ($\lambda_0 = 3.7$ m) and in a large discharge region, e.g., in stripe laser systems in which the electrodes and the discharge between them have the shape of a long stripe. This fact should not be ignored, because wave effects may deteriorate the discharge homogeneity.

1.3.7 Electromagnetic wave penetration into the plasma and the field potentiality criteria

Clearly, a treatment of oscillating field interaction with an RF discharge plasma in terms of the 'wave' model does not have much sense, for we deal here with a limiting case of quasipotential field, which is opposite to the case of wave field. On the other hand, the 'wave' approach, which allows for the effects of displacement current and of electromagnetic induction, is more general and may throw light on new aspects of electrodynamic phenomena. For instance, a consideration of the interaction between an electromagnetic wave and a plasma may provide an insight into the physics of the above criteria for the field quasipotentiality in RF discharges.

We would like to stress that in the framework of the field quasipotentiality model, one may not use the concept of displacement current. Oscillating current can be treated only with electrostatic (1.12) and charge conservation (1.15) equations, from which it follows that the sum of the charge currents j_t and the vector $(1/4\pi)\dot{E}$ has no sources, i.e., its divergence is zero. This is sufficient for consideration of current in a discontinuous circuit of conductors or for description of current in a conducting medium, such as plasma, capable of polarization.

It is well known (see, for example, [1.1]) that Maxwell's equations (1.11) and (1.13) yield wave equations for the fields E and H which permit solutions in the form of a plane running wave $E, H \sim \exp[i(\omega t - kx)]$, where $k = \sqrt{\varepsilon'}\omega/c$ is the wave vector and $\varepsilon' = \varepsilon - i4\pi\sigma/\omega$ is a complex dielectric permittivity which should be introduced instead of the complex conductivity of (1.17). After we have identified the real and imaginary parts of the wave vector by finding $\sqrt{\varepsilon'} = n + i\varkappa$, established the relationship between the complex amplitudes of E and H with equations (1.11) and (1.13) and calculated n and $\varkappa$ with the expression for ε', we can now represent the fields in the wave as

$$H = \sqrt{\varepsilon'}E \sim \exp[i\omega(t - nx/c) - \varkappa(\omega/c)x] \qquad (1.30)$$

$$n = \sqrt{\frac{\varepsilon + \sqrt{\varepsilon^2 + (4\pi\sigma/\omega)^2}}{2}} \qquad \varkappa = \sqrt{\frac{-\varepsilon + \sqrt{\varepsilon^2 + (4\pi\sigma/\omega)^2}}{2}} \qquad (1.31)$$

which permits a clear interpretation.

It follows from the original equations that at $\varepsilon' \neq 0$ (a wave is 'electromagnetic' only under this condition), the vectors E, H and k are perpendicular to each other.[2] In equation (1.30), n defines the phase velocity c/n and the wavelength $\lambda = 2\pi c/\omega n$ in the medium, and $\varkappa$ describes the damping of the wave amplitude. In an ideal dielectric medium with $\sigma = 0$, there is no damping ($\varkappa = 0$), and $n = \sqrt{\varepsilon}$ is a conventional refractive index.

A typical 'wave' case is observed when the displacement current dominates over the conduction current: $4\pi\sigma/\omega|\varepsilon| \ll 1$ (the medium is a 'weak conductor') and $\varepsilon > 0$. According to (1.26) and (1.20), this happens when ν_m^2, $\omega_p^2 \ll \omega^2$ and the medium is sufficiently large. One can speak of propagation if the medium size is $R \gg \lambda = \lambda_0/n \approx \lambda_0$. RF discharges are, however, too far from this ideal model, which is valid in the RF range only for fairly short waves propagating through the ionosphere.

If the medium is a good conductor in the sense that $4\pi\sigma/\omega|\varepsilon| \gg 1$, i.e., the conduction current dominates, $n \approx \varkappa \approx \sqrt{2\pi\sigma/\omega}$ from the formulas of (1.31). The wave is strongly damped at a distance of about a medium wavelength or at $\mathchar'26\mkern-10mu\lambda = \lambda/2\pi = c/n\omega$. Its amplitude is reduced by a factor of e at a depth $\delta = \mathchar'26\mkern-10mu\lambda = c/\sqrt{2\pi\sigma\omega}$, which was termed as skin-layer thickness we mentioned in the previous section. It often happens in a good conductor, e.g., in a metal, that a wave entering this medium is nearly totally reflected and only partly penetrates and is absorbed. At normal incidence, the portion of the incident energy flux that is absorbed is $1/n$ [1.1].

If a medium is 'nonconductive' in the sense that $4\pi\sigma/\omega|\varepsilon| \ll 1$ and $\varepsilon < 0$, which happens when $\nu_m \ll \omega \ll \omega_p$, then from (1.31) $n \approx 0$, $\varkappa \approx \sqrt{|\varepsilon|} \approx \omega_p/\omega$. As in the case of a good conductor, the wave cannot penetrate into this medium but for a different reason. The phase velocity and the wavelength tend to infinity at $\sigma \to 0$, and the field oscillates only in time, decreasing in amplitude with distance from the surface without energy dissipation. This is similar to total internal reflection. This effect is widely used for laboratory plasma diagnostics by microwave radiation and for ionospheric probing by radiation in the intermediate radio-frequency range ($f \sim 1$ MHz, $\lambda_0 \sim 300$ m), when a wave either travels through a plasma or is reflected, depending on whether $\omega > \omega_p$ or $\omega < \omega_p$ ($\varepsilon > 0$ or $\varepsilon < 0$). The field penetration depth in a nonabsorbing medium, or the distance at which the amplitude reduces e-fold, is equal to the critical wavelength we mentioned above for a similar case: $c/\omega\varkappa = c/\omega_p = \mathchar'26\mkern-10mu\lambda_{cr}$.

Thus, the conditions for field potentiality in an RF discharge are similar to those for a weak electromagnetic wave damping along the whole plasma length. They are valid for situations with dominant conduction currents (weak skin-effect) as well as with dominant displacement currents (collisionless, 'reflecting' plasma).

---

[2]If $\varepsilon' = 0$, in a 'collisionless' plasma ($\sigma = 0$) and at $\varepsilon = 0$, Maxwell's equations permit the existence of longitudinal waves of the electric field and space charge; here $H = 0$. Infinitely long waves correspond to oscillations of the electron gas as a whole with a frequency $\omega = \omega_p$, which turns ε to zero in accordance with equation (1.20).

Indeed, strong damping results from a strong effect of electromagnetic induction: the vortex electric field induced by an alternating magnetic field of the conduction currents in a good conductor or of the displacement currents in a nonconductive medium cancels the electric field of the incident wave.

In the RF range, the skin-effect is well manifested in atmospheric inductive discharges. For example, an equilibrium strongly ionized plasma with a temperature $T \approx 10,000$ K (for both the electrons and the gas) is created in the air at the same frequency of 13.56 MHz. In this case $\sigma \approx 25 \ \Omega^{-1} \, \text{cm}^{-1}$ and $\delta \approx 0.27$ cm, which is much less than the plasma size (1–10 cm, as in RF capacitive discharges [1.1]).

The 'wave,' or rather, 'field' model of the Joule heat release by RF discharge currents is as follows. An electromagnetic energy flux, $S_r = -c\langle E_z H_\varphi \rangle / 4\pi$, is radially incident on the outer side surface of a conducting plasma cylinder (Figure 1.2). The flux S is dissipated according to the exact equation $\text{div} S = \langle jE \rangle$. Here E is nearly independent of the cylinder radius owing to the field quasipotentiality, since in equation (1.13) $\text{rot} E \approx 0$, and the field H inside the conductor is $H \sim r$ from (1.11). The energy flux density $S_r = S_0(r/R)$ decreases towards the axis due to the geometry. The flux at the surface S_0 makes up only a small portion of the 'incident wave,' since most of it is 'reflected' and 'returns' to the generator. The field model is more suitable for the microwave range and describes adequately the situation in a real microwave discharge, with the only difference that the wave reflected from the plasma column cannot return to the generator because of the arrangement design. On the other hand, the reflected energy is much smaller, providing a high efficiency of the input power. For details see [1.1].

1.4   Electron production and losses: Plasma maintenance

A steady-state RF discharge is capable of maintaining strictly periodic processes for a long time, thus maintaining a steady plasma. This does not mean, of course, that its density n_e, let alone its temperature T_e, remain constant with time. They are constant only in average, being modulated by the double field frequency. The double frequency results from the electrons being produced primarily at maximum field values, regardless of its direction, while during the rest of each halfcycle they are mainly lost. Below, we consider briefly the mechanisms of electron production and loss in RF discharges. For a detailed description of these processes the reader is referred to [1.1], while here we will focus on effects associated with field oscillations.

1.4.1   Electron impact ionization

The basic mechanism of electron production in discharges is ionization of un-excited atoms and molecules by electron impact. The rate of this process is characterized by ionization frequency ν_i, which is the number of ionization events per electron per second. Ionization frequency is determined by the energy spectrum of electrons or by the distribution function $\varphi(v)$ over the absolute velocities. If φ is normalized to the density n_e, then

$$\nu_i = N \int_{\sqrt{2I/m}}^{\infty} \varphi(v) v \sigma_i(v) dv / n_e \qquad (1.32)$$

where $\sigma_i(v)$ is the ionization cross section. The lower integral limit is a velocity corresponding to the electron energy ε equal to the ionization potential $\varepsilon = mv^2/2 = I$; N is the density of atoms. At moderate mean energies, when $kT_e \ll I$, as is usually the case with the RF discharge plasma, the ionization cross section is approximated, with a good accuracy, by the formula $\sigma_i = C(\varepsilon - I), (\varepsilon \geq I)$, where C is a constant varying with the gas, e.g., for argon $C \approx 2 \times 10^{-17}$ cm^2 eV^{-1}. Despite the unlimited growth of σ_i in this formula, the integral of equation (1.32) rapidly converges because of a sharp drop of $\varphi(v)$ at $\varepsilon > I$. For the Maxwellian distribution

$$\nu_i \approx N\bar{v}C(I + 2kT_e)\exp(-I/kT_e) \qquad (1.33)$$

The temperature dependence of ν_i has a Boltzmann character.

Equation (1.33) is fairly approximate, since the energy spectrum of electrons in the field often differs from the Maxwellian one. For this reason, in order to find the distribution function, we have to solve a kinetic equation for the electrons in the field (for details, see [1.1]). These kinds of calculations have been made for many gases but usually for the case of steady field. So the question arises whether these results could be used for finding the ionization frequency in RF fields. This has proved to be possible in some limiting cases that quite often occur in RF capacitive discharges.

The energy spectrum of electrons is established at a rate characterized by the frequency of energy loss by the electron in its collisions with atoms or molecules: $\nu_u = \nu_m \delta$, where δ is the energy fraction transferred to the atom on collision [see equation (1.9)]. For elastic collisions, $\delta = 2m/M$ is very small. In collisions with molecules, there is a high probability of exciting molecular vibrations, hence, δ is much larger. For instance, in nitrogen $\delta \approx 2.1 \times 10^{-3}$ at $\bar{\varepsilon} \approx 1.5$ eV, whereas $2m/M = 3.9 \times 10^{-5}$; for nitrogen $\nu_m \approx 4.2 \times 10^9$ p [Torr] s^{-1} and $\nu_u \approx 10^7$ p [Torr] s^{-1}. If $\omega \gg \nu_u$, the spectrum has no time to respond to the field oscillations. The spectrum and the mean energy remain nearly constant over a cycle and almost coincide with the spectrum and $\bar{\varepsilon}$ that are established in a steady effective field equal to

$$E_{\text{eff}} = \frac{E_a}{\sqrt{2}} \frac{\nu_m}{\sqrt{\omega^2 + \nu_m^2}} = E \frac{\nu_m}{\sqrt{\omega^2 + \nu_m^2}} \qquad E = \frac{E_a}{\sqrt{2}} \qquad (1.34)$$

As for $\bar{\varepsilon}$, the foregoing follows directly from equation (1.10), because in a steady field $\bar{\varepsilon} = e^2 E^2 / m \nu_m^2 \delta$. For the spectrum, this follows from the form of the kinetic equation in which the term $(\omega^2 + \nu_m^2)$ enters naturally at $\nu_m = \mathrm{const}$ [1.1]. Hence, if $\omega \gg \nu_u$, the ionization frequency ν_{iRF} in an RF field of amplitude E_a is approximately the same as the ionization frequency ν_i in a steady field E equal to the effective field

$$\nu_{iRF}(E_a) = \nu_i(E_{eff}) \quad \text{at} \quad \omega \gg \nu_u = \nu_m \delta \tag{1.35}$$

In particular, if $\nu_m \gg \omega \gg \nu_m \delta$, as it often happens, ν_{iRF} approximately equals the ionization frequency in a steady field equal to the rms field.

In the limiting case of low frequencies $\omega \ll \nu_u$, the energy spectrum and the ionization frequency are not defined by these quantities, although the cycle mean energy is still described by the same formula (1.10) and corresponds to the effective field (1.34). The spectrum and the mean energy considerably vary within a cycle. They are strongly modulated and at each moment approximately correspond to the instantaneous field as if it were steady-state. The ionization frequency ν_{iRF} follows closely the slow field variation and coincides with the ionization frequency ν_i in a steady field equal to the instantaneous field. In this case, one should average directly ν_i corresponding to the instantaneous field, $\nu_i(E_a \sin \omega t)$.

The averaging may be made analytically if the function $\nu_i(E)$ is given in the Townsend form, $\nu_i \sim \exp(-Bp/E)$, conventionally employed for dc discharges. Here B is a characteristic constant of the gas [1.1].[3] Making use of the fact that the exponent index is a very large number for typical discharge plasma values of E/p and, hence, the slope of the $\nu_i(E)$ curve is very steep, we can expand $(\sin \omega t)^{-1}$ around the values of $\omega t = \pm \pi/2$ and evaluate the ωt integral. Eventually, we get [1.2]

$$\nu_{iRF} = (2E_a/\pi Bp)^{1/2} \nu_i(E_a) \quad \text{at} \quad \omega \ll \nu_u \tag{1.36}$$

This result has a clear physical meaning. At low frequencies, when the mean energy and the fraction of energetic electrons capable of ionization are strongly modulated, the ionization occurs only at maximum field values due to the steep slope of the Townsend curve $\nu_i(E)$. For this reason, the ionization frequency is determined by the amplitude E_a, and the pre-exponential factor in equation (1.36) characterizes the effective time during which ionization occurs in each cycle. This time is short due to the large exponent index of Bp/E_a for the discharge plasma. Since there is no ionization during most of the cycle, the electrons are only lost between the moments of peak field. In a steady-state discharge the short bursts of electron production compensate, on the average, their continuous losses. We may assume that in the general case of an arbitrary ratio between ω and ν_u, ionization

---

[3] Generally, the Townsend function is approximational, though it can be given a clear physical sense. If the spectrum is Maxwellian and the electron temperature $T_e \sim E$, as in the case of constant electron path length approximation (1.10*a*), the Townsend function $\nu_i(E)$ immediately follows from (1.33).

frequency lies within the range of values given by equations (1.35) and (1.36), in particular, between $\nu_i(E_a/\sqrt{2})$ and $\nu_i(E_a)$ at $\nu_m \gg \omega \gg \nu_u$. Formulas (1.35) and (1.36) permit evaluation of ionization frequencies from numerous data borrowed from theory, numerical calculations and experiments concerning steady electric fields and discharges.

1.4.2 Stepwise ionization and other mechanisms

At low plasma densities, when the gas is weakly ionized, the ionization of unexcited atoms occurs through electron impact. We have already discussed this mechanism in the previous section. At high plasma densities, stepwise ionization may become significant. Electrons excite metastable atomic levels, and when the electron density in a discharge is high, the concentration of long-living metastable atoms may become substantial. The ionization potentials of metastables are several times lower than those of unexcited atoms, so the cross sections of ionization by electron impact are quite large. Naturally, such atoms are readily ionized by electrons. There is also so-called associative ionization when two excited metastable atoms unite to form a molecular ion, e.g., $He(2^3S) + He(2^3S) \rightarrow He_2^+ + e$. The excitation energy of the atoms is expended for the electron detachment. Below (Section 2.3.4), we will demonstrate in what way these mechanisms permit interpretation of some important features of the RF discharge in helium. In gas mixtures, one may also observe the Penning effect, or the ionization of atom B on collision with excited atom A if the excitation energy of A exceeds the ionization potential of B, as it happens in a mixture of neon (A) and argon (B) at $E_{Ne}^* = 16.6$ eV and $I_{Ar} = 15.8$ eV. This two-step process goes on more rapidly than direct ionization of neon atoms that requires the energy $I_{Ne} = 21.6$ eV.

1.4.3 Electron loss mechanisms

The mechanisms of electron loss include diffusional escape of charges towards the discharge chamber walls where they become mutually neutralized, bulk recombination and, in electronegative gases, attachment to atoms and molecules to form negative ions. The latter are not better charge carriers than positive ions. Except for the electrode sheaths (see below), charge diffusion has an ambipolar character, and its coefficient in a nonequilibrium plasma, where $T_e \gg T$ (T is the gas temperature), is $D_a = \mu_+ T_e$ if T_e is expressed in volts. The frequency of diffusion losses, inverse to the characteristic time of the charge loss to the discharge space, is $\nu_d = D_a/\Lambda^2$, where Λ is the characteristic diffusion length. The latter parameter depends on the geometry: for a cylinder of radius R and length L, $\Lambda = [(2.4/R)^2 + (\pi/L)^2]^{-1/2}$, while for a plane gap of length L and large transverse dimensions $\Lambda = L/\pi$. For example, in argon at $p = 30$ Torr, $T_e = 1$ eV and $\mu_+ = 1.5 \times 10^3$ cm^2 Torr V^{-1} s^{-1}, we have $D_a = 50$ cm^2 s^{-1}; if $R = \infty$, $L = 2$ cm, $\nu_d = 120$ s^{-1}.

Recombination occurs primarily through a dissociative mechanism in the following way: $A_2^+ + e \rightarrow A + A$. This is always so in molecular gases, but in inert gases ions of the type He^+ transform to He_2^+ via the conversion reaction $He^+ + He + He \rightarrow He_2^+ + He$, and He_2^+ ions then recombine through a dissociative mechanism. The coefficients of electron-ion recombination in molecular gases are of the order $\beta \sim 10^{-7}$ cm^3 s^{-1} but smaller for inert gases. The recombination frequency, or the inverse lifetime of the electron, is $\nu_r = \beta n_e$. For instance, at the plasma density $n = n_e = 10^{10}$ cm^{-3}, $\nu_r \approx 10^3$ s^{-1}.

Electron affinity is exhibited by oxygen, by the mixture $CO_2 + N_2 + He$ used in CO_2 lasers in which the dissociative attachment $CO_2 + e \rightarrow CO + O^-$ occurs, and by halogen-containing compounds commonly used in plasma and ion technologies. The frequency of dissociative electron attachment ν_a, as the ionization frequency ν_i, varies with the field but more smoothly. Apart from attachment, there is the reverse process of electron detachment from negative ions by impact of excited molecules which accumulate in steady-state discharges. Dynamic equilibrium is often established between the amounts of electrons and negative ions, so that the detachment compensates for the attachment, and the negative ions recombine with the positive ions. In this case it is sometimes possible to describe electron losses in terms of recombination with an effective coefficient β_{eff} an order of magnitude larger than the real β [1.1].

The details of electron loss mechanisms and the rate constants of the reactions for some processes can be found in [1.1]. We will refer to these, whenever necessary, during the discussion of particular experiments and situations.

1.4.4 Weakly ionized nonequilibrium plasma maintenance by the field

The complex process of plasma maintenance by the field will more than once be discussed throughout this book. Here, we will consider a simple case which may give a general notion of the subject in order to facilitate further discussion. Suppose a gas in a certain space region is ionized by electrons produced and oscillating in the same region (but not coming from the electrode sheaths), and the charges are lost due to recombination. This often happens in a plasma located in the middle of a long discharge gap and also at elevated pressures. Such homogeneous electrically neutral plasma is called a positive column, a term borrowed from dc glow discharge physics. At $\nu_r > \nu_d$, recombination dominates over diffusional losses. In the example of the previous section (nitrogen, $p = 30$ Torr, $L = 2$ cm, $\beta = 10^{-7}$ cm^3 s^{-1}), this situation is observed at $n > D_a/\Lambda^2\beta \approx 1.2 \times 10^9$ cm^{-3}.

The percentage modulation of the plasma density in the RF discharge is usually very small. Over a half-cycle, between two successive moments of intense ionization, n decreases by $\Delta n = \beta n^2/2f$, e.g. only by $\Delta n/n = 0.037\%$ at $\beta = 10^{-7}$ cm^3 s^{-1}, $n = 10^{10}$ cm^{-3}, and $f = 13.56$ MHz. The conditions for the charge number balance are then equivalent to the condition for ionization-

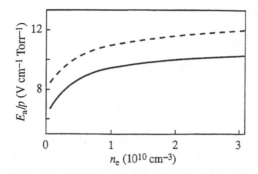

FIGURE 1.3
RF field for plasma maintenance (heating ignored). Laser mixture CO_2:N_2:He=1:6:12, $p = 30$ Torr, effective recombination coefficient $\beta_{\text{eff}} = 10^{-6}$ cm^3 s^{-1}. Electron spectrum following closely field oscillations (solid curve), ionization determined by rms values of the field (dashed curve). 'True' dependence is close to the solid curve at 13.56 MHz.

recombination equilibrium

$$\nu_{i\,\text{RF}}(E_a) = \nu_r = \beta n \tag{1.37}$$

where $\nu_{i\,\text{RF}}$ is determined by the RF field amplitude from equation (1.35) or (1.36), depending on the ratio between ω and ν_u. Equation (1.37) defines the field amplitude E_a necessary for the maintenance of a plasma of density n, or actually the current-voltage characteristic (CVC) of the positive column, because the current density amplitude (the displacement current then is insignificant) is proportional to n

$$j_a = \left[(e\mu_e n E_a)^2 + (\omega E_a/4\pi)^2\right]^{1/2} \approx e\mu_e n E_a \tag{1.38}$$

Figure 1.3 illustrates the results of calculation with equations (1.37) and (1.38). The calculation, however, neglected the gas heating due to the Joule heat release (for the account of heating and stepwise ionization, see Section 2.3). Because the pressure is constant and $p \sim NT$, the gas heating reduces molecular density N and, hence, the necessary field. In fact, the ionization frequency depends on E/N rather than on E/p, i.e., on ET.

Thus, the chain of cause-effect relationships that provide the maintenance of a steady-state plasma is this. The charge number balance indicates what ionization frequency can compensate for the losses, or, conventionally, how high the electron temperature should be for the spectrum to contain a sufficient number of energetic electrons. The electron energy balance, i.e., the compensation of electron energy losses in collisions with molecules through the field energy gain, shows what field value is required.

1.5   A simplified model of the RF discharge

1.5.1   Electrode space charge sheaths

Suppose a discharge operates between two plane electrodes and its transverse dimensions $2R$ are much larger than the interelectrode distance L, so that the process may be considered as one-dimensional. The x coordinate is counted off from the left-hand electrode while the electric potential is counted off from the right-hand electrode which is grounded. RF voltage is applied to the left-hand electrode.

At the moment of initial ignition, the high voltage applied to the electrodes causes breakdown, creating a plasma. Leaving aside the discussion of the breakdown mechanism and the transition process, we will focus on a well-formed discharge stationary in the sense that all processes in it occur regularly with time-independent amplitudes. Even at a fairly low electron density $n_e = 10^8$ cm^{-3} and a characteristic electron temperature $T_e = 1$ eV, the Debye radius $\lambda_D = 0.05$ cm is much smaller than typical electrode spacings $L \sim 1-5$ cm, so the plasma at the spacing center is electrically neutral. At the electrodes, however, the gas oscillates relative to the slow ions, periodically 'flooding' or exposing the positive ions, like a sea wave that recedes, exposing the coastal sand. This is the primary reason for the appearance of positive space charge sheaths at the electrodes. We will use the term 'electrode space charge sheath' also for the case when metallic electrodes are insulated from the plasma with a dielectric coating. For simplicity, in this section we will discuss the case of naked metallic electrodes.

1.5.2   Phenomenological description of the evolution of charge density, field, and potential

Assume, for simplicity, that the density of 'immobile' ions is constant in space, being the same in the plasma and in the electrode sheaths (further referred to just as sheaths). This assumption, which is in fact quite essential quantitatively, does not distort the picture, making it simple and clearly evident. The electric field in a homogeneous plasma must be uniform, therefore the electron gas as a whole oscillates around a middle point with the same amplitude A. The electrons, which at the moment of crossing this point were at a distance smaller than A from the electrodes, hit the metal during the first oscillations and are lost in it for good. (If the electrode is coated with a dielectric, the electrons become irreversibly attached to its surface.) In further oscillations, the electrons only slightly touch the solid surfaces. As a result, the electron gas passes its equilibrium position, leaving, on each side of the plasma, uncompensated positively charged layers of thickness A. The gas as a whole becomes positively charged, too. This model was suggested by S. M. Levitsky in 1957 [1.3], and since then it has often been employed for qualitative interpretation and evaluation of discharge processes. The electron gas oscillation is illustrated in Figure 1.4 for every quarter-cycle on the assumption that

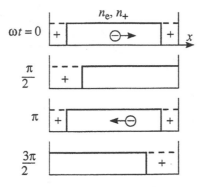

FIGURE 1.4
Electron gas oscillation with ion density taken to be constant (dashed curve) and $n_e(x, t)$
distributions for each quarter cycle (solid curves).

there are no diffusion charge flows towards the electrodes or diffusive smearing of
the boundary between the plasma and the sheaths. It follows from the electrostatic
equation

$$\frac{\partial E}{\partial x} = 4\pi e(n_+ - n_e) \tag{1.39}$$

that inside the sheaths, where $n_e = 0$ and $n_+ = $ const, the instantaneous field
depends linearly on x and the respective potential $\varphi = -\int E dx$ varies with x
parabolically. In a plasma with E independent of x, the instantaneous potential
varies in space linearly (Figure 1.5). The plasma current is mostly conduction
current, at least in moderate-pressure discharges; therefore, the discharge current
j flows along the plasma field E_p during most of the cycle. This is indicated by
arrows in Figure 1.5. There are no charge currents towards the electrodes in the
immobile ion approximation for the case of plasma momentary contact with them
and in the absence of electron diffusion, or thermal motion.

The time average harmonic potential of the left-hand electrode, as well as of
the grounded electrode, is zero, while the plasma potential is always positive with
respect to the electrodes. Hence, the field in the sheaths is in average directed to the
electrodes. This is of primary importance for applications and can be accounted
for by the fact that the gas in the gap is in general positively charged and, for this
reason, possesses a steady positive potential $\overline{V}$. In contrast with this simplified
model, in a real plasma there is always a continuous, relatively small thermal
ion flow into the sheath. In low pressure discharges, the ions cross the sheath
almost without collisions, gaining from the field, energy of the order of a constant
plasma potential. This energy may be as high as a few hundred electron volts.
Technological applications of low-pressure RF capacitive discharges are based just
on the bombardment of the electrode or of the target material by energetic ions.

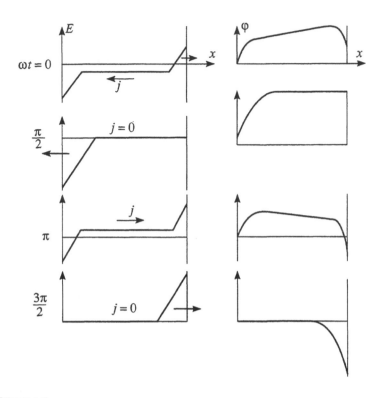

FIGURE 1.5
Field and potential distributions in the gap between plane electrodes for the n_+ and n_e distributions of Figure 1.4. Arrows indicate the direction of current j and field E. The left-hand electrode is grounded.

1.5.3   Equations for discharge parameters

Let us consider the above picture, employing the equations of motion for electrons in a field and the electrostatic equations. We will denote the left-hand and the right-hand instantaneous sheath thicknesses as d_1 and d_2, respectively. At zero charge currents at the electrodes, the total charge and thickness of both sheaths remain constant:

$$en_+d_1 + en_+d_2 = \text{const} \qquad d_1 + d_2 = 2A \qquad (1.40)$$

From equation (1.39), the field distributions in the left-hand (E') and in the right-hand (E'') sheaths are

$$E' = E_p - 4\pi en(d_1 - x) \qquad E'' = E_p + 4\pi en[x - (L - d_2)]$$

The plasma potentials relative to the electrodes, or the instantaneous voltage falls in the sheaths, are

$$V_1 = 2\pi e n d_1^2 \qquad V_2 = 2\pi e n d_2^2 \tag{1.41}$$

The electrode voltage V equal to the left-hand electrode potential is expressed as

$$V = \int_0^L E dx = \int_0^{d_1} E' dx + \int_{d_1}^{L-d_2} E_p dx + \int_{L-d_2}^L E'' dx = E_p L + 2\pi e n \left(d_2^2 - d_1^2\right)$$

Substituting here $d_2 = 2A - d_1$ according to (1.40) and solving the equality relative to E_p, we get the following equation for the plasma field

$$E_p = V/L + 8\pi e n(A/L)y \qquad y = d_1 - A \tag{1.42}$$

Equation (1.42) relates parametrically E_p to the displacement y of the left-hand boundary of the plasma (y is also the electron displacement from the equilibrium position). On the other hand, the displacement obeys the general equation of motion for the electron (1.1). In this equation $v = \dot{y}$, and by $E_a \sin \omega t$ we mean the plasma field E_p.

Substituting E_p from equation (1.42) into equation (1.1) and using expression (1.21) for the plasma frequency, we have the equation for the electron displacement in the plasma:

$$\ddot{y} + \omega_p^2(2A/L)y + \nu_m \dot{y} = -eV/mL \tag{1.43}$$

It is furthermore convenient to deal with harmonic quantities in a complex form. With the voltage $V = V_a e^{i\omega t}$ applied to the electrodes, the steady-state solution of equation (1.43) is

$$y = \frac{eV_a e^{i\omega t}}{mL(\omega^2 - \omega_p^2 2A/L - i\omega\nu_m)} \tag{1.44}$$

By equating the real displacement amplitude y to A, to which y is equal by definition, we have

$$A^2 \left[(\omega^2 - \omega_p^2 2A/L)^2 + \omega^2 \nu_m^2\right] = (eV_a/mL)^2 \tag{1.45}$$

This is the fourth-power algebraic equation with respect to A. The root that has a physical sense defines the electron oscillation amplitude as a function of the voltage amplitude V_a, the frequency ω_p and the plasma density n. The latter enters into equation (1.45) through ω_p as a tentatively unknown parameter.

We have shown in Section 1.4.5 that the plasma density is related to the field amplitude E_{pa} through the charge number balance equation, e.g., by the condition of ionization-recombination equilibrium (1.37). From equation (1.42), the plasma field is

$$E_p = \frac{V_a e^{i\omega t}}{L} \frac{\omega^2 - i\omega\nu_m}{\omega^2 - \omega_p^2 2A/L - i\omega\nu_m} \tag{1.46}$$

Thus, the set of equations in the model relating all discharge parameters becomes closed. Equations (1.45) and (1.37), containing amplitude E_{pa} found from equation (1.46), form a set of two equations for finding A and n as a function of the applied voltage (V_a and ω). With their values found, we can evaluate any other quantity, e.g., the current, the discharge impedance and the current-voltage characteristic. All the equations can be easily generalized for the case of dielectric-coated electrodes. Then the voltage drop in the dielectric should be added to the total voltage drop in the sheaths and the plasma.

1.5.4 Discharge current

The density of discharge current which flows through the external circuit, in particular through the electrodes, is, by definition, equal to the variation rate of the surface charge density q at the left-hand electrode (chosen with respect to the x-axis direction). The electrode may be assumed to be an ideal conductor. There is no field and, hence, no displacement current $\dot{E}/4\pi$ in it. If the plasma were also an ideal conductor, the negative charge density q at the electrode would coincide in the absolute value with the amount of positive charge in the left-hand sheath per unit area end_1. But unlike a metal, plasma is a poor conductor, so it has a field E_p which must maintain the current and, possibly, ionization. In electrostatic laws, it is related to the surface charge densities as follows:

$$E_p = 4\pi(q + end_1) \tag{1.47}$$

By differentiating this equation with respect to time, we can find the discharge current density in the form

$$j = \dot{q} = -en\dot{y} + \dot{E}_p/4\pi \tag{1.48}$$

The value of j is independent of x, being constant for all cross sections of the gap. The first term in equation (1.48) represents the electron current density (conduction and polarization currents), the second is the vacuum displacement current. Due to the poor plasma conduction, the latter appears in the expression for the quantity j conserved along the x-axis. It is for this reason that the sheath charge differs in magnitude from q, being unable to respond instantly to the variation in the electrode charge which comes from or goes to the external circuit under the action of the power supply emf. In fact, the above analysis clarifies the physical sense of the conventional notion of vacuum displacement current that we have successfully discarded by writing equation (1.48) in terms of the charge motion and electrostatic laws.

After substitution of E_p into equation (1.48) with equation (1.42), the expression for the current density takes the form

$$j = -en\dot{y}(1 - 2A/L) + i\omega V/4\pi L \tag{1.49}$$

In the absence of discharge (at $n = 0$), there is only the second term, which

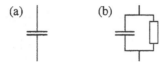

FIGURE 1.6
Electric circuits (a) for the gap in absence of discharge and (b) for discharge current corresponding to equation (1.49).

is nothing else but the reactive current through a 'vacuum' capacitor formed by two electrode plates; $1/4\pi L$ is the capacitance per unit area. Ignition of a discharge and the appearance of the first term in the electrical sense is similar to connecting a complex resistance parallel to the capacitor (Figure 1.6), because the first term is also proportional, according to equation (1.44), to voltage V. Plasma production inside the capacitor changes the electric system so radically that an equivalent electric circuit with a series connection of elements would seem to be more appropriate. This we are now going to demonstrate.

1.5.5 'Impedance' of the discharge plasma space

Let us substitute equation (1.44) into equation (1.49). With a slight transformation, we can write

$$j = \frac{V}{Z} \qquad Z = \frac{Lm(\nu_m + i\omega)/e^2 n + 4\pi 2A/i\omega}{1 - \omega^2/\omega_p^2 - i\omega\nu_m/\omega_p^2} \qquad (1.50)$$

The quantity Z has the sense of impedance (complex resistance) per unit area of the gap containing a plasma. The electric system, however, is nonlinear: A and n in Z depend on the voltage amplitude as they are defined by equation (1.45) and by the charge number balance equation. If we find and substitute the functions $A(V_a)$ and $n(V_a)$ into equation (1.50), we will get the dependence of j_a on V_a, or the discharge current-voltage characteristic. Although A and n in (1.50) depend on V, it is useful to interpret the expression for Z as if A and n were fixed values, i.e., as if Z were the real impedance of the electric system. In the general case, such an interpretation is not easy to find, but in the limit of strong ionization with $\omega_p^2 \gg \omega^2$ and $\omega_p^2 \gg \omega\nu_m$, $(\omega_p^2 \sim n)$ characteristic of RF discharges makes the interpretation easy to grasp. In this limiting case,

$$Z = L/\sigma_t + 4\pi 2A/i\omega \qquad \sigma_t = \frac{e^2 n}{m(\nu_m + i\omega)} \qquad (1.51)$$

where σ_t is the complex plasma conductivity given by formulas (1.17)–(1.19), $2A$ is the time-independent total thickness of the two electrode sheaths, and $1/4\pi 2A$ is the equivalent capacitance (per unit area) of two capacitances connected in series and corresponding to the sheaths. Each of them varies with time, but the

FIGURE 1.7
Electric circuits for RF discharge corresponding to the impedance of equation (1.51):
(a) sheaths represented as individual capacitances, (b) combined sheath capacitances.

equivalent capacitance of the two sheaths remains constant. So, in accordance
with equation (1.51), the electric system may be considered as a series connection
of the sheath capacitances and the complex plasma resistance (Figure 1.7).

Expression (1.51) for the RF discharge impedance was deduced in [1.4] directly
from this physical consideration. Here, it has been derived as a limiting case of
formula (1.50) 'exact' within the model discussed. The work [1.4] was based on
the same quantitative model, but one of its governing equations corresponding to
equation (1.45) was based on entirely different grounds with which we disagree.
This issue is of fundamental importance because it concerns basic qualitative
aspects of a discharge—what defines what. We will return to this issue in Section 3.2. A similar model with a closing equation of the type (1.45) was analyzed
in [1.5], assuming the electron motion to be purely drift motion. So terms with $\ddot{y}$
and ω^2 dropped out of equations (1.43)–(1.46) and (1.50).

1.6   Constant positive plasma potential

It has long been known that the RF discharge plasma possesses a constant potential positive relative to the electrodes. This causes electrode sputtering similar
to cathode sputtering in a glow discharge. Ions escaping the plasma due to thermal motion are accelerated away by the steady field towards the electrodes and
bombard them. All available ion–plasma technologies with RF discharges are
based on this effect: a target sample to be treated is mounted on an electrode and
bombarded by ions. This is a subject of our subsequent consideration.

The positive plasma potential relative to the electrode $\overline{V}_p$ is found by averaging
over a cycle the instantaneous voltage fall in sheath V_1 given by equation (1.41).
Let us substitute into it the expression $d_1 = A + y$. Noting that in harmonic
displacement $\overline{y} = 0$ and $\overline{y^2} = A^2/2$, we have

$$\overline{V}_p = \overline{V}_1 = \overline{V}_2 = 3\pi e n A^2 \tag{1.52}$$

Evaluate now $\overline{V}_p$ in two limiting cases in which we can find explicitly the solution
of fourth-power equation (1.45) for A.

1.6.1 Oscillation amplitude and positive plasma potential at moderate pressures

Let us assume that $\nu_m \gg \omega$ and $\omega \nu_m \gg \omega_p^2 2A/L$. Such inequalities arise at elevated pressures and for relatively large gaps when the oscillation amplitude of electrons is much smaller than L. For this case we find from equation (1.45) that $A \approx eV_a/Lm\omega\nu_m$. Electrons perform drift oscillations in a field of amplitude $E_a \approx V_a/L$, which is obtained if, in a first approximation, all the voltage applied to the electrodes contributes to the plasma, and the voltage fall in the sheaths is comparatively small. The constant plasma potential, small as compared to V_a and evaluated in the next approximation from the sheath voltage fall, is

$$\overline{V}_p = 3\pi e n \mu_e^2 E_a^2 / \omega^2 \qquad \mu_e = e/m\nu_m \qquad (1.53)$$

where μ_e is the electron mobility in the plasma and E_a is actually the field amplitude in it, $E_a \approx E_{pa}$. A formula of the type (1.53) was first derived in [1.3] (with an error 6 instead of 3). Thus, at $E_a/p \approx 10 \text{ V cm}^{-1} \text{ Torr}^{-1}$, $A \approx 0.1$ cm; if $n = 10^{10} \text{ cm}^{-3}$, $\overline{V}_p \approx 140$ V, which is more or less consistent with experiment. If $p = 30$ Torr, $L = 3$ cm, $f = 13.56$ MHz, the assumptions of $\nu_m \gg \omega$ and $\omega\nu_m \gg \omega_p^2 2A/L$ are fulfilled with a fairly good accuracy and $V_a \approx 900$ V.

1.6.2 Low pressures

Suppose ω^2, $\omega\nu_m \ll \omega_p^2 2A/L$. This happens in discharges of not very high frequencies and at low pressures, when the plasma oscillation amplitude is nearly as large as the electrode spacing, so that $2A/L \lesssim 1$. At moments of maximum displacement, the thickness of one of the sheaths becomes comparable to the electrode spacing, and nearly all applied voltage falls in one sheath. As it follows from equation (1.45), $A \approx \sqrt{V_a/8\pi e n}$, which in fact follows from initial formulas (1.41) with allowance for $V_{1\,max} \leq V_a$ and $d_{1\,max} = 2A$.

From equation (1.52), the constant plasma potential $\overline{V}_p = (3/8)V_a$ makes up a substantial fraction of the applied voltage amplitude and is proportional to it. A similar result can be easily obtained without much theory but just from a simple reasoning. Assume that the oscillation amplitude of electrons is nearly as large as the electrode spacing and that nearly all the voltage falls in one sheath at maximum displacement over a half-cycle. In the next half-cycle the plasma is held against the electrode so the voltage fall here is small. Then

$$\overline{V}_p \approx \frac{1}{2\pi} \int_0^\pi V_a \sin \omega t \, \mathrm{d}(\omega t) \approx \frac{1}{\pi} V_a \qquad (1.54)$$

and this is different from the previous result only in the replacement of $8/3 \approx 2.7$ by π. The result in equation (1.54) was obtained in [1.4] as a limiting case of a more detailed theory and generally agrees with experiment (Figure 1.8).

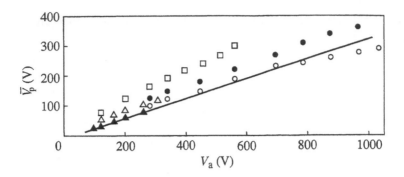

FIGURE 1.8
Constant plasma potential versus electrode voltage amplitude at low pressure [1.6].
Calculation with equation (1.54) (solid curve), measurements in He and H_2 (points).

1.6.3   Dielectric-coated electrodes

An important point in the RF discharge model we are discussing is the statement that some electrons escape from the plasma to be lost in the electrode metal, leaving the gap positively charged. There are reasons to believe that when the electrodes are separated from the plasma by dielectrics, nothing principally new occurs in the gap. The electrons that have accumulated along the boundary with the dielectric are pressed by the field to the dielectric surface during oscillations and they become attached to it 'for good.' The only difference with a metallic surface is that the electrons may incorporate into the metal bulk, while in a dielectric they are deposited on its surface. If this is so, the above theory, including equation (1.53) for the constant plasma potential, can be automatically extended to dielectric-coated electrodes. Since the field in a dielectric is harmonic, the cycle average constant voltage fall in a dielectric is absent, and $\overline{V}_p$ is, as previously, a constant potential relative to the electrode.

It is interesting to see what kind of constant potential a plasma would have, if any, if no electrons were attached to the dielectric surface and if the gas as a whole were electrically neutral. Evidently, during oscillations the electrons would expose the ion sheath at one electrode and press themselves to the other. The ion sheath on each side would exist only for a half-cycle, as is illustrated in Figure 1.9. Then in the case of moderate pressures described by equation (1.53), the voltage fall in the sheath during a half-cycle could be described by the same expression (1.41) but with $d_1 = A \sin \omega t$ instead of $d_1 = A(1 + \sin \omega t)$ as in the case of a naked metallic electrode (in both cases $t = 0$ was taken arbitrarily). During the other half-cycle, when the electrons are pressed against the dielectric, the voltage fall in the negative space charge sheath is close to zero because of the small thickness of this sheath (at the same maximum field in the dielectric as in the first half-cycle). Then the average constant potential of the plasma relative to

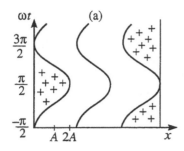
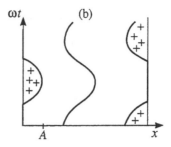

FIGURE 1.9
Schematic trajectories of the plasma boundaries and a bulk electron in a gap between
coated electrodes: (a) electrons assumed to attach themselves to the dielectric as if
entering a metal, (b) electrons assumed to be only pressed against the dielectric without
attachment.

the electrode would be equal to

$$\overline{V}_p = \overline{V}_1 = (2\pi e n \overline{d_1^2} + 0)/2 = (2\pi e n A^2/2 + 0)/2 = \pi e n A^2/2 \qquad (1.55)$$

It would be 6 times smaller than for purely naked electrodes (but larger than zero!).
Experiment does not show much difference in the constant plasma potentials for
naked or dielectric-coated electrodes, though such a great difference could hardly
remain unnoticed. This fact may be taken as indirect evidence for the electron
'permanent' attachment to dielectrics.

1.7 Stochastic heating of electrons

It has been pointed out above (Section 1.2.2 and further) that electrons become
heated in a discharge due to collisions with atoms, when the energy of regular
oscillations acquired from the field after the previous collision transforms to
the energy of random motion on the next collision. At low pressures, when
$\omega^2 \gg \nu_m^2$, the conduction and Joule heat (see Section 1.3.4) are proportional to
the collision frequency ν_m. Hence, at extremely low pressure there must be very
low power released in the discharge. Contrary to that, experiment demonstrates
something quite different: the released power exceeds, by an order of magnitude
and higher, the estimates taking into account only electron collisions with atoms.
This discrepancy is due to another electron heating mechanism, which acts even
in a 'collisionless' plasma. It has been termed by different workers as stochastic
heating or as the mode of electron collision with sheaths. To the RF discharge, it
was first applied by V. A. Godyak [1.7].

1.7.1 The physical mechanism

The mechanism of this kind of electron heating is briefly as follows. At the electrode, the plasma borders on a positive space charge sheath, in which the field increases with distance from the plasma to the electrode during most of the cycle and throughout most of the space and is directed towards the electrode (Figure 1.5). This field pushes out an electron that has occasionally penetrated into the sheath back into the plasma. The boundary between the plasma and the sheath oscillates at velocity u_s, moving alternatively towards the plasma or the electrode. Having in mind the case of low pressures, when collisions of electrons with atoms are quite rare, consider in the laboratory coordinate system collisionless motion of an electron that has accidentally entered the sheath with the velocity component v_x normal to its boundary. Since the mean random velocity of an electron in a plasma $\bar{v}$ is usually much larger than its oscillation velocity, the penetration of electrons into the sheath happens quite frequently.

As the electron goes deeper into the sheath, it is slowed down by the retarding space charge field. After it has entirely lost its velocity v_x, it turns back and increases its velocity under the action of an accelerating field, eventually returning to the plasma. In a coordinate system related to the moving plasma boundary (the boundary acts as a potential barrier or a 'wall'), the electron experiences purely elastic recoil. In the laboratory coordinates, the velocity of the recoil electron is $-v_x + 2u_s$, and its kinetic energy on recoil changes to

$$\Delta \varepsilon = \frac{m(-v_x + 2u_s)^2}{2} - \frac{mv_x^2}{2} = -2mv_x u_s + 2mu_s^2 = -2mu_s v_x'$$

where $v_x' = v_x - u_s$ is the relative incident velocity of the electron. Naturally, the electron hits the 'wall' only if $v_x' > 0$ or $v_x > u_s$, where both quantities are algebraic. The kinetic energy of the lateral motion does not change on recoil.

At given values of v_x and $|u_s|$, the recoil electron gains energy if the wall moves to meet the electron ($v_x > 0 > u_s$) or tries to overtake it ($u_s < v_x < 0$). If, on the contrary, the electron overtakes the running wall ($v_x > u_s > 0$), it loses its energy on recoil. In an average of two such acts, the electron inevitably gains energy, though not much: $2mu_s^2$. Typically, when $v_x \gg |u_s|$, this average gain is by a factor of $v_x/|u_s| \gg 1$ less than the real gains and losses in individual acts, which are identical with an accuracy of a small value of $|u_s|/v_x \ll 1$ and are close to $2mv_x|u_s|$. The ensemble-average effect of collisions also depends on the recoil frequencies of electrons with the velocity v_x, which also slightly vary with the direction of the wall movement, as they are proportional to the relative velocity $v_x' = v_x - u_s$.

1.7.2 Evaluation of dissipated power and electrical resistance

Let $f(v_x, t)\mathrm{d}v_x$ be the number of electrons with velocities from v_x to $v_x + \mathrm{d}v_x$ per cubic centimeter at the plasma boundary at time t. Over the time $\mathrm{d}t$, the number

$v'_x f(v_x,t)dv_x dt$ of such electrons is incident on 1 cm^2 of the boundary surface, and their energy changes by $-2mu_s v'_x$. The total energy gain by electrons from 1 cm^2 per second is

$$P_1 = \left\langle -2mu_s \int_0^\infty v_x'^2 f(v_x,t)dv'_x \right\rangle \qquad (1.56)$$

This formula takes into account $dv'_x = dv_x$ and makes the time averaging over an oscillation cycle.

The velocity distribution of electrons is time-dependent because of the electron oscillations that superimpose on their random motion. The oscillation velocity u in a general case does not necessarily coincide with the plasma boundary velocity u_s (see below). Since $|u|$, $|u_s| \ll \overline{v}$, the function $f(v_x,t)$ is only slightly time-modulated with the oscillation frequency as compared to stationary $f(v_x)$. We can neglect this modulation in the integral evaluation of (1.56). Taking into account that for the majority of electrons v_x, $v'_x \gg u$, u_s, we put

$$f(v_x,t) \to f(v_x) = f(v'_x + u_s) \approx f(v'_x) + u_s \left(\frac{df}{dv_x} \right)_{v_x = v'_x} \qquad (1.57)$$

Substituting the expansion into equation (1.56), integrating by parts and assuming that $f \to 0$ at $v_x \to \infty$ and $\overline{u_s} \equiv \langle u_s \rangle = 0$, we find

$$P_1 = 4m\overline{u_s^2}\Gamma \qquad \Gamma = \int_0^\infty v'_x f(v'_x)dv'_x = \frac{1}{4}n_e\overline{v} \qquad (1.58)$$

Here Γ is the one-directional electron flux density and n_e is the electron density at the plasma boundary. The average energy gain per collision with the wall $\overline{\Delta\varepsilon} = 4m\overline{u_s^2}$ is twice as large as the estimated value, because the electron more often collides with the wall when the latter moves to meet the electron than when the wall runs away.

Assume that there is no correlation between the electron recoils from the two opposite plasma boundaries and that these events are entirely independent of one another. This is likely to be true if the electron performs many oscillations along its path L during its 'collisionless' flight from 'wall' to 'wall' (at a velocity close to the mean random velocity), i.e., if $\omega \gg \Omega = \overline{v}/L$. Then the powers gained from recoils from both boundaries are just summed up, and the power dissipated in the discharge through a stochastic mechanism per 1 cm^2 cross section will be

$$P_s = 2P_1 = 2n_e\overline{v}m\overline{u_s^2} = n_e\overline{v}mu_a^2 \qquad (1.59)$$

where u_a is the velocity amplitude of the plasma boundary oscillations. Now assume the field and the electron density at the plasma boundary to be the same as in the bulk. By summing up P_s and the power dissipated in collisions with atoms using formulas (1.8) or (1.23), (1.18), and also (1.3) for u_a, we can find the power that is released in the discharge column per unit cross section due to both the mechanism under consideration and the collisions with atoms:

$$P = Ln_e \frac{e^2 E_a^2}{2m\omega^2}(2\Omega + \nu_m) = n_e \bar{v} m u_a^2 \left(1 + \frac{L}{2l}\right) \tag{1.60}$$

Their relative contribution is determined by the number of electron path lengths l along the half-length of the plasma column $L/2$. The lower the pressure and the smaller the discharge length, the smaller the contribution of collisions with atoms as compared to stochastic heating. Thus at $L = 6$ cm and $l = 0.03/p$ [Torr] cm, stochastic heating dominates at $p < 10^{-2}$ Torr. The total power may be obtained formally by adding the double frequency of electron collisions with the plasma boundary $\Omega = \bar{v}/L$ to the frequency of collisions with atoms ν_m.

Since $e^2 n_e^2 \bar{u}_s^2$ represents the mean square of the conduction current j_{cond}, the quantity

$$R_s = \frac{P_s}{\langle j_{cond}^2 \rangle} = \frac{2m\bar{v}}{e^2 n_e} = \frac{4\pi\bar{v}}{\omega_p} \approx 4\pi\lambda_D$$

may be considered as the active resistance per 1 cm^2 discharge column due to the stochastic mechanism. Resistance R_s should be added to the impedance (1.51)

$$Z = L\frac{m}{e^2 n_e}\left[(\nu_m + 2\Omega) + i\omega\right] + \frac{4\pi 2A}{i\omega} \tag{1.61}$$

These results were obtained in [1.7, 1.4].

1.7.3 Absence of effect for steady-state distribution of relative electron velocities

Suppose that the electron distribution function related to the oscillating plasma boundary is time-independent, or that the relative velocity distribution is steady

$$f(v_x', t) = F[v_x - u_s(t)] = F(v_x')$$

In this case, the integral of (1.56) is independent of u_s. When the boundary moves to meet the electrons ($u_s < 0$), the electron gas gains as much energy as it gives off in the opposite situation ($u_s > 0$). Hence the time-average effect is absent: $P_1 \sim \overline{u_s} = 0$.

If we look at this matter just formally, this situation may seem somewhat similar to that considered in Section 1.5 within the simplified RE discharge model which takes ionic density to be identical throughout. The oscillation velocity of electrons in the plasma $u(t)$ is then the same as the boundary velocity $u_s(t)$. Indeed, for collisionless motion of electrons through a homogeneous plasma in a uniform oscillating field $E = E_a \sin \omega t$, their distribution function obeys the kinetic equation [1.1]

$$\frac{\partial f}{\partial t} - \frac{eE_a}{m}\sin\omega t\frac{\partial f}{\partial v_x} = 0$$

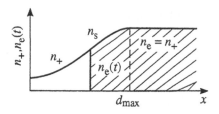

FIGURE 1.10
More realistic (as compared to Figure 1.4) ion and electron density distributions from the side of the left-hand electrode [$n_e(t)$ for a fixed moment].

Its solution has the form we discussed above:

$$f(v_x, t) = F[v_x - u(t)] \qquad u(t) = u_s(t) = (eE_a/m\omega)\cos\omega t$$

where u is the electron oscillation velocity satisfying the equation of motion (1.1) at $\nu_m = 0$. The absence of stochastic heating in the discharge model with $n_+(x) = $ const was pointed out by M. Lieberman [1.8].

1.7.4 Electron recoil from low ion density sheaths

In contrast with the simplified model we have discussed, real RF discharges, in which plasma only momentarily touches the electrode and the current is closed on the electrode exclusively by the displacement current, have a nonuniform ion density in the gap. Within a sheath, it decreases several-fold from the plasma toward the electrode. When the oscillating electron gas alternatively floods and exposes the ionic gas, the plasma density in the sheath also decreases as the boundary approaches the electrode (Figure 1.10). The details of available sheath models will be discussed in Sections 2.1 and 3.8. Here we will describe phenomenologically the way the real sheath structure affects stochastic heating of electrons. A quantitative treatment of this problem was offered in [1.8].

The distribution function of electrons coming to the plasma boundary, which enters into the integral of (1.56), is determined by processes occurring in a homogeneous plasma, because the electrons come just from this plasma. One may assume, as in [1.8], that the distribution function has the form

$$f(v_x, t) = \frac{n_s(t)}{n_e} F[v_x - u(t)] \tag{1.62}$$

where F is the distribution function for a homogeneous plasma normalized to the plasma density n_e in the bulk. The factor $n_s(t)/n_e$ allows for the fact that the function f at the boundary must yield the electron density at this boundary at time t, $n_s(t)$. The latter is equal to the ion density at this point of the sheath.

By expanding the function $F[v_x - u(t)]$ in the vicinity of the point of the relative velocity of the electron and the sheath $v'_x = v_x - u_s$, as in deriving (1.58)

$$F(v_x - u) \approx F(v'_x) + (u_s - u) \left(\frac{\mathrm{d}F}{\mathrm{d}v_x} \right)_{v'_x}$$

and then integrating exactly (1.56) by parts, one obtains an expression generalizing and refining equation (1.58) [1.8]

$$P_1 = 4m \left\langle \frac{n_s}{n_e} u_s(u_s - u) \right\rangle \Gamma = 4m \langle u(u_s - u) \rangle \Gamma \qquad (1.63)$$

The last expression of (1.63) follows from the previous one, since $n_s u_s = n_e u$ due to the conservation of the electron flux or electric current.

Neglecting the oscillation effect on the distribution function of electrons, as was done in deducing equation (1.58), and putting $n_s = n_e$ and $u = 0$ in the first part of (1.63), we will return to equation (1.58). If we put $n_s = n_e$, $u = u_s$, strictly following the concepts of the model with $n_+ = \text{const}$, we will get $P_1 = 0$, as in Section 1.7.3. In the general case, when the ion density in the sheath is lower than the bulk plasma density, $n_s < n_e$, and hence, the sheath boundary velocity $|u_s| > |u|$, the power in (1.63) is smaller than in (1.58). One may say that the easily deduced expression in (1.58) yields the upper estimate of stochastic heating power. In [1.8], the calculation of (1.63) was made with an actual anharmonic law of boundary motion $u_s(t)$ that follows from consideration of a sheath with the calculated density distribution $n_+(x)$ (Section 3.8).

Note a curious implication from the general expression (1.63) that is consistent with the process physics. If, on the contrary, the sheath ion density is higher than in a homogeneous plasma, and hence, the boundary oscillation is slower than the electron oscillation in the plasma bulk, the electron gas may become cooled rather than heated owing to the electron recoil from the sheath and plasma boundaries. Thus if we put $u = \beta u_s$ with $\beta(t) = \text{const}$, then $\langle u(u_s - u) \rangle = \overline{u_s^2}\beta(1 - \beta)$. At $\beta < 1$, $P_1 > 0$ as above. But if $\beta > 1$, $P_1 < 0$. It will be shown further that there is an RF discharge in which the sheath ion density is indeed higher than in the plasma bulk, but this kind of discharge operates at not very low pressures, when stochastic heating is not pronounced as compared to the effect of electron-atom collisions. In principle, a cooling effect seems to be possible, and it would be interesting to find the appropriate experimental conditions.

1.8   RF discharge modes

It has been found experimentally that RF capacitive discharges usually operate in one of the two strongly differing modes. Externally, these modes differ in the intensity and luminosity distribution along the discharge length, but they essentially differ in the ionization processes in the electrode sheaths and in the mechanisms of current closure on the electrodes. The first observation of these facts was made in 1957 by S. M. Levitsky [1.3], who described the transition of one mode to the other and offered a basically correct interpretation of this phenomenon. In 1978, one of the authors of this book started a systematic experimental study of the properties of both modes at moderate pressures and of the conditions under which one kind of discharge transforms to the other. As a result, we have today a fairly clear understanding of the underlying mechanisms. Interest in these processes was largely simulated by the promise they held (and that has mostly been realized) for RF moderate-pressure discharge application for creating the active medium in CO_2-lasers. About the same time, interest in low-pressure RF discharges considerably increased primarily due to their successful application in plasma technology. These discharges also exhibit two operation modes, but at low pressures their differences are not so much pronounced. For this reason, in this chapter we will first discuss moderate-pressure discharges, leaving a more detailed consideration for Chapters 2 and 3.

1.8.1 The current-voltage characteristic and ignition of a moderate-pressure α-discharge

The properties and specific features of the two discharge modes will become clearer if we first consider the experimental current-voltage characteristic (CVC) of a discharge. The CVC relates the current and voltage amplitudes, i_a and V_a, or their root-mean-square values, i and V, for the electrodes in a steady-state discharge. The measurements we are going to discuss were made in a large vessel of 60 l in volume with a pair of parallel disc water-cooled electrodes of 10 cm diameter at its center [1.9]. The distance L between the electrodes could be varied from 1 mm to 10 cm. Sometimes, they were insulated from the plasma with a glass sheet or other dielectric material. The discharge was excited by a 3 kW generator operating at 13.56 MHz. Let the pressure and the electrode gap be not too large so that their product pL is smaller than a certain critical value $(pL)_{cr}$ characteristic of each gas; e.g., for air it is 40 Torr cm. The point is that both discharge modes can be observed only at $pL < (pL)_{cr}$, otherwise only one of them can be operative (see Chapter 2).

As the voltage and the current are gradually increased starting with small values, they are at first proportional to each other. This indicates that a reactive current (displacement current) flows through the capacitor formed by the two electrodes. At a large voltage—which depends on the gas, pressure and the electrode spacing—breakdown occurs, initiating a discharge. The current immediately rises and the voltage drops. At first, the discharge may not cover all the electrode area; if the arrangement is fairly symmetric, it forms a circle at the disk center. With further growth of the discharge current, the electrode potentials remain constant, while the area on the electrodes occupied by the discharge column grows in such a way that the electrode current density does not change. There is the effect of normal current density, as in a dc glow discharge. The normal current density

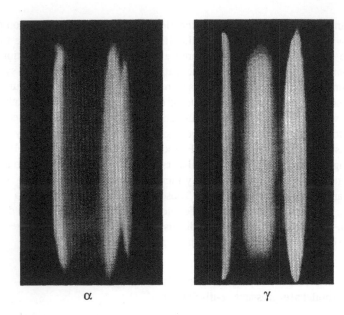

α   γ

FIGURE 1.11
Photographs of RF α- and γ-discharges with air at $p = 10$ Torr, $f = 13.56$ MHz, and
$L = 2$ cm. rms value of $V = 320$ V for both modes, rms $j_\alpha = 7$ mA cm^{-2} and
$j_\gamma = 120$ mA cm^{-2}.

is not large, about $1-10$ mA cm^{-2}, and the plasma at the gap center is not very
bright; there are brighter layers closer to the electrodes, but the gas between these
layers and the electrodes is dark (Figure 1.11(α) and curve α in Figure 1.12).

Levitsky termed this type of discharge as α-discharge (it is sometimes called
a low-current discharge, meaning that the current density is less than in the other
mode called the γ- or high-current discharge). It occurs in much the same way
as was described in Section 1.5 with the simplified model. If the gap is large
enough, a more or less homogeneous plasma forms at the gap center and is
maintained by a more or less uniform RF field. In reality, the field amplitude
and the rate of electron production, as well as the excitation rate of atoms, have
higher values at the boundary between the plasma and the space charge sheaths.
This is shown by the luminosity distributions in Figures 1.11(α) and 1.12. The
sheaths practically conduct no current, and the conduction current through the
plasma is closed on the electrodes exclusively by the displacement currents. The
electrodes are entirely indifferent to the electric process and behave as impassively
as dielectrics. Naturally, if the electrodes are insulated from the plasma with a
dielectric, the state of the plasma in the gap at same p and L is actually the same
as with naked electrodes. But of course, the electrode voltages are higher and
increase with the dielectric thickness, since some of the voltage falls on them.

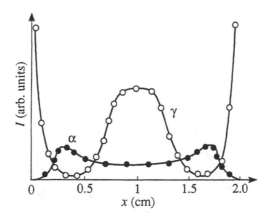

FIGURE 1.12
Glow intensity distribution in the gap of α- and γ-discharges with air, $p = 10$ Torr, $f = 13.56$ MHz, and $L = 2$ cm between brass electrodes. rms voltage is close to 300 V in both modes.

The space charge sheaths (dark zones near the solid boundaries of the gap) in the α-discharge are quite thick, a few millimeters in air. The fact that the sheaths are very poor conductors was supported experimentally by passing a dc current or a low-frequency current (50 Hz) through the discharge [1.9]. The electrical resistance of the sheaths in an α-discharge was found to be very large for steady currents. Of course, there is some ionic current flowing from the sheath to the electrode, but it is hundreds of times less than the discharge current. Therefore, at the moment the plasma 'touches' the electrode, a much stronger electron current passes momentarily to the electrode, because on the average the amounts of positive and negative charges falling on the electrode over a cycle are equal in a steady discharge.

The current variation within which the α-discharge is operative in the normal mode (at normal current density) is not large, and this is clear from the CVCs in Figure 1.13. Under some conditions, for example in the case described by curve 7 in this figure or in the experiments [1.10], no normal mode is observed in α-discharges: immediately after initiation, the discharge spreads all over the electrode. This, in particular, happens when the generator fails to produce a very weak current at a given discharge load and the electrodes are small. Then even a weak current is capable to fill them up. (This also depends on the pressure and the gap, but we will discuss this issue in Chapter 2.) If an α-discharge operates in the normal mode under experimental conditions (with a given kind of gas, p, L, and electrode area) then, as the current rises, the discharge expands and eventually embraces all the electrode. With further current increase, the current density and the electrode voltage grow (Figure 1.13).

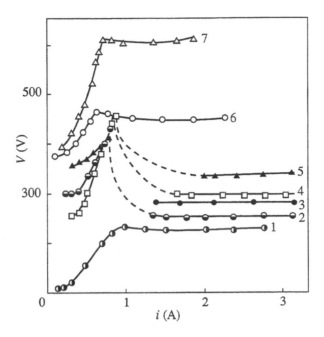

FIGURE 1.13
CVC rms values of RF discharge at 13.6 MHz: 1, helium, $p = 30$ Torr, $L = 0.9$ cm; 2 air,
$p = 30$ Torr, $L = 0.9$ cm; 3 air, $p = 30$ Torr, $L = 3$ cm; 4, CO_2, $p = 30$ Torr,
$L = 0.9$ cm; 5, CO_2, $p = 15$ Torr, $L = 3$ cm; 6, air, $p = 7.5$ Torr, $L = 1$ cm,
glass-coated electrodes; 7, air, $p = 7.5$ Torr, $L = 1$ cm, Teflon-coated electrodes [1.9].

1.8.2   Ignition of a γ-discharge

If the discharge current and voltage continue to grow, there comes a moment when
their values become such that an abrupt restructuring of the discharge occurs. It
contracts so that the current density at the electrodes rises by an order, the voltage
somewhat decreases and the discharge current increases.[4] The discharge column
changes its shape radically at each electrode, where new sheaths similar to cathode
sheaths of a dc glow discharge appear. One can see well-defined bright regions
of 'negative glow' and the 'Faraday dark space.' The positive column at the
gap center is much brighter than the α-discharge [Figure 1.11(γ) and γ-curve

---

[4]The current and voltage values that set in after the column transformation lie in the CVCs like in
Figure 1.13; they are also determined by the characteristics of the generator and external circuit. Indeed,
the values of i and V satisfy not only the discharge CVC but the electrical equation for the circuit.
This can be illustrated with reference to the initiation and transformation of a dc discharge, when i
and V are determined by the interception points of the CVC and so-called load curve $\mathcal{E} = V + i\Omega_R$,
where $\mathcal{E}$ is the emf of the power source and Ω_R is the circuit resistance [1.1]. The situation with the
RF discharge is very similar to this one, but the equation of the circuit may be more complex primarily
owing to a greater complexity of the process in an RF generator, whose emf may depend on the load.

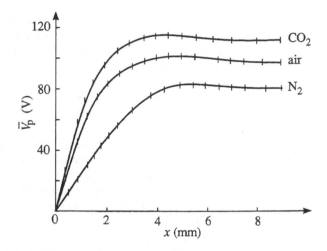

FIGURE 1.14
Measured constant potential distributions in the left-hand electrode vicinity (with a symmetrical right-hand side) in α-discharges; $f = 13.56$ MHz, $p = 7.5$ Torr, $L = 2$ cm, rms value of $V = 305$ V.

in Figure 1.12]. In air at $p = 15$ Torr, the total thickness of the 'near-cathode' regions, i.e., of the negative glow and Faraday space, is 1.2 cm. However, the thickness of the electrode space charge sheath $d_\gamma \approx 0.03$ cm is by an order less than in the α-discharge, $d_\alpha = 0.4$ cm. These values have been obtained by probe measurement of the constant potential distribution as a function of the distance x from the electrode, $\overline{V}(x)$. Typical distributions are shown in Figure 1.14: the time average (constant) potential grows from zero at the electrodes to a constant value in the homogeneous plasma region. The regions of increasing $\overline{V}(x)$ correspond to the space charge sheaths at the electrodes. Indeed, $\mathrm{d}^2\overline{V}/\mathrm{d}x^2 = -4\pi(n_+ - \bar{n}_e)$, where n_+ is the ion density and $\bar{n}_e$ is the cycle average electron density which is lower than n_+ because the ions are 'exposed' for some time in each cycle. In the plasma bulk, $n_+ \approx \bar{n}_e$ and $\overline{V} \approx$ const.

This form of RF discharge has much in common with a dc glow discharge, and this similarity is by no means accidental: its both electrodes behave like the glow discharge cathode during most of the cycle. A positive space charge sheath appears around the 'cathode.' It has an ion density and electric field much larger than in the α-discharge. When the ions reach the electrode, they produce electron emission, and the emitted electrons multiply in the 'cathode' sheath like in the glow discharge. Because of this effect, Levitsky called this form of discharge as γ-discharge (it is sometimes referred to as high-current discharge). The letter γ symbolizes the contribution of secondary emission (γ-processes). Direct measurements [1.9] have shown that the γ-discharge sheaths possess conductivity, and the sheath resistance to direct or low-frequency current is two orders of magnitude less than in the α-discharge. The sheath thickness in the γ-discharge, d_γ, is close to the cathode sheath thickness in the normal glow discharge, d_n. The constant voltage fall in the sheath, which is equal to the constant plasma potential relative to the electrode $\overline{V}_p$, is close to the normal cathode fall of the glow discharge V_n. As in the glow discharge, $d_\gamma \approx d_n \sim p^{-1}$, $\overline{V}_p \approx V_n$ is independent of p and both quantities are frequency-independent.

The γ-mode is also characterized by a normal current density $j_{n\gamma}$, but its value is much larger than in the α-discharge and may be substantially larger than in the glow discharge. In air at $p = 30$ Torr, the root-mean-square value of $j_{n\gamma} \approx 400$ mA cm^{-2}, which exceeds by a factor of 50 $j_{n\alpha} \approx 8$ mA cm^{-2} and is twice as large as j_n of the glow discharge. The ion current reaching the electrode in the γ-discharge has a density close to the normal current density of the glow discharge, since the processes occurring in the γ-sheath are similar to those in the cathode sheath. The normal current density in the γ-discharge is, however, larger due to the displacement current which is added to the ion current in the sheath. The relative contribution of the displacement current increases as the pressure decreases, but this relationship will be discussed in the next chapter.

With increasing current in the γ-discharge, the current spots on the electrodes grow, while the current density and voltage at the electrodes remain constant (the latter fact is well illustrated with the CVCs in Figure 1.13). When the electrode is filled up with the current, an abnormal regime arises, in which the current density and voltage grow with growing current. The abnormal regime does not, however, occur very often: the expanding plasma column may start contracting to a thin filament, very similar to that in the glow discharge, before all the electrode surface is covered with the current.

1.8.3   The α–γ transition as a result of the α-sheath breakdown

This is exactly what happens. The ion sheath of the α-discharge may be considered as a gas gap containing no electrons (not 'broken down') and having a variable thickness, which passes twice the values from zero to maximum $d_{\alpha\,max}$ over a cycle. As the discharge current density rises, the ion density in the sheath increases, which, according to equation (1.41), leads to a larger voltage amplitude in the sheath, $V_{s\,max}$. This voltage amplitude and the cycle average voltage in the sheath $\overline{V}_p$ eventually reach values corresponding to the breakdown threshold in the gas gap nearly equal to the average thickness d_α. The values of $\overline{V}_p$ and pd_α approximately correspond to the point $V_t(pd)$ in the Paschen curve for the breakdown threshold (Figure 1.15). The condition for breakdown is approximately the cycle average Townsend condition for self-sustained charge current in the sheath. The current is generally self-sustained by ion-electron emission from the electrode followed by charge multiplication during the electron flight through the sheath (Section 2.4).

Following breakdown of the gap-sheath boundary, an ionization wave travels

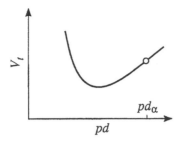

FIGURE 1.15
The Paschen curve for the breakdown voltage of a gap between plane electrodes and
α-sheath breakdown.

towards the electrode, leaving behind a plasma and reducing the initial sheath
thickness d_α to d_γ, a value corresponding to optimal conditions for charge multiplication through secondary emission. These conditions correspond to minimum
average sheath voltage. This process is quite similar to the formation of a normal
glow discharge with a characteristic cathode fall following the gap breakdown.
On initiation of a glow discharge, the sheath reduces to a thickness d_n, and the
rest of the gas space, except for a thin anode sheath, is filled by electrically neutral plasma. In a similar way, the γ-discharge is formed from the α-discharge at
moderate pressures. The γ-discharge has two 'cathodes,' and the plasma filling
the middle of the electrode spacing acts as the 'anode.'

This process is characteristic of moderate pressures, when the value of pd_α
corresponds to the right-hand branch of the Paschen curve. At low pressures,
when pd_α corresponds to the left-hand branch of this curve, the formation of a
'normal cathode sheath' would require expansion rather than contraction of the
sheath. But this, as we will see in Chapter 2, is impossible. For this reason,
the $\alpha-\gamma$ transition in low pressure discharges occurs smoothly, without abrupt
changes in the sheath thickness or in other parameters. The $\alpha-\gamma$ transition as
a process of breakdown of the α-discharge sheaths will be discussed in detail in
Chapter 2; the references will be given there.

1.8.4 Moderate and low pressures

RF discharges in the two pressure ranges are fairly different. The boundary
between these ranges is, of course, quite rough and depends on a particular feature
in which the distinction is drawn as well as on specific parameters, such as
size, geometry, gas and frequency. In general, this boundary may be drawn at
$p \sim 1$ Torr. Here are some of the obvious distinctions.

(i) At moderate pressures the α-discharge exhibits the effect of normal current
density, while at low pressures it does not. This means that in the first case the
discharge does not occupy all the area of a large electrode, but in the second case

it does. If the plane-parallel electrodes differ in area, in the first case this does not affect the size of the current spots on each electrode, because the discharge is inevitably symmetrical relative to the mid-plane between the electrodes. At low pressures, however, the discharge tends to occupy the whole electrode area and is asymmetric in an asymmetric system.

(ii) In a moderate-pressure discharge, the α–γ transition occurs abruptly, changing radically the whole discharge structure. The electrode current density and the CVC change abruptly, too. A low-pressure discharge does not exhibit abrupt changes, and only the CVC continuous curve shows a bend.

(iii) In a moderate-pressure γ-discharge, the bulk plasma possesses a fairly high electron temperature close to that in the α-discharge. At low pressures the electron temperature in the γ-discharge plasma is extremely low—much lower than in the α-discharge.

The differences we have listed are naturally associated with specific processes involved in each type of discharge. Because of these differences and in view of a systematized presentation of material in this book, we think it more convenient to analyze moderate-pressure and low-pressure discharges separately. Moderate-pressure RF discharges will be considered in Chapter 2. The abrupt changes in discharge behavior on transition to low pressures will be explained. Low-pressure RF discharges will be discussed in Chapter 3.

2

Moderate-Pressure RF Discharges

2.1 Space charge sheaths

We have pointed out above that space charge sheaths arise at the electrodes or at their dielectric coatings. The sheaths are produced by plasma electron oscillations relative to slow ions which, in contrast to the electrons, always fill up the space between the electrodes. In the simplified discharge model (Section 1.5), the ion density n_+ is taken to be constant throughout the gap; the electron gas or the plasma boundary performs strictly harmonic oscillations. Now we will discard our earlier assumption of $n_+ = $ const and turn to a more detailed discussion of the sheath structure in α- and γ-discharges.

2.1.1 Formulation of a simplified problem for discharge simulation

We will consider a discharge between plane electrodes in a one-dimensional approximation without taking account of charge diffusion to the electrodes. The gas is assumed to be incapable of electron attachment, i.e., there are no negative ions in it. The distribution and time variation of the charge densities n_e and n_+ and of the field E in the gap are described by a set of continuity equations and electrostatic equation

$$\frac{\partial n_e}{\partial t} + \frac{\partial \Gamma_e}{\partial x} = \nu_i n_e - \beta n_e n_+ \qquad \Gamma_e = -n_e \mu_e E$$

$$\frac{\partial n_e}{\partial t} + \frac{\partial \Gamma_+}{\partial x} = \nu_i n_e - \beta n_e n_+ \qquad \Gamma_+ = n_+ \mu_+ E \qquad (2.1)$$

$$\frac{\partial E}{\partial x} = 4\pi e(n_+ - n_e)$$

where Γ_e and Γ_+ are the densities of drift charge flows. Having in mind not too high frequencies and not very low pressures, when $\nu_u > \omega$ in molecular gases (see Section 1.4), we will take the ionization frequency ν_i to be dependent on the instantaneous field value. It may be conveniently expressed through the Townsend ionization coefficient α in accordance with the relations $\nu_i n_e = \alpha |\Gamma_e|$ and $\nu_i = \alpha v_d$, where $v_d = \mu_e |E|$ is the instantaneous drift velocity of the electrons. Since we are primarily considering the electrode sheaths rather than the positive plasma column (for the latter, see Section 2.3), we will not complicate the set of equations (2.1) with the gas heating due to the Joule heat release or with stepwise ionization processes. In this approximation, ν_i and α are the known functions of $|E|$ and the set of equations (2.1) is closed.

Let us formulate the boundary conditions at the surface of an electrode or a dielectric coating. When the field is directed to a body, it may induce secondary electron emission. When a positive ion hits a metallic surface, it is neutralized by a metal electron and, with a small probability γ, knocks out another electron. A similar process occurs on a dielectric surface with the only difference being that the ion interacts with the gas electrons attached to the surface during the first oscillations rather than with the body electrons. In any case, the boundary condition states that $\Gamma_e = -\gamma\Gamma_+$. When the field is directed away from the body, there is no ion flux near it, since the body ions do not escape into the gas. In this situation $\Gamma_+ = 0$. We will seek a steady, strictly periodic solution of equations (2.1) with a given harmonic electrode voltage or discharge current density. (One of these quantities is an independent variable, while the other, the CVC, is found by solving the equations.) A steady solution can be found by a rough numerical calculation of the unsteady stage beginning with an appropriate initial state of the gap plasma. A similar set of equations was used in [2.1] to calculate an ac discharge at a frequency $f \sim 10$ kHz and in [2.2] to calculate a non-self-sustained RF discharge with the gas ionization induced by an external source (the α-mode). Here we are discussing a self-sustained RF discharge in the α- and γ-modes.

Numerical integration of equations (2.1) shows that a uniform electrically neutral positive plasma column with a harmonic field $E = E_a \cos \omega t$ is formed in the gap middle if the gap is sufficiently long. The ionization–recombination equilibrium over a cycle is maintained in the plasma with $\langle \alpha\mu_e E \rangle = \beta n$ where $n = n_e = n_+$. Since $\alpha = \alpha(E)$, this expression relates E_a to n and, through formula (1.38), to the density amplitude of discharge current j_a. When analyzing the sheath structure, it is more convenient to deal with given current density $j = j_a \cos \omega t$ rather than electrode voltage. This allows us to limit the integration range to the vicinity of one (left-hand) electrode only. The quantities n and E in the positive column far from the electrode are then known, and this is a new boundary condition for the right-hand boundary, i.e., for the plasma.

The results of the calculation [2.3] for nitrogen at $f = 13.56$ MHz and $p = 15$ Torr are presented below. The following set of parameters often used for nitrogen was employed (here p is measured in Torr)

$$\alpha = 12\, p \exp\left(-342 p/E\right) \text{cm}^{-1} \quad \text{at} \quad E/p > 100 \text{ V cm}^{-1} \text{Torr}^{-1}$$

$$\alpha = 2.4\, p \exp\left(-155 p/E\right) \text{cm}^{-1} \quad \text{at} \quad E/p < 100 \text{ V cm}^{-1} \text{Torr}^{-1}$$

$$\mu_e = 4.4 \times 10^5/p \qquad \mu_+ = 10^3/p \text{ cm}^2 \text{ V}^{-1} \text{ s}^{-1}$$
$$\beta = 2 \times 10^{-7} \text{ cm}^3 \text{ s}^{-1} \quad \gamma = 0.01$$

The values of $n_e = n_+ = n$ and E_a corresponding to a uniform positive column were fully specified as an initial condition. A strictly periodic solution was reached after $10^2 - 10^3$ cycles.

2.1.2 The sheath structure and current balance in steady α- and γ-discharges

The distribution of n_+, n_e and E characteristic of the α- and γ-modes are automatically found as a function of the given values of n or j_a by solving the above set of equations. The structural alteration, as the $\alpha-\gamma$ transition, is abrupt and occurs at $n_{tr} = 8.7 \times 10^9$ cm^{-3} and $j_{atr} = 9.2$ mA cm^{-2}.[1] In the experiment in air at $p = 30$ Torr, the transition was registered at $j_{atr} = 17$ mA cm^{-2} [2.4]. But at $p = 15$ Torr, as in the calculation [2.5], j_{atr} was 1.5 times less, i.e., 11.3 mA cm^{-2}; in other words, the agreement between the calculated and experimental values is quite satisfactory. Below (Section 2.4), the critical transition parameters n_{tr} and $j_{a,tr}$ will be evaluated analytically in terms of the theoretical concepts of transition as an α-sheath breakdown. We will now continue the discussion of the results of numerical simulation. This sequence of presentation is quite consistent with conventional interpretation of physical phenomena, in which theory follows a description of experimental facts. As for electrode sheaths, detailed experimental measurements are actually lacking and mathematical simulation should be regarded as a numerical experiment. As in a real experiment, the mechanisms governing a physical phenomenon are implicitly present in the final calculations but can be understood only from their analysis in terms of adequate theoretical concepts.

Figure 2.1 shows the results of the calculation of a typical α-mode with $j_a = 4$ mA cm^{-2} and with $n = 3.7 \times 10^9$ cm^{-3} and $E_a/p = 14$ V cm^{-1} Torr^{-1} in the plasma. The ion density in the sheath decreases by about a factor of 4 from the positive column towards the electrode. It, no doubt, pulsates in time but so slightly that this could not be represented on the given scale. The value of n_+ decreases towards the electrode because of a regular, though weak, ion flux into the electrode under the influence of the mean dc field $\bar{E}$ in this direction. The ion flux $\Gamma_+ = n_+\mu_+\bar{E}$ grows by a factor of 7 from the point $x = 0.3$ cm near the boundary with the positive column (PC) to the point $x = 0$ at the electrode. This

---

[1]The respective value of $(E_a/p)_{tr}$ for the plasma, 15 V cm^{-1} Torr^{-1}, is not quite representative, because the real CVC of the positive column may significantly differ from that in the calculation due to the low reliability of the Townsend coefficients α at such low E/p values, to the neglected effect of gas heating and, possibly, stepwise ionization (see below). However, these factors do not affect the sheath structure, so their neglect seems to be justified in this case. Still less representative is the measured electrode voltage at which the transition occurs. It depends on the gap and the dielectric thickness, if the electrodes are coated, as well as on the gas cooling procedure.

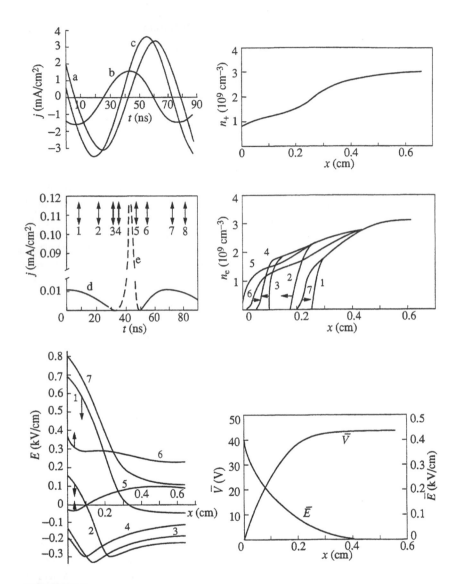

FIGURE 2.1
The left-hand electrode vicinity in the α-mode: (a) conduction plasma current,
(b) displacement plasma current, (c) displacement current on the electrode, and (d) ion and
(e) electron currents to the electrode. The current and field toward the electrode are taken
to be positive and the initial time arbitrary. Vertical arrows with digits indicate the times of
the given distributions of n_e and E; the respective curves have the same digits. Arrows for
n_e indicate the direction of plasma boundary movement. Upward arrows for E indicate
field rise. There are also cycle average (constant) field $\bar{E}$ and potential $\bar{V}$ distributions.

results from ionization occurring at each sheath point when there are available electrons. The ion current to the electrode and the 'cathode' stage last over $3/4$ of a cycle. The ion current is very weak and makes up $10^{-3}-10^{-4}$ of the total electrode current which, therefore, is almost exclusively displacement current. Since the charge flow to the electrode is in average absent in a strictly periodic process, the amounts of positive and negative charges are equal. The electron current flows to the electrode during the short 'anode' stage when the plasma is pressed against the electrode. The average electron current is by a factor of 4–5 larger than the average ionic current at the cathode stage (the factor 4–5 is the duration ratio of the stages). The plasma boundary motion is not strictly periodic, because the electron gas flows across a region of space-variable ion density but in such a fashion that n_e coincides with the local density $n_+(x)$ at any point and any moment of time.

The maximum sheath thickness of 0.3 cm agrees well with the value of 0.35 cm measured in air [2.5] and is twice the value calculated in the model with $n_+ = $ const and $2A = 2\mu_e E_a/\omega = 0.14$ cm. The constant plasma potential $\overline{V}_p = 45$ V is 1.7 times that calculated from equation (1.53) in the same model. These results appear to be theoretically hopeful. They indicate that the simplified discharge model with $n_+ = $ const, applicable to analytical or semi-analytical interpretations of many effects, is quite acceptable, providing a reasonable accuracy.

Figure 2.2 shows results, analogous to those in Figure 2.1, for a typical γ-mode with $j_a = 0.12$ A cm^{-2} and with $n = 9 \times 10^{10}$ cm^{-3} and $E_a/p = 18.6$ V cm^{-1} Torr^{-1} in the positive column. The last value should not be regarded as being very important for the reason explained in the previous footnote. Such current densities were measured in a γ-discharge with air at 15 Torr and found to be 0.16 A cm^{-2} [2.5] and 0.11 A cm^{-2} [2.6]. The ion current to the electrode in the γ-mode is considerable but less than the displacement current. It is 3–4 times less than the total current, which agrees with measurements [2.6] (see Section 2.5). At the anode stage, all current flowing to the electrode is electron current. The space charge sheath has a smaller thickness than in the α-mode with a maximum of 0.05 cm. Unlike the α-sheath, in which n_+ monotonically decreases from the positive column towards the electrode, the ion density in the γ-sheath has a maximum and is generally higher by several-fold than the ion density in the column. The average sheath field and the positive plasma potential $\overline{V}_p = 270$ V are much higher than in the α-discharge illustrated in Figure 2.1 in spite of the smaller sheath thickness. The reason is that n_+, i.e., the space charge density, is two orders of magnitude greater. Oscillations of the sheath ion density are, in contrast with the α-discharge, large enough to be depicted on the plot scale of Figure 2.2.

The discharge structure for the case of coated electrodes is the same as for naked electrodes. Figure 2.3 shows the results of calculation of a γ-discharge with the interelectrode distance $L = 2$ cm and the coating thickness $l = 1$ mm (the dielectric permittivity $\varepsilon = 5$). Unlike Figures 2.1 and 2.2 showing the calculations for the vicinity of one electrode at a given current, Figure 2.3 presents

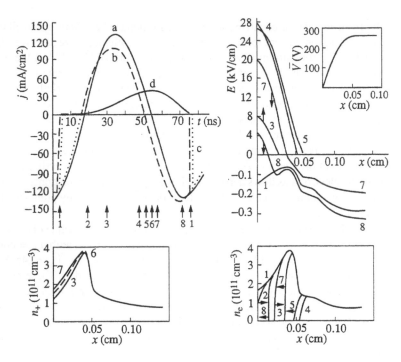

FIGURE 2.2
The left-hand electrode vicinity in the γ-mode: (a) conduction plasma current,
(b) displacement electrode current, and (c) electron and (d) ion currents to the electrode.
The rest is as in Figure 2.1, except for different scales for $E > 0$ and $E < 0$.

calculations for the whole gap at a given electrode voltage, $V_a = 1.3$ kV. For
the dielectrics, γ was taken to be 0.01. One can see that the distributions of
the values near the two electrodes are phase-shifted relative to one another. The
maximum discharge current density was found to be $j_{max} = 55$ mA cm^{-2} and
the maximum gap voltage, except for the voltage fall at the dielectrics, $V_{g\,max} =$
0.7 kV. Strictly, these quantities cannot be interpreted as 'amplitudes,' because
when the electrode voltage is harmonic, the current, and hence the voltage falls in
the dielectrics, are not quite harmonic. This is due to the anharmonic oscillations
of the sheath boundaries and, therefore, of the total voltage fall in the sheaths. In
terms of electrical engineering, current anharmonicity results from the nonlinearity
of sheath resistance.

Figure 2.3 also demonstrates the evolution of the field at the dielectric boundaries $E(0, t)$ and $E(L, t)$, and of the charge densities $q_1(t)$ and $q_2(t)$ on the
left-hand and right-hand dielectric coatings, respectively. The surface charges are
found by solving the equation $\dot{q}_{1,2} = e(\Gamma_+ - \Gamma_e)_{1,2}$. The integration constant
is evaluated from the equalities $4\pi\bar{q}_1 = 4\pi\bar{q}_2 = \bar{E}(0) = -\bar{E}(L)$ for the cycle-

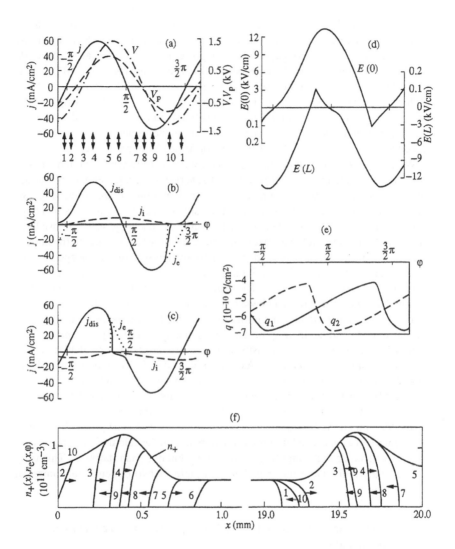

FIGURE 2.3
Calculations for the γ-discharge between coated electrodes with a given wave-shaped voltage: (a) electrode voltage V, plasma voltage V_p, discharge current density j; (b) displacement currents j_{dis}, ion current j_i, electron current j_e on the left-hand electrode; (c) the same for the right-hand electrode; (d) fields at the left-hand $E(0)$ and the right-hand $E(L)$ dielectrics; (e) deposited charge densities on the left-hand q_1 and on the right-hand q_2 dielectrics; (f) evolution of the $n_e(x,t)$ distribution in the gap. Digits at the curves correspond to the time arrows in (a); arrows indicate the movement of the plasma boundary. Upper curve above the $n_e(x,t)$ is the distribution $n_+(x)$ slightly varying in time.

average values. The latter two equalities follow from the condition of electrical neutrality of the gas–dielectric system. There are always excess electrons on the dielectric surface, and the gas is charged positively, as in the naked electrode discharge. The least mean distance between attached electrons, 6×10^{-5} cm, is much larger than the atomic size, so the electrons do not densely populate the dielectric surface. The field in the dielectric $E_{d1} = \varepsilon^{-1}[E(0) - 4\pi q_1]$, being cycle average, is zero, as in a uniform positive column, therefore the constant plasma potential relative to a coated electrode is the same as in the respective naked electrode discharge. This has been confirmed by experiment [2.4].

We would like to conclude this discussion with the following comment. It was pointed out in Section 1.8 that α- and γ-discharges are characterized by normal current density. In the one-dimensional approximation we are discussing, we cannot naturally obtain a normal current density as in experiment or as in a two-dimensional approximation. For an illustration of the discharge structure, we have taken discharges in which the current densities are more or less close to the normal density, experimental or predicted. Qualitatively, the structure is independent of a particular value of j_a or V_a, so Figures 2.1–2.3 give an adequate picture of the structure of α- and γ-discharges. The same refers to the process of establishing a steady discharge structure, which we will discuss in the next section.

Two-dimensional mathematical modeling of the RF discharge with equations similar to (2.1) and neglecting diffusion was attempted in [2.7]. A tendency for establishing a normal current density was revealed, although the calculation was not brought up to a steady state. As was explained in [2.8] on two-dimensional modeling of a dc glow discharge the normal current density is, in fact, stabilized by charge diffusion transverse to the current. In [2.7] and glow discharge calculations made earlier by the same workers, the role of diffusion ignored in (2.1) was played by the 'computational' diffusion that arises inevitably when one proceeds to finite-difference equations. The latter diffusion depends on the calculation step and can, for this reason, describe an effect qualitatively but not quantitatively. This fully refers to the RF discharge and should not be ignored in two-dimensional numerical simulations.

2.1.3   Formation of the α- and γ-discharge patterns from an arbitrary initial state

The jump-like pattern of the α–γ transition at moderate pressures is clearly supported by numerical simulation of the α- and γ-structure formation near the electrode from a qualitatively identical uniform plasma state [2.9]. There is a critical current density, below which the initially uniform plasma evolves into the α-pattern and above this value into the γ-pattern. These situations were illustrated in Figures 2.1 and 2.2.

Figures 2.4 and 2.5 show the calculations made for the same conditions: nitrogen, $f = 13.56$ MHz, $p = 15$ Torr. Initially uniform plasma had a given density $n = n_+ = n_e$ and a field $E = E_a \cos \omega t$ providing ionization–recombination

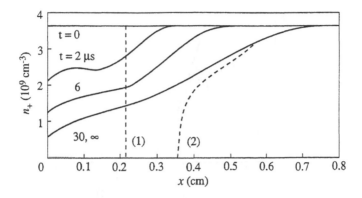

FIGURE 2.4
Formation of the α-structure at the left-hand electrode as a uniform RF field is switched on in half-gap with uniform nitrogen plasma, $p = 15$ Torr, $f = 13.6$ MHz, $E_a/p = 14.1$ V cm^{-1} Torr^{-1}, current amplitude $j_a = 3.55$ mA cm^{-2}. Vertical dashed line (1) is for the plasma boundary at maximum distance from the electrode after the primary shift. Dashed line (2) is for n_e at maximum shift in a steady mode.

equilibrium, which means that the current density was actually given

$$j = en\mu_e E_a \cos \omega t - \frac{\omega}{4\pi} E_a \sin \omega t = j_a \cos(\omega t + \Delta\phi) \qquad (2.2)$$

with an amplitude j_a defined by equation (1.38).

Constant plasma density and harmonic field amplitude were maintained at the right-hand boundary set up in the positive column bulk. By virtue of the conservation law, the current density further remained constant in space and harmonic in time.

The formation of a discharge pattern at the left-hand electrode depends on the value of one of the three interrelated quantities, n, E_a and j_a, although n is physically crucial. During the first few cycles, while the ion density remains constant, the electrons at the electrode oscillate with the same amplitude $A = \mu_e E_a/\omega$ as in the column and escape from the sheath into the electrode. In the α-discharge (Figure 2.4), the sheath becomes inevitably depleted of ions which escape into the electrode under the influence of an immediately arising dc field. This process goes on until the ion loss is balanced by ionization at the moments when the ion sheath is flooded by electrons. Then it ceases, leaving a weak ionic current into the electrode. The ion density at the electrode decreases several-fold as compared to the plasma density, and the sheath thickness becomes nearly an order of magnitude larger than A. The process of stabilization lasts for several hundreds of cycles ($f^{-1} = 74$ ns), almost as long as is necessary for the ions to drift through the sheath. As the current rises, the sheath thickness increases. For instance, at $j = 1.65$ mA cm^{-2}, the maximum thickness is $d_{max} \approx 0.25$ cm and at $j \approx 3.55$ mA cm^{-2}, $d_{max} = 0.35$ cm.

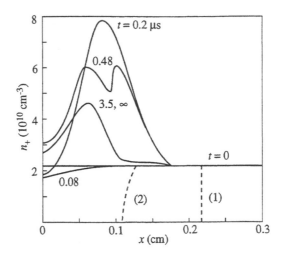

FIGURE 2.5
Formation of the γ-structure, $E_a/p = 16.3$ V cm^{-1} Torr^{-1}, $j_a = 24.4$ mA cm^{-2}. The rest is as in Figure 2.4.

The point, which separates the directions of evolution of the α- and γ-patterns, corresponds approximately to $j_{a\,tr} \approx 10$ mA cm^{-2}. The γ-discharge starts exactly as the α-discharge, but a sheath breakdown occurs right in the second cycle, when the plasma shifts far away from the electrode. Electron avalanche develops in the sheath owing to the emission from the electrode. An abrupt rise of the charge density at the end of the strong field region and at the end of the avalanche transforms this region into a plasma, which joins the initial plasma space. A continuous process of this kind is equivalent to the propagation of an ionization wave that moves towards an electrode, leaving behind a plasma and a thinner space charge sheath. This goes on until the sheath acquires a thickness and other parameters corresponding to the cathode sheath of a dc glow discharge. The process of sheath formation has much in common with the formation of a cathode sheath in a glow discharge after the gap breakdown. Since the thicknesses are smaller, this process occurs more rapidly than in the α-discharge. In the latter case, a steady d_{max} is larger than the amplitude of initial electron displacement $2A$, while in the γ-discharge, $d_{max} < 2A$. In contrast to the α-discharge, the thickness of the γ-cathode sheath decreases with growing current density, as in the glow discharge.

Our conception of the moderate pressure $\alpha-\gamma$ transition as being due to breakdown of the α-sheaths is not the only model currently proposed by researchers. For example, some workers [2.10, 2.11] believe that the α- and γ-discharges differ in which current (displacement or conduction current) dominates and that the transition corresponds to the equality of both components. Other workers think that the $\alpha-\gamma$ transition occurs when the rate of gas ionization by fast electron fluxes, formed in the sheaths due to electrode emission, becomes higher than the rate of ionization by oscillating plasma electrons. Clearly, the last two mechanisms should lead to a smooth transition pattern, whereas the first one makes the transition abrupt, producing a rapid restructuring of the discharge and a radical charge in its external appearance. The jump-like transition conception is supported by a moderate pressure experiment (Figure 1.12) and by the numerical experiment discussed above.

2.1.4 Analytical and semi-analytical sheath models based on the cycle average ion density

Within the approximations of equations (2.1), a numerical solution can provide a full description of the longitudinal structure of both RF discharge modes and of the transition between them. However, a search for a steady, strictly periodic solution consumes much computer time and is not always reasonable or practical. Therefore, of interest are less rigorous theoretical models of electrode sheaths, which ignore ion oscillations and assume the ion density n_+ to be constant over a cycle. It is considered that the ions slowly drift in an average constant field and diffuse due to spatial density gradients.

Such an 'averaged' sheath model was, probably, first suggested in [2.14]. It may be called averaged only in the sense that it deals with a cycle-average ion density. This model did not take into account ionization in the sheath, therefore, it analyzed the α-discharge. Ion mobility in a collisional sheath was assumed to be determined by resonance charge exchange. No allowance was made for the diffusion. A flux of 'thermal' ions flows from the plasma and enters the sheath at a Bohm velocity (see Chapter 3). Because of the absence of sources, the flux persists throughout the sheath, thus simplifying the integration of equations. Computations yielded a distribution pattern of $n_+(x)$ qualitatively similar to that in Figure 2.1. A more comprehensive and self-consistent model of collisional and collisionless sheaths based on similar premises was developed later [2.15] (see Chapter 3).

Production and loss of charges changes the flux of ions moving in an average field. Such a situation was analyzed in [2.16] for a non-self-maintained discharge with an external uniform ionization source. As previously, the flux was assumed to be of drift nature and the diffusion was neglected. The authors of [2.17] took account of ambipolar charge diffusion and ambipolar drift (for the latter, see [2.18]). Since ionization by plasma electrons and electron multiplication through secondary emission were taken into consideration, this model provided the distributions for both the α- and γ-patterns. These distributions are consistent with the computational results [2.3] presented in Figures 2.1 and 2.2. This model also describes the transition between the α- and γ-discharge modes.

2.2   Experimental current-voltage characteristics

Experimental data on the current-voltage characteristics (CVC) are not numerous, but together with available theoretical models they are sufficient to provide an understanding of discharge behavior and its basic features. Some of the measurements were discussed in Section 1.8. It was noted that moderate-pressure α-discharges exhibit the effect of normal current density. When the current strength is not too high, a discharge does not cover the whole electrode; but as the current rises, the discharge area grows in such a way that the current density on the electrode remains constant. The value of normal current density $j_{n\alpha}$ varies with the gas, pressure and, which is curious, with the interelectrode distance L. These relationships are illustrated in Figure 2.6 [2.19]. The measurements were made using an installation described in Section 1.8.1. At the same values of L, current densities of an order, say $j_{n\alpha} \approx 5-10$ mA cm^{-2}, were obtained in helium at pressures an order of magnitude higher than in a molecular gas, air.

The CVC measurements in RF nitrogen discharges made at several frequencies in the range $f = 15-60$ MHz and over the range of $p = 25-55$ Torr [2.20] may give an idea of respective relationships. A discharge was excited in the gap between water-cooled aluminum disks of radius 0.85 cm and with rounded edges. The radius of the plane part of the disks was $R = 0.7$ cm. The gap was varied, but most of the measurements were made at $L = 0.59$ cm. Right after the ignition, the α-discharge spread practically all over the plane electrode area $S = 1.5$ cm^2 and persisted until the $\alpha-\gamma$ transition when it constricted to an area of 0.1 cm^2

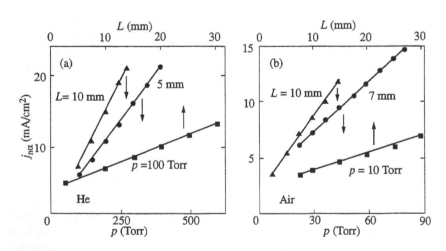

FIGURE 2.6
Plots for the normal α-current density on the electrodes versus pressure and gap size in (a) helium and (b) air at $f = 13.56$ MHz.

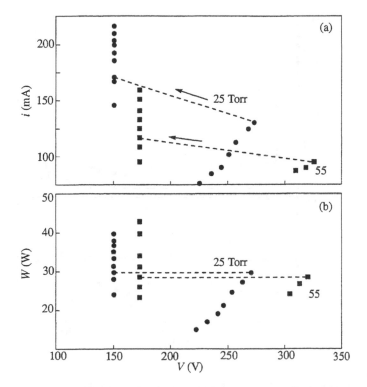

FIGURE 2.7
Root-mean-square values of the (a) current–voltage and (b) power–voltage characteristics
for a nitrogen discharge at $L = 0.59$ cm, $f = 29.25$ MHz for two pressure values.
Dashed lines with arrows are the $\alpha - \gamma$ transition due to increasing current and voltage. As
the current in the γ-mode decreases, the current, prior to the reversal $\gamma - \alpha$, drops below
the end points of the dashed lines, producing a hysteresis [2.20].

corresponding to a normal γ-discharge at transition current.

The measurements are shown in Figures 2.7 and 2.8. Qualitatively, they fit well
the data in Figure 1.13 [2.4] with the only difference that no normal α-mode was
observed in the experiments we are discussing. We consider this as being due to
the small electrode area, 1.5 cm^2, in contrast to the 80 cm^2 area in [2.4]. At the
lowest currents of 50–70 mA necessary to start an α-discharge, current densities
of $30-50$ mA cm^{-2} were already above normal; in other words, the α-discharge
operated in an abnormal mode from the very beginning. Figures 2.7 and 2.8
also show powers released in the discharge. The power is known to be equal to
the product of the rms values of current and voltage and the phase angle cosine.
Transition voltage increases with increasing pressure and decreasing frequency.
Transition current slightly reduces as the pressure grows and strongly rises with

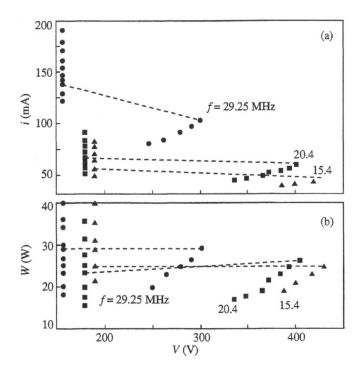

FIGURE 2.8
The same as in Figure 2.7 at $p = 35$ Torr for three frequency values.

frequency. These relationships are illustrated in the table in Section 2.4 which compares experimental and theoretical parameters.

In the $\alpha-\gamma$ transition region (Figures 2.7 and 2.8), one can observe some hysteretic effects. As the current and voltage are increased, the α-mode transforms to the γ-mode with a current strength corresponding to a constant power which does not change at the transition. If the γ-discharge current is reduced, the back-transition is delayed and occurs at a lower current and power than those achieved in the γ-discharge as a result of direct $\alpha-\gamma$ transition.

Experiments [2.4] have shown that the α-discharge cannot operate in any combination of the gap and pressure values. If L and p are too large, or, more exactly, if the product of L and p has a certain critical value $(pL)_{cr}$ varying with the gas, the α-mode becomes unsteady. In this case it either transforms to the γ-mode or dies out. At $pL > (pL)_{cr}$, one generally fails to excite an α-discharge. At $pL < (pL)_{cr}$, both modes may exist. The α-mode boundaries are shown in Figure 2.9 [2.4]. The critical values for molecular gases are essentially smaller than for atomic gases. For example, in air $(pL)_{cr} \approx 40$ Torr cm and in helium $(pL)_{cr} \approx 150$ Torr cm. In the mixtures $CO_2 + N_2 + He$ used in CO_2 lasers, the dis-

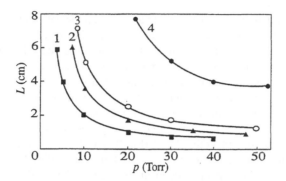

FIGURE 2.9
The boundaries of α-regions at $f = 13.56$ MHz with the γ-mode only above and on the right of the curves: nitrogen (curve 1), air (curve 2), CO_2 (curve 3), and helium (curve 4).

charge often operates in the γ-mode at L of a few centimeters, as in power lasers, and at characteristic laser pressures of tens of Torr. However, small waveguide CO_2 lasers with $L \approx 1$ mm generally operate in the α-mode. As for other gases, the α-mode in argon and xenon was found to be steady at $p \lesssim 200$ Torr, but then it constricted to a filament [2.20]. Moreover, at $p < 200$ Torr and $L = 0.3-3$ cm, with the corresponding $pL \lesssim 60-600$ Torr cm, no $\alpha-\gamma$ transition was observed, and the γ-mode could not be started.

A comparison of temperatures in α- and γ-discharges operating in a neutral gas was made in [2.20]. Measurements were taken with a mercury thermometer outside the discharge and RF field regions at a distance $r = 2$ cm from the glowing edge. At this point, the temperature was found to be 150 °C in nitrogen at $f = 30$ MHz and $p = 35$ Torr in an α-discharge of the power of 30 W and bulk power input of 33 W cm^{-3}. In a γ-discharge of the same power and the bulk power input of 423 W cm^{-3}, the temperature was 230 °C. With the thermal conductivity equation, one can calculate the temperature in the heat source, whose center is at $R + r = 2.7$ cm from the position of the thermometer [2.21]. It has turned out that the temperature in the α-discharge prior to the transition is $T = 800$ K, which will be shown to fit direct calculations (see Section 2.4.3).

2.3 CVC and normal current densities in the α-discharge (theory and numerical simulation)

2.3.1 Physical premises for normal current density and limits in p and L

The effect of normal current density for a dc glow discharge has long been known. The discharge occupies such an area of the cathode that the cathode current density has a definite, 'normal' value varying with the gas, pressure, and cathode material. As far back as the 1930s, von Engel and Steenbeck found that this effect was associated with a minimum in the CVC for the cathode sheath, $V_c(j)$. If the cathode is only partly filled with current, the experimental situation corresponds to a minimal (normal) cathode fall of the potential $V_c \approx V_{min} = V_n$, which, in turn, corresponds to the normal current density j_n. For a glow discharge, $j_n \sim p^2$, and V_n is independent of p. Interpretation of the physical reason for nature's selection of the state with $V_c \approx V_{min}$ has a long history, and the solution to the problem has been found quite recently [2.8, 2.18]. Briefly, the reason lies in the unstable states with the cathode fall $V_c > V_{min}$. The left-hand, falling branch of the $V_c(j)$ curve represents the unstable state inside the cathode layer. This was clear to von Engel and Steenbeck. The right-hand, rising branch of the curve, where all states are generally stable, unstable is the state at the sheath edge, i.e., at the boundary between the current and currentless zones of the cathode. This has not been clear until very recently. If the cathode is completely filled with current, there is no current–currentless boundary on the cathode. Since the layer on the right-hand branch is stable, it reveals itself in experiment as an abnormal glow discharge ($V_c > V_n$, $j > j_n$).

Something similar to that occurs in the α-discharge. The effect of normal current density is associated with the presence of a minimum in the CVC, but in contrast with the glow discharge, it is the CVC for the whole α-discharge [2.22]. It has been shown experimentally that the CVC for an RF positive column $V_{PC}(j)$ goes down, while that for the electrode sheath containing displacement current goes up. Indeed, $j \approx \omega V_s/4\pi d_\alpha$, where V_s is the voltage fall on the sheaths, and d_α is their thickness determined by the oscillation amplitude of the plasma electrons and is only slightly dependent on the current density. Therefore, the electrode voltage V, which is the sum of the sheath and plasma voltages, must have a minimum as a function of j.

Since E/p in the positive column has a more or less fixed value, let the falling CVC in it have the form $V_{PC} = C p L j^{-m}$, where C and $m > 0$ are constant. With account of the $\pi/2$ phase shift between V_s and V_{PC} with the respective reactive and active currents, the electrode voltage is

$$V = (V_s^2 + V_{PC}^2)^{1/2} = \left[\left(\frac{4\pi d_\alpha j}{\omega} \right)^2 + (CpL)^2 j^{-2m} \right]^{1/2} \tag{2.3}$$

The function $V(j)$ has a minimum at

$$j = j_{n\alpha} = \left(\frac{\omega C p L \sqrt{m}}{4\pi d_\alpha} \right)^{1/(m+1)} \tag{2.4}$$

The normal current density (2.4) corresponding to minimum V grows with pressure (d_α weakly depends on p) and the gap size, in full qualitative agreement with experiment. The value of $j_{n\alpha}$ also grows with frequency, which means that, other things being equal, one can achieve higher current densities at higher frequencies. This is essential for RF discharge application in laser technology.

It becomes clear, too, why the α-discharge has an upper limit in p and L. As pL increases, the values of $j_{n\alpha}$, of the field and voltage fall in the sheaths become higher. At a certain critical value of $(pL)_{cr}$, the sheath voltage of a normal α-discharge becomes so large that breakdown occurs resulting in the transition to the γ-mode. For this reason, there is no α-discharge at $pL > (pL)_{cr}$.

2.3.2 Heating and vibrational relaxation in a molecular gas

One of the reasons why the CVC for the positive column has a falling appearance is the gas heating. We will illustrate this effect with reference to nitrogen in order to be able to use the results further in theoretical treatment of the $\alpha-\gamma$ transition. It was for nitrogen that the transition parameters were measured in [2.20] at various frequencies and pressures, permitting a comparison between experiment and theory. In our estimations, we will have in mind the experimental conditions of this work described in Section 2.2.

Let us start with a simple α-discharge model in which there is a sufficiently long positive column (PC) between two plane electrodes covered with sheaths. The PC plasma has the density $n = n_e = n_+$ and the gas has the temperature T. Due to the release of Joule current heat, T is higher than the initial gas and electrode temperature T_0, which can be taken to be equal to room temperature, $T_0 = 293$ K. The ion density in the sheaths is assumed to be constant and equal to n. Stepwise and other ionization mechanisms are ignored. Note that ionization mechanisms in the PC plasma of a nitrogen glow discharge are very complicated and are still poorly understood [2.18]. In addition to the trivial electron impact ionization of unexcited molecules, there are electron impact ionization of excited metastable molecules and associative ionization. In the latter case, two highly excited molecules with a possible combination of electronic and strong vibrational excitations unite to produce a complex N_4^+ ion and a free electron. Electron-excited molecules may be produced by collisions of vibrationally excited molecules.

In discharge experiments [2.4, 2.20], the plasma usually occupies a small portion of working gas space. A certain pressure $p = NkT = p_0$ with N and T as local molecular density and gas pressure is maintained in this space. Since the pressure is spatially uniform, the heated gas density is lower than the density N_0 of the cold gas far from the discharge and possessing the same temperature $T_0 = 293$ K as the cooled electrodes. If the interelectrode distance L is smaller than the column transverse diameter $2R$ and the gas flow is not too fast, the Joule heat is largely transported by thermal conduction to the electrodes. The ionization frequency in the RF field ν_{iRF} and the Townsend ionization coefficient α are proportional, by virtue of a similarity law, to N and to a certain function E/N, or ET. Therefore, the equation of ionization–recombination equilibrium (1.37) defining the CVC of a uniform PC must be supplemented, on the gas heating, by the balance equations of heat (thermal conduction) and of constant pressure that determine T and N.

The respective set of equations has the form

$$\nu_{i\,RF}(E_a, N) = \beta n \qquad E = E_a \cos \omega t$$

$$-\frac{d}{dx}\left(\lambda \frac{dT}{dx}\right) = \eta \langle jE \rangle \qquad j \approx en\mu_e E \qquad \mu_e \sim N^{-1} \qquad (2.5)$$

$$p = NkT = N_0 k T_0$$

where λ is thermal conductivity and η is the portion of current energy actually expended for the gas heating. Most of the energy gained by electrons from the field in a molecular gas is spent for the excitation of molecular vibrations, and only through vibrational relaxation does it transform to heat. A certain portion of the energy, $1 - \eta$, can be lost to the discharge space for good due to the escape of vibrationally excited molecules.

In the calculation of $\nu_{i\,RF}$, we will deal with a gas temperature averaged over the whole discharge space. Keeping in mind that in nitrogen $\lambda \approx 2.4 \times 10^{-4}(0.25 + 0.75\, T[K]/300)\, W\, cm^{-1}\, K^{-1}$, introduce the heat flow potential $\Theta(T) = \int_0^T \lambda dT$ and integrate the thermal conductivity in the limits $0 < x < L$ on the assumption of constant energy release $\langle jE \rangle$ over the space and of the conditions $T(0) = T(L) = T_0$. By averaging the obtained parabolic distribution $\Theta(x)$ over x, we get an equation for the average potential Θ which will be regarded as the function of the average temperature T

$$\Theta(T) - \Theta_0 = \frac{\eta \langle jE \rangle L^2}{12} \qquad \Theta_0 \equiv \Theta(T_0) \qquad (2.6)$$

It is not an easy matter to evaluate reliably the coefficient η, especially in nitrogen, in which vibrational relaxation represents a very complex process. Only a small portion of the energy, in fact a few percent, gained by the electrons from the field is contributed to the translational degrees of freedom of nitrogen molecules. The rest of the energy is spent for the excitation of vibrations. The probability of vibration damping due to collisions of vibrationally excited N_2 molecules with metallic and dielectric electrode surfaces is small, $10^{-3}-10^{-4}$.[2] If the discharge transverse size $2R$ is not much larger than L as in [2.20], an excited N_2 molecule has no time for multiple collisions with the electrodes before its diffusional loss to the discharge space, so this mechanism of vibrational energy losses is unessential.

The diffusion time of a molecule is

$$\tau_{dif} = 2\Lambda^2/D$$

where $D = 116\,(T[K]/300)^{3/2}/p\, cm^2\, s^{-1}$ is the diffusion coefficient in nitrogen and $\Lambda \approx R/2.4$ is the diffusion length. The factor 2 roughly takes into account

---

[2]The probability for vibrationally excited H_2 molecules is also very small. Vibrationally excited CO_2 molecules, on the contrary, lose their energy in collisions with metallic and dielectric electrode surfaces with a probability 0.2–0.4 [2.23].

that the side edge of the discharge cylinder does not absorb, as is assumed in deriving Λ, so a molecule that has escaped from the space has a chance to return. At $R = 0.7$ cm, $p = 35$ Torr, and $T \approx 300{-}1000$ K, $\tau_{\text{dif}} \approx 10^{-3}{-}5 \times 10^{-2}$ s. Even at a high temperature $T = 1000$ K, the time of vibrational (VT) relaxation of the lower vibrational level of an N_2 molecule [2.24]

$$\tau_{\text{VT}}^{(10)} = 4.9 \times 10^{-6} \frac{\exp\left[137(T[\text{K}])^{-1/3}\right]}{p} \text{ s} \tag{2.7}$$

greatly exceeds τ_{dif} (at $p = 35$ Torr, $\tau_{\text{VT}}^{(10)} = 0.12$ s).

This does not mean, however, that all vibrational energy is transported out of the discharge space. Under intense vibration pumping, the relaxation goes on in another, much faster way through excitation of high vibration states by VV-exchange in collisions of two vibrationally excited molecules. Highly excited molecules become deactivated very rapidly because of the small size of vibration quanta due to anharmonicity. In nitrogen, for instance, at $T = 500$ K, vibration temperature $T_V = 3500$ K and $p = 35$ Torr, the time of VT relaxation $\tau_{\text{VT}}^{(v \gg 1)} \approx 1.6 \times 10^{-4}$ s [2.25] is much less than the diffusional escape time. In such complex kinetics, the relaxation rate is limited by the transfer of vibrational excitation from the lower to the upper levels with a flux of quanta along the vibration energy axis Π cm^{-3} s^{-1}.

When the VT relaxation from the lower levels is very slow as compared to the loss rate of vibrational quanta due to molecular diffusion but the VT relaxation from the upper levels is very fast, there is a competition between the diffusion and the upward excitation transfer. If $W \approx \langle jE \rangle / \hbar \omega_0$ is the number of quanta $\hbar \omega_0$ at the lower levels produced per 1 cm^3 per second and $N \epsilon_V$ is the total number of quanta per 1 cm^3, the number balance of quanta under the steady-state conditions is described by the equation

$$W = \frac{N \epsilon_V}{\tau_{\text{dif}}} + \Pi \tag{2.8}$$

The coefficient of transformation of vibration energy to thermal energy is then $\eta \approx \Pi / W$. The average number of quanta per molecule ϵ_V can be conveniently replaced by effective vibration temperature T_V defined by the expression

$$\epsilon_V = \left[\exp\left(\frac{\hbar \omega_0}{k T_V}\right) - 1 \right]^{-1} \tag{2.9}$$

This is the real number of quanta per oscillator in a steady-state system of harmonic oscillators at a given temperature T_V which is now unified temperature.

Without going into the details of vibration kinetics [2.26], we will write some useful formulas with which we can find the flux Π as a function of T and T_V that may become necessary for analysis of various situations

$$\Pi = \tilde{Q} N_0^2 v^{*2} \exp\left[-\left(\frac{2\kappa \hbar \omega_0 v^{*2}}{kT}\right) + 1 \right]$$

TABLE 2.1
Characteristic parameters of vibrational relaxation of nitrogen calculated in [2.21] for the
experimental conditions of [2.20]: $p = 35$ Torr, $R = 0.7$ cm, and $L = 0.59$ cm

$\langle jE \rangle = 3\,\text{W cm}^{-3}, W = 6.42 \times 10^{19}\,\text{cm}^{-3}\,\text{s}^{-1}$								
T (K)	T_V (K)	N (cm^{-3})	$\tau_{VT}^{(10)}$ (s)	v^*	ϵ_V	τ_{dif} (s)	τ^* (s)	η
300	2600	1.15×10^{18}	108	10	0.37	5.1×10^{-2}	6.6×10^{-3}	≈ 1
500	3500	9.3×10^{17}	4.37	12	0.28	2.5×10^{-2}	4×10^{-3}	≈ 1
$\langle jE \rangle = 30\,\text{W cm}^{-3}, W = 6.42 \times 10^{20}\,\text{cm}^{-3}\,\text{s}^{-1}, (\alpha - \gamma\ \text{transition})$								
500	4500	9.3×10^{17}	4.37	10	0.66	2.5×10^{-2}	9.7×10^{-4}	≈ 1
1000	6000	3.5×10^{17}	0.12	14	1.31	8.9×10^{-4}	7.1×10^{-4}	≈ 0.5

$$\tilde{Q} = \frac{4\kappa\hbar\omega_0 Q}{kT\Delta^3} \qquad N_0 = N\left[1 - \exp\left(-\frac{\hbar\omega_0}{kT_V}\right)\right] \qquad v^* = \left(\frac{T}{2\kappa T_V}\right) + \frac{1}{2} \quad (2.10)$$

Here κ is the anharmonic constant in the formula $\epsilon_V = \hbar\omega_0 v - \kappa\hbar\omega_0 v^2$ for the
vibration energy of an anharmonic oscillator on the v-th level, v^* is the vibration
number beyond which the vibration level distribution of molecules n_v shows the
so-called 'plateau' $n_v \sim v^{-1}$, N_0 is the density of unexcited molecules with $v = 0$,
Q is a constant velocity of the VV exchange, Δ is the so-called inverse radius of the
VV exchange, though it is a dimensionless value. For nitrogen $\hbar\omega_0 = 0.292$ eV,
$\kappa = 6.13 \times 10^{-3}$, $Q \approx 2.5 \times 10^{-14}(T/300)^{3/2}$ cm^3 s^{-1}, $\Delta \approx 6.8/\sqrt{T[\text{K}]}$.
Equations (2.8)–(2.10) define T_V, ϵ_V as well as Π and η through the energy release
in the discharge $\langle jE \rangle$ and the gas temperature T.

The calculations [2.21] for the experimental conditions [2.20] have shown
that in most cases where vibrational relaxation does take place, the time of the
upward excitation transfer $\tau^* = N\epsilon_V/\Pi$ is significantly smaller than τ_{dif}, and the
coefficient η is, therefore, close to unity. But in the general case, when the loss
of vibration quanta due to molecular diffusion or to flow transport is essential, the
quantity $\eta \approx \Pi/W$ is to be evaluated by solving the general set of equations (2.5)
together with the number balance equation for vibration quanta (2.8), as well
as (2.9) and (2.10). Table 2.1 illustrates the calculated values of some quantities
related to vibrational relaxation for some of the above experimental conditions.

The flow transport can be easily entered into (2.8) by adding to the right-hand
side the term $N\epsilon_V/\tau_F$, where $\tau_F \approx 2R/u$ is the characteristic transport time, $2R$
is the discharge length along the flow and u is the flow velocity. If $2R \approx 1$ cm,
even a weak flow of velocity $u \approx 1$ m s^{-1} will produce an effect comparable to
the diffusion effect, $\tau_F \approx 10^{-2}$ s, and at $u \sim 100$ m s^{-1}, η and the heating will be
considerably smaller.

2.3.3 Gas heating effect on the CVC and normal current density

Figures 2.9 and 2.10 present the results from solving equations (2.5). For most of the experimental conditions described in [2.20], ω is much smaller or at least smaller than ν_u, hence we use (1.36) to find ν_{iRF}. Still, in order to emphasize the difference in the results derived from approximation (1.35) corresponding to $\omega \gg \nu_u$, Figure 2.10(b) compares the curves obtained in these extreme approximations. The difference is visible only for low plasma and current densities. In the calculations, η was taken to be unity. Test calculations with reasonably smaller η values did not show much change in the results. The plots for E_a/p versus n, in fact, characterize the PC CVC, because the PC voltage $V_{PC} = \mathrm{const}\, E_a$ and the current density $j_a = \mathrm{const}\, T E_a n$ is nearly proportional to n, because the product of T and E_a weakly varies with n. The CVC for the PC generally goes down; and, only at very low densities of plasma and current, when the gas is not heated at all, the recombination CVC smoothly goes up [Figure 2.10(b)]. Thus, the PC CVC has a maximum in the range of very weak currents.

Let us now find the CVC for the whole α-discharge. Within the model considered, the total voltage amplitude on both sheaths coincides with the maximum voltage on one sheath and according to (1.42) is

$$V_{sa} = 8\pi e n A^2 = \frac{8\pi e n \mu_e^2 E_a^2}{\omega^2} = \frac{2e}{\pi}\mu_e^0 \left(\frac{E_a}{p}\right)^2 \left(\frac{T}{T_0}\right)^2 \frac{n}{f^2} \qquad (2.11)$$

where μ_e^0 is mobility at $p = 1$ Torr and $T = T_0 = 293$ K, and p is the real gas

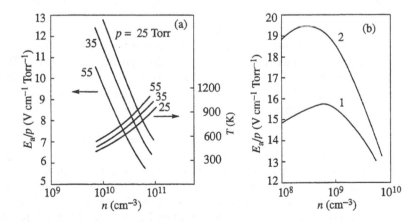

FIGURE 2.10
Field amplitude-to-pressure ratios at initial pressure p and gas temperature T (a) at large plasma densities, nitrogen, $L = 0.59$ cm, $\eta = 1$; ionization frequency $\nu_i(E)$ averaged over a cycle. (b) E_a/p at moderate and low plasma densities; ν_i averaged over a cycle (curve 1), ν_{iRF} corresponding to ν_i in the rms field (curve 2).

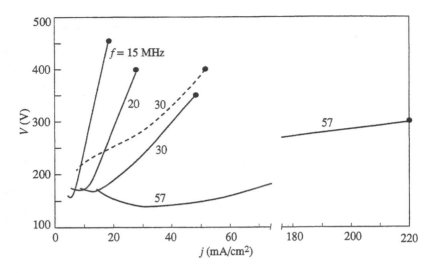

FIGURE 2.11
The CVC of the α-discharge at various frequencies in nitrogen, $p = 35$ Torr,
$L = 0.59$ cm, $\eta = 1$. The upper end points of the curves indicate the α–γ transition.
Dashed line is for $\eta = 0.05$ (weak relaxation or fast gas flow); there is no minimum, the
transition point lies higher.

pressure. If the sheaths do not occupy much of the gap ($2A \ll L$), the voltage
amplitude in the PC is $V_{\text{PC}\,a} \approx E_a L$. The discharge CVCs calculated from (2.3)
are shown in Figure 2.11. All the curves possess minima that correspond to the
normal current density j_n (see Section 2.3.1). On the side of large j values, the
curves extend as far as the α–γ transition points (see Section 2.4). The normal
current densities and the respective plasma densities grow with frequency. This
is associated with the shift of the V_a minimum, because $V_{s\,a}$ decreases with rising
frequency and V_{PC} is independent of f. For this reason, high frequencies are more
profitable for lasers, permitting an increase in the values of n and j and in the
input discharge energy. The values of n and j_n also grow with pressure and L, but
only to a certain limit because there is a limit in pL (see Sections 2.2 and 2.3.1).

In any one-dimensional analytical theory that considers effects responsible for
a falling PC CVC, the total discharge CVC may have a minimum and, hence, two
possible values of the current density corresponding to one value of the electrode
voltage. So, in order to select a real value of j, one has to resort to additional
reasoning concerning the steady states (Section 2.3.1). The left-hand, falling
branch of the CVC must not exist at all because of the instability of even a one-
dimensional state, while the right-hand branch may be stable only if there is no free
boundary between the current and currentless zones of the electrode, i.e., when
all its surface is occupied by current. The instability of the right-hand curve for
a partially occupied electrode may be revealed only by a two-dimensional theory.

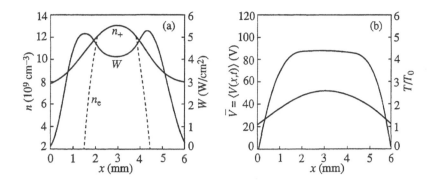

FIGURE 2.12
Calculated distributions of (a) ion densities n_+ and energy release W, (b) the gas temperature T and constant potential for the normal α-discharge in nitrogen, $p = 35$ Torr, $L = 0.59$ cm, $f = 30$ MHz, $\eta = 1$, $T_0 = 300$ K; electrode voltage amplitude $V_a = 350$ V, current amplitude $j_a = 15.6$ mA cm^{-2}.

A CVC minimum, which according to this conception may be realized for a partly occupied electrode, also indicates a minimum voltage necessary for α-discharge operation. A steady discharge cannot be maintained if the electrode voltage is lower than this minimum. In that case, the gap would behave as a capacitor in an ac circuit, passing only the displacement current.

It is remarkable that in numerical simulation of a one-dimensional process, the entirely unstable left-hand branch of the CVC is automatically discarded during the evolution of an arbitrary initial state. The evolution eventually leads to a steady discharge that corresponds only to the right-hand branch. The numerical 'experiment,' like the laboratory experiment, indicates that a steady discharge operation is impossible with electrode voltage smaller than a certain value. Computations based on equations (2.1) and the thermal conduction equation [the second one in (2.5)] literally simulate the natural process.

Figure 2.12 illustrates the results of simulation [2.9] for one set of experimental conditions [2.20]. These results clearly demonstrate the distributions of the ion density, gas temperature, energy output and of the constant potential in the gap at the minimum electrode voltage, $V_a = 350$ V and $j_a = 15.6$ mA cm^{-2} (η is taken to be unity). Note the two maxima of the energy output W at the boundaries of the space charge sheaths. Energy output maxima in an α-discharge were also obtained from calculations in [2.11] and [2.3] without account of the gas heating. The physical reason for such maxima is as follows. In the PC plasma bulk, the active current component and the plasma density are large, but the electric field strength is relatively low. The sheaths, on the contrary, have a strong field but a low cycle-average electron density; in order words, the current is primarily reactive, being out of phase with the field. For this reason, the energy output $W = \langle jE \rangle \sim \langle n_e E^2 \rangle$ has a maximum on each side since it is a product of the falling and rising functions with respect to x. The density of the Joule heat release is somewhat proportional to the energy gain rate of the electron gas, and hence, to the ionization and gas excitation rates. This, in particular, is the reason for the two maxima in the glow distribution in the α-discharge (Figure 1.11). The computed rising CVC branch is in good qualitative agreement with the branch calculated in the simple model (Figure 2.11) but naturally differs from it quantitatively (about 1.5-fold).

2.3.4   Disappearance of the normal current density effect at low frequency or pressure

It follows from (2.3), (2.11), $V_{PCa} = E_a L$ and from the real PC CVC which descends only at sufficiently large n and j, i.e., on noticeable gas heating, that the effect of normal current density disappears at fairly low frequencies. Indeed, the sheath CVC rises steeply with decreasing f, while the PC CVC is independent of f. Therefore, as the frequency decreases, the V_a minimum is shifted towards lower plasma and current densities. At a certain frequency, the values of n and j become so small that they go beyond the PC CVC maximum [Figure 2.10(b)] where this curve, as the sheath CVC, goes up. In other words, with decreasing frequency, the CVC minimum in the α-discharge becomes increasingly more shallow, degenerates and finally disappears. Mathematically, this follows from the differentiation of the $V_a(n)$ curve and from solving the equation for the extremum $dV_a/dn = 0$ with respect to f, which relates the plasma density n_n, corresponding to the normal current density, to f [2.27, 2.21].

Figure 2.13 shows the respective $n_n(f)$ plot for the experimental conditions of [2.20]. The curve stops at a minimum frequency of about $f_{min} = 10$ MHz. At still lower frequencies, the CVC has no minimum, and there must be no effect of the normal current density at the same parameters. At larger p and L values, this effect persists down to lower frequencies. There seem to be no published data to test these conclusions. Most measurements at moderate pressures have been made for the frequencies $f = 13.56$ MHz and higher. We have already pointed out that the absence of normal current density in experiments [2.20] should be associated with the small electrode size, so the α-discharge was excited as initially abnormal. This is supported by the growing CVC measured experimentally.

The above effect also disappears at rather low densities, although for a different reason. As the pressure is decreased, the PC voltage fall, $V_{PCa} = E_a L = (E_a/p)pL$, makes an increasingly smaller contribution to the total fall on the electrodes, since the value of E_a/p for the PC lies within narrow limits. For this reason, nearly all applied voltage falls on the sheaths possessing a rising CVC. Hence, the total discharge CVC acquires a rising character. In this case at a given current, a state with minimum voltage is also established in the discharge owing to the influence of the factors associated with stable states. If, however, the minimum voltage corresponds to the minimum current density, the discharge completely covers the electrode. The absence of normal current density and

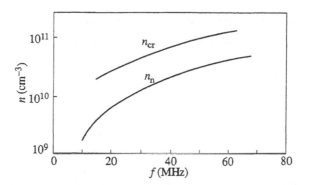

FIGURE 2.13
Plasma density n_n corresponding to the normal current density in the nitrogen α-discharge and critical plasma density n_{cr} for the $\alpha-\gamma$ transition versus frequency, $p = 35$ Torr, $L = 0.59$ cm.

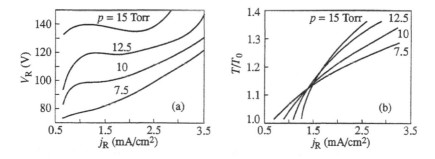

FIGURE 2.14
Degeneration of the CVC minimum in a nitrogen α-discharge at decreasing pressure. Plots for the rms voltage and average gas temperature versus the rms current density, $T_0 = 300$ K, $f = 13.56$ MHz, $L = 1$ cm.

completely occupied electrodes at low pressures is well known from experiment. This is one of the distinctions between low- and moderate-pressure discharges. The above qualitative arguments are illustrated by calculations [2.28] presented in Figure 2.14. The calculation was made within the simplified α-discharge model. One can see that as the pressure is decreased, the minimum in the CVC becomes less pronounced and completely disappears at $p < 10$ Torr if $L = 1$ cm. At larger L this effect disappears at lower pressures because V_{PC} is nearly proportional to pL.

2.3.5   Stepwise ionization and normal current density in light gases

In light gases, such as helium possessing high thermal conduction, the gas heating is insignificant due to the fast heat transport. The thermal effect considered above is too small for the PC CVC to show such a steep fall that could account for the total CVC minimum at the pL values sufficient for supporting normal current density. It seems likely that this effect is due to stepwise ionization leading also to the PC voltage fall at higher current densities. Within the simplified model, the CVC with account of stepwise ionization follows from equations of the kind (2.5) and (2.6) but with different equations for the number balance of electrons n_e.

Let us take into account the basic processes of production and loss of electrons in helium: electron impact ionization of unexcited atoms, ionization of the metastables He(2^1S) and He(2^3S) (with their densities denoted as N_1 and N_2 and ionization rate constants as k_{i1} and k_{i2}), associative ionization in which two He(2^3S) atoms combine to give a molecular ion He$_2^+$ and a free electron (with a rate constant k_2), and ambipolar diffusion loss of charges (recombination in helium goes on with a smaller effective coefficient $\beta \approx 4 \times 10^{-9}$ cm^3 s^{-1} than, for example, in nitrogen with $\beta \approx 2 \times 10^{-7}$ cm^3 s^{-1}). The number balance equation for electrons can be written as

$$(\nu_{iRF} + k_{i1}N_1 + k_{i2}N_2)n_e + k_2N_2^2 = \beta n_e^2 + D_a \left(\frac{\pi}{L}\right)^2 \qquad (2.12)$$

where D_a is the ambipolar diffusion coefficient. Equation (2.12) should be supplemented with the number balance equations for metastable atoms

$$\nu_1^* n_e = (\nu_d^* + \tau_1^{-1} + k_{12}n_e + k_{i1}n_e)N_1$$

$$\nu_2^* + k_{12}n_eN_1 = (\nu_d^* + k_{i2}n_e + k_{20}n_e)N_2 - 2k_2N_2^2 \qquad (2.13)$$

to include the basic processes of their production and loss. Here ν_1^* and ν_2^* are the excitation frequencies of the metastable states in collisions of electrons with unexcited atoms, $k_{12} \approx 3.5 \times 10^{-7}$ cm^3 s^{-1} is the rate constant of H$_2$(2^1S) transformation to He(2^3S) with release of 0.79 eV, $k_{20} \approx 1.5 \times 10^{-9}$ cm^3 s^{-1} is the rate constant of He(2^1S) transformation to the ground state, $\tau_1 \approx 0.02$ s is the lifetime of He(2^1S) relative to the radiative transition, and $\nu_d^* = D^*(\pi/L)^2$ is the diffusion loss frequency of metastables with the diffusion coefficient $D^* = 420/p$ cm^2 s^{-1}. The process of electron impact deactivation of He(2^1S) is much slower than the almost resonant transformation to He(2^3S), so it has been neglected together with the slow radiative transition of He(2^3S). To write equations (2.12) and (2.13) and to choose the constants, we have made use of the detailed analysis of phenomena occurring in the glow discharge PC [2.29]. (These equations were used in [2.31] to solve the problem for the structure of glow discharge cathode parts and the transition region between the Faraday dark space and the PC for helium.)

In atomic helium, the frequency of establishing the mean electron energy $\nu_u = \delta\nu_m$ is much lower than in nitrogen, since $\delta = 2m/M$ is defined only

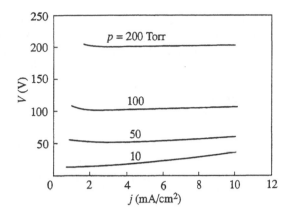

FIGURE 2.15
The CVC for a helium α-discharge at various pressures, $f = 13.56$ MHz, $L = 1$ cm (rms V and j).

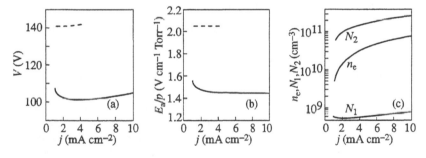

FIGURE 2.16
Helium α-discharge parameters at $f = 13.56$ MHz, $p = 100$ Torr, $L = 1$ cm: (a) the CVC (enlarged), (b) the amplitude E_a/p in the positive column, and (c) the densities of electrons and metastable atoms (rms V and j). Dashed line—neglecting ionization involving metastable atoms.

by elastic collisions. So the inequality $\nu_m \gg \omega \gg \nu_u$ for the frequencies commonly used in RF discharges is fulfilled. The mean electron energy and the ionization frequency ν_{iRF} are determined by the rms field (Sections 1.2.2 and 1.4.1). Superelastic collisions of electrons with the metastables, and especially associative ionization producing energetic electrons with $\epsilon \approx 15-17.6$ eV, greatly distort the distribution function of electrons in the high energy range, a fact that should be taken into account in the description of ionization and excitation rates of unexcited helium atoms. Another factor that influences the electronic spectrum is the reduction of the amount of energetic electrons due to the energy losses for ionization and excitation.

The results of computation of the α-discharge CVC in helium with account of the above effects [2.30] are given in Figures 2.15 and 2.16. At sufficiently high pressures, the CVCs show minima but are not very deep. In accordance with measurements [2.19], the normal current densities corresponding to these minima are nearly proportional to the pressure, $j_n \sim p$. The calculated values of j_n are by a factor of 1.5–2 smaller than the experimental values. The CVC minima degenerate and entirely disappear with decreasing pressure. For $f = 13.56\,\text{MHz}$ and $L = 1$ cm, this occurs at $p \approx 10$ Torr. Below this pressure there is no normal current density in helium, so the conclusion (Section 2.3.4) that this effect disappears at sufficiently low frequencies as well remains valid. There should be no normal current density effect in a fast gas flow, because the metastable atoms responsible for this effect are carried away by the flow. In this case even a weak current will fill up the entire electrode.

2.4   The $\alpha-\gamma$ transition parameters

The jump-like character of the $\alpha-\gamma$ transition at moderate pressures was clearly observed in the numerical simulation (see Section 2.1). The transition parameters (e.g., current density) showed a fairly good fit with the experimental data. This indicates that the physical mechanism of this effect can be described quite well by the appropriate equations which contain nothing else but the conventional laws of drift motion, of electron production and loss, of secondary emission and electrostatic laws. Nevertheless, the nature of this effect is masked by the computations in much the same way as it was masked by experimental data that needed a theoretical analysis based on certain assumptions. In order to understand the transition mechanism, we will evaluate its parameters in terms of the above hypothesis. Briefly, we will try to show that the calculations made in terms of the α-sheath breakdown model fit well both the laboratory experiment and the numerical simulation.

2.4.1   A simple analytical evaluation of the transition parameters

A gap of size d between two plane electrodes can be broken down if the applied constant voltage exceeds a certain threshold value V_t that provides charge reproduction in the gap and current self-maintenance. This means that each electron emitted by the cathode multiplies in the field $E = V_t/d$ to generate as many ions as is necessary to produce at the cathode a new electron in secondary emission with the γ-coefficient. The condition for charge multiplication is expressed by the well-known Townsend equation

$$\gamma\left(e^{\alpha d} - 1\right) = 1 \qquad \alpha = \alpha(E) \qquad E = \frac{V_t}{d} \qquad (2.14)$$

By substituting into (2.14) Townsend's expression for the ionization coefficient $\alpha = A_1 p \exp(-Bp/E)$ for V_t, we get the formula

$$V_t = \frac{Bpd}{C + \ln(pd)} \qquad C = \ln \frac{A_1}{\ln(1 + \gamma^{-1})} \qquad (2.15)$$

which describes well the Paschen experimental curves $V_t(pd)$ [2.18].

Assume an α-sheath to be a discharge gap. The plasma will then play the role of the anode, and the naked or coated electrode will act as the cathode. The plasma has a positive potential relative to the electrode and absorbs electrons emitted by the electrode and multiplied in the sheath. Due to the strong dependence of α on E, intense charge multiplication occurs, in fact, during a short period within a cycle when the sheath thickness and the voltage fall in it are close to the maximum values. In the simplified α-discharge model with $n_+(x) = n = \text{const}$, these maximum values are

$$d_{\max} = 2A \qquad V_{\max} = 8\pi e n A^2 \qquad (2.16)$$

where $A = \mu_e E_a/\omega$ is the oscillation amplitude of the plasma boundary. The critical values of the plasma density n and of the current density $j_a = e n \mu_e E_a$ at which the sheath breakdown and the α−γ transition occur, can be evaluated by equating $V_{\max}$ in (2.16) to the breakdown voltage V_t in (2.15) with $d = d_{\max}$ from (2.16). This yields

$$n_{cr} = \frac{Bp}{4\pi e A[C + \ln(2pA)]} \qquad (2.17)$$

The plasma field amplitude E_a which defines A is related to n by the condition of ionization–recombination equilibrium; so, generally speaking, equality (2.17) is not yet solved relative to n_{cr}. To find n_{cr}, it is, however, sufficient to substitute into (2.17) a fixed value of A, since E_a/p and A are only weakly dependent on n. Suppose $f = 13.56$ MHz, $p = 15$ Torr, $\gamma = 0.01$, $A_1 = 12$ cm^{-1} Torr^{-1} and $B = 342$ V cm^{-1} Torr^{-1}, as for the nitrogen discharge simulation (Section 2.1.1). Take $(E_a/p)_{cr} = 15$ V cm^{-1} Torr^{-1} with a corresponding $A = 0.077$ cm. This value of $(E_a/p)_{cr}$ was obtained from numerical simulation without account of the gas heating. From (2.17) we find $n_{cr} = 2 \times 10^{10}$ cm^{-3}. This value agrees with $n_{cr} = 0.87 \times 10^{10}$ cm^{-3}, which separates the α- and γ-regions of the discharge, indicating an agreement with experimental data.

This may be taken as an evidence for the validity of the above α−γ transition conception, in which the transition is due to the α-sheath breakdown. This conception is also supported by the twofold increase in the critical plasma density calculated from (2.17) as compared to the respective simulation result. In the latter case, the constant plasma potential equal to the cycle-average voltage in the sheath is about twice as large as in the $n_+ = \text{const}$ model. In this model, therefore, the breakdown voltage in the sheath would require a double plasma density.

From our estimations, a sheath of 'thickness' $pd_{\max} = 2.3$ Torr cm will be broken down at $V_{\max} = 440$ V and $E_{\max}/p = V_{\max}/(pd_{\max}) = 190$ V cm^{-1} Torr^{-1}.

($E_{max}/p > 100$ V cm^{-1} Torr^{-1}, i.e., the choice of A_1 and B values corresponding to the first formula for α in Section 2.1.1 is justified.) The breakdown voltage and the sheath thickness exceed the respective characteristics for the Paschen curve minimum, which are

$$V_{min} = \frac{2.72B}{A_1} \ln(1 + \gamma^{-1}) = 357 \text{ V}$$

$$(pd)_{min} = \frac{2.72}{A_1} \ln(1 + \gamma^{-1}) = 1.0 \text{ Torr cm} \tag{2.18}$$

Thus, the α–γ transition parameters correspond to the right-hand rising branch of the Paschen curve $V_t(pd)$, which is typical of moderate-pressure discharges and distinguishes them from low-pressure discharges (see below). It is clear from (2.17) that n_{cr} and $j_{a\,cr}$ grow with frequency, since $A \sim \omega^{-1}$.

2.4.2 Extension of the Townsend breakdown criterion to an oscillating sheath

One can hardly expect something qualitatively new from the allowance for a variable sheath thickness and field; but a revision may become useful, in particular, for the understanding of the charge multiplication process in sheath breakdown. Within the α-discharge model with $n_+ = $ const, the sheath thickness and the absolute field in the left-hand sheath are represented by the functions (see Section 1.5.3)

$$d(t) = A(1 + \sin \omega t) \quad A = \frac{\mu_e E_a}{\omega}$$

$$E(x, t) = 4\pi e n A \left(1 + \sin \omega t - \frac{x}{A}\right) \tag{2.19}$$

In the expression for E, the relatively small term for the plasma field E_p may be dropped, because the amplitude E_a is at least an order of magnitude smaller than the maximum field at the electrode.

When considering electron multiplication in a sheath with a field variable in time and space, one should bear in mind that in the α-mode, as distinct from the γ-mode (see below), the time of the electron drift through the sheath is comparable with a cycle [2.3]. Over the time of propagation of an electron avalanche initiated by emission of a single electron from the 'cathode,' the field phase changes remarkably. This significantly affects the ionization coefficient $\alpha(E)$, which is very sensitive to the field value. Let $x = x(t, t_0)$ describe the avalanche propagation starting with the emission of an electron at time t_0. This formula can be easily deduced by integrating the equation $dx/dt = \mu_e E(x, t)$, assuming $t = t_0$ at $x = 0$ (Figure 2.17). The avalanche dies out at the time t_1 when it meets the plasma boundary. The value of t_1 can be found from the equation $x(t_1, t_0) = d(t_1) \equiv d_0$ with d_0 as the final avalanche length. The number

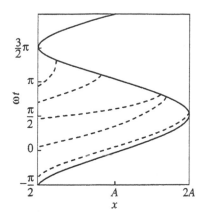

FIGURE 2.17
The dynamics of the boundary between the plasma and the left-hand sheath. Dashed
lines—cathode electron trajectories along which electron avalanche grows, $A = \mu_e E_a / \omega$
is the oscillation amplitude for a plasma electron.

of ion pairs formed in this avalanche is

$$Z(t_0) = \exp \left(\int_0^{d_0(t_0)} \alpha \left\{ E\left[x, t\left(x, t_0\right)\right] \right\} \mathrm{d}x \right) - 1$$

where $t(x, t_0)$ is the result of inversion of the function $x = x(t, t_0)$. These
cumbersome functions will not be written here.

Unlike electrons that can go through the sheath very fast, ions need many cycles $T_\omega = f^{-1}$ in order to arrive at the electrode from their generation point
somewhere in the middle or at the end of the sheath. So, the ions do not 'remem-
ber' their original phase. Since there is practically no correlation between the
times of the ion production and arrival at the electrode, only a cycle-average
number balance of ions in the sheath should be valid during multiplication.
The number of ions that hit 1 cm² of the electrode surface over the time T_ω
is $\int_0^{T_\omega} n_+ \mu_+ E(0, t) \mathrm{d}t$. The same number of ions is produced over a cycle in the
sheath avalanches that are generated from 1 cm² of the electrode. This number is
equal to $\int_0^{T_\omega} \gamma n_+ \mu_+ E(0, t_0) Z(t_0) \mathrm{d}t_0$. If the sheath voltage is insufficient for a
breakdown, the drift ion flux into the electrode is compensated by the gas ioniza-
tion via the plasma electrons rather than in avalanches. This creates the number
balance of ions in the α-discharge. By equating the above expression and reducing
them by μ_+ and the constant density n_+, we obtain an equation that generalized
the Townsend condition (2.14)

$$\gamma \int_0^{T_\omega} E(0, t_0) \left\{ \exp \left(\int_0^{d_0(t_0)} \alpha \left\{ E\left[x, t(x, t_0)\right] \right\} \mathrm{d}x \right) - 1 \right\} \mathrm{d}t_0 = \int_0^{T_\omega} E(0, t_0) \mathrm{d}t_0$$

$$(2.20)$$

TABLE 2.2
Root-mean-square values of voltage and current in the α-discharge and the gas
temperature at $\alpha-\gamma$ transition in the experimental conditions [2.20] in nitrogen

			$p = 35$ Torr		
	Calculation [2.21]			Experiment [2.20]	
f (MHz)	V (V)	I (mA)	T (K)	V (V)	I (mA)
15	460	29	620	440	50
20	400	42	740	405	60
30	350	72	910	300	108
57[a]	300	234	1250	220	350
			$f = 30$ MHz[b]		
p (Torr)	V (V)	I (mA)	T (K)	V (V)	I (mA)
25	330	66	790	270	134
35	350	72	910	300	108
55	390	90	1080	320	95

[a] $\nu_{i\,RF}$ was derived from (1.35) with $E_{eff} = E_a/\sqrt{2}$, since $\omega > \nu_u$.
[b] The use of $\nu_{i\,RF}$ derived from (1.36) and (1.35) with $E_{eff} = E_a/\sqrt{2}$ gives nearly the same values.

Equation (2.20), together with the relationship $E_a(n)$, defines n_{cr}. Calculation
with the same parameters as those used for solving (2.17) gives a close value of
$n_{cr} = 2.2 \times 10^{10}$ cm^{-3}.

2.4.3   Comparison of calculations and experimental data

Figure 2.13 showing a plot of the plasma density in a normal α-discharge versus
the field frequency also presents the plot for $n_{cr}(f)$. The calculation of the α-
sheath breakdown was based on the Townsend condition (2.20) with account of
the gas heating that affects the PC CVC and $E_a(n)$. In the calculation, η was taken
to be 1. The current $I = jS$ was calculated with $j(n_{cr})$ and $S = 1.5$ cm^2. The
parameters of the $\alpha-\gamma$ transition and the respective gas temperatures calculated
for the experimental conditions of [2.20] are presented in Table 2.2 together with
the measurements. The calculated voltages of the transition are indicated by
circles in the CVC of Figure 2.11. The circles are the end points of the $V(j)$
curves for the abnormal α-discharge. If the conditions for heat removal become
worse, for example, due to a larger interelectrode distance L, the plasma density
in the normal mode n_n may turn out to be higher than n_{cr}, and the arrangement of
the n_n and n_{cr} curves in Figure 2.13 will be reverse. The α-discharge will then
not be able to operate at all. In this way the discharge becomes limited in L (see
Section 2.3.1).

One can see from Table 2.2 that the calculated values for the frequency dependence of voltage V and current I of the $\alpha-\gamma$ transition are in reasonable agreement with the respective measurements. One may hardly expect a better fit, considering the complicated character of the phenomenon and the simplicity of the theoretical model. With increasing frequency, the transition voltage drops while the transition current grows. The physical reason for this relationship is that the increasing frequency leads to a reduction in the electron oscillation amplitude and sheath thickness, so a higher charge density in the sheaths and plasma and a higher current density are required in order to achieve a breakdown field. In this case the voltage drops because of the greater gas heating, inducing a voltage drop in the PC. Note that prior to the transition the gas is heated to very high temperatures. We illustrated this in Section 2.2: at $p = 35$ Torr and $f = 30$ MHz, the discharge temperature deduced from indirect measurements [2.20] is 800 K; the calculated temperature given in Table 2.2 for the same conditions is 910 K. The calculation adequately reflects the growth of the transition voltage with pressure, as the voltage on the PC becomes greater. However, the calculated transition current grows with p while the measured values show a slight decrease. This discrepancy may be interpreted as being due to imperfections of the model and to the effect of many neglected factors.

2.5 The γ-discharge

2.5.1 Specific features of the γ-discharge: Normal current density

Experimental data and numerical simulation evidence that the sheaths at both electrodes of a moderate-pressure γ-discharge are similar in their basic behavior and average characteristics to the cathode sheath of a glow discharge. In the latter type of discharge, the cathode sheath plays a primary role in the generation of electric current. Within a certain approximation assuming the ionization frequency at a given point x (in ionization of atoms by electrons) to be dependent on the local field only $\alpha(x) = \alpha[E(x)]$, the cathode sheath operates quite independently: it maintains the ionization process and the current through the gap. Moreover, the cathode sheath of a normal glow discharge creates optimal conditions for ionization approximately corresponding to the voltage minimum in the Paschen curve. The normal cathode fall V_n and the sheath thickness d_n are approximately defined by equations (2.18) with $d_n \sim p^{-1}$ and V_n independent of p.

The normal values of d_n and V_n are close to the experimental and calculated values of the average sheath thickness d and the constant plasma potential $\overline{V}_p$ of the γ-discharge. Like d_n and V_n, $d \sim p^{-1}$ and $\overline{V}_p$ is independent of p, both d and $\overline{V}_p$ being independent of frequency. We will show below that the ionic component j_{ni} of the normal electrode current density in a γ-discharge is, on the average, close to that of a glow discharge and is proportional to the squared pressure, $j_{ni} \sim p^2$. From this analogy also naturally follows the resultant normal current density j_n which contains, in addition to j_{ni}, the electrode displacement current. These features of the γ-discharge are absolutely alien to the α-mode in which the electron oscillation amplitude A in the plasma field serves as the measure of the sheath thickness: the thickness $d \sim A \sim \omega^{-1}$ is inversely proportional to frequency, nearly independent of pressure and an order of magnitude larger than d_n.

All characteristic features of the two RF discharge modes are determined by whether electron multiplication through secondary emission is self-sustained and whether the Townsend condition for charge multiplication (2.14) and (2.20) is fulfilled; in order words, whether there is a sheath breakdown or not. A breakdown does occur in the γ-discharge but does not in the α-mode. The ion flux from the sheath into the electrode of a γ-discharge is quite intense and is compensated by ion production in avalanches that develop in the sheath; in this sense, the ion flux is self-sustained. In the α-discharge, the ion flux into the electrode is very weak and is compensated by ion production in the plasma.[3] The avalanches sometimes arising at the electrode due to electron emission die out without providing charge multiplication.

The difference between the γ- and α-modes does not lie in the different proportions of the conduction and displacement currents on the electrode, as is sometimes believed. The dominant electrode current in γ-discharges of moderately high pressures is the displacement current, too. The principal result of a sheath breakdown is not the appearance of a more intense ion flux than in the α-mode but a sharp reduction in the sheath thickness d to a value optimal for multiplication of electrons and corresponding to the Paschen curve minimum. This is the reason for the sharp rise of the field at the electrode $E \sim V/d$ with the same order of voltage magnitude V on the sheath. The displacement current density may then become very high, $j \sim (4\pi)^{-1}\omega V/d$. A thin γ-sheath with a strong field passes a high displacement current, whereas a thick α-sheath does not. This fact may be interpreted as follows. Owing to the self-maintenance of charge multiplication in the γ-sheath, the gas becomes ionized to a far greater charge density, providing a high current in the sheath (the higher the ion density, the stronger the field) and a high conduction current in the plasma; this is the reason for strong currents in the γ-mode.

This model of normal current density has been directly supported by experiment [2.6]. Figure 2.18 shows normal current density measurements in a γ-discharge in air at various pressures and their interpretation in terms of the above model. The measurements were made using an installation described in Section 1.8.1. The ion component of the current can be represented, as in a glow discharge, in the form $j_n = C_1 p^2$, where C_1 is a constant specific to each gas-electrode metal pair. In the experiments to be described, these were

---

[3]It will be shown below that in the γ-mode, too, a large portion of ions flowing into the electrode is produced in the plasma, but this effect has quite a different nature being due to the nonlocality of the electron spectrum and ionization coefficient α.

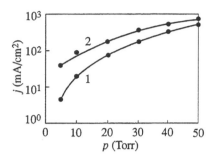

FIGURE 2.18
Plots for the electrode normal current density in the γ-mode versus pressure in air; brass electrodes, $f = 13.56$ MHz. j_n in a dc normal glow discharge (curve 1), calculations with equation (2.21) and displacement current (curve 2). Points near curve 2 are measured values.

air and brass for which $C_1 = 2 \times 10^{-4}$ A cm^{-2} Torr^{-2}. If the sheath thickness $d = C_2/p$, where $C_2 = 0.23$ cm Torr for this pair, the displacement current density $j_{dis} \approx (\omega V/4\pi C_2)p \sim p$. Due to the difference in the pressure dependence of the conduction current j_n and the displacement current j_{dis}, their relative contributions to the resultant current j depend on p. Since the active current j_n and the reactive current j_{dis} are $\pi/2$ phase-shifted relative to one another, the rms density of the resultant current is

$$j = \left(j_n^2 + j_{dis}^2\right)^{1/2} = C_1 p^2 \left[1 + \left(\frac{\omega V}{4\pi C_1 C_2 p}\right)^2\right]^{1/2} \qquad (2.21)$$

Figure 2.18 compares the results of calculation using (2.21) and measurements made at $f = 13.56$ MHz. The sheath voltage is taken to be $V = 320$ V; it represents an rms RF electrode voltage measured at the moment of the nearest approach of the electrodes ($L \approx 1$–2 cm), so that the positive column is practically absent. There are only the sheaths and, partly, the Faraday space present (see below). Most of the applied voltage falls on the sheaths. The experimental rms values fit theoretical curve 2 well. One can see that at a low pressure of $p \approx 5$ Torr, j_{dis} is an order of magnitude larger than j_n. Only at relatively large pressures, $p \approx 50$ Torr, both current components become comparable. No visible changes in the discharge structure or CVC are observed. The whole $j(p)$ curve in this experiment and in Figure 2.18 corresponds exclusively to the γ-mode.

We have pointed out above that the scale of the α-sheath thickness d_α is the plasma electron displacement amplitude $d_\alpha \approx A = v_d/\omega$, where $v_d = \mu_e E_a$ is the amplitude of electron drift velocity. The plasma–sheath boundary oscillates with a velocity of the same order of magnitude. The relationship $d_\alpha \approx v_{d\,\alpha}/\omega$ remains valid in the γ-mode, too. In the α-mode, however, the field E_a in the plasma is established at a value high enough to maintain its quasi-stationary operation;

the primary magnitude is $v_{d\,\alpha}$ while $d_\alpha \approx v_{d\,\alpha}/\omega$ is a derivative. In a normal moderate-pressure γ-discharge, the primary magnitude is the sheath thickness d_γ, which is established at a value close to d_n corresponding to the Paschen minimum. The field E_p and the electron drift velocity at the plasma–sheath boundary $v_{d\,\gamma} = \mu_e E_p \approx d_n\omega$ 'get adjusted' to d_n, being much lower than in the α-mode. The plasma maintenance conditions are not violated in this case: the γ-sheath does not border on the positive column as in the α-discharge, but it borders on a negative glow plasma maintained in a different way (Section 2.5.4).

2.5.2 Analytical model of the electrode sheath

For simplicity, assume the ion density in a sheath n_s to be constant but not coinciding with the plasma density n in the positive column; n_s is to be evaluated. Neglecting the relatively weak plasma field E_p (as compared to the sheath field) and the low density of rapidly escaping electrons (as compared to n_s), write down the familiar expressions (Section 1.5.3) for the absolute field in the left-hand sheath

$$E(x,t) = E_0(t)\left(1 - \frac{x}{x_s}\right) \qquad E_0(t) = 4\pi e n_s x_s(t) \qquad (2.22)$$

where E_0 is the field at the electrode and x_s is the sheath thickness. By taking the discharge current j to be harmonic and neglecting the plasma displacement current on the assumption $E_p \ll E_0$, we have the following charge balance equation for the sheath

$$e n_s \frac{dx_s}{dt} = j_a \cos\omega t - (1 + \gamma)e n_s \mu_+ E_0 \qquad (2.23)$$

This equation also expresses the total current (conduction and displacement currents) conservation law for the sheath. The sign, i.e., the initial phase, of j is chosen such that the moment $\phi \equiv \omega t = -\pi/2$ and $j = 0$, the plasma touches the left-hand electrode ($x_s = 0$) and then moves away from it. By integrating (2.23) with E_0 with respect to (2.22) and the initial conditions just specified, we find the expression for the plasma–sheath boundary motion

$$x_s = \frac{d_0}{1 + \xi^2}\left\{\xi\cos\phi + \sin\phi + \exp\left[-\xi\left(\phi + \frac{\pi}{2}\right)\right]\right\} \qquad (2.24)$$

The length $d_0 = j_a/e\omega n_s$ serves as the sheath thickness scale, and the dimensionless parameter $\xi = (1 + \gamma)4\pi e n_s \mu_+/\omega$ characterizes the conduction-to-displacement current ratio at the electrode. The oscillation amplitude of plasma electrons $A = j_a/e\omega n$ does not coincide with d_0 because $n \neq n_s$. In reality, $n < n_s$ and $d_0 < A$.

After the plasma boundary has passed the point of maximum distance $x_{s\,\mathrm{max}}$ from the electrode, it returns to the electrode at a certain moment $\phi_1 \equiv \omega t_1 < 3\pi/2$ and remains pressed against it until the cycle is over $\phi = 3\pi/2$. Indeed, the electron current into the electrode (in our approximation it is equal to the discharge current j) must compensate, at this 'anode' stage, for the positive charge that vanishes at the 'cathode' stage $-\pi/2 < \phi < \phi_1$. This restriction, equivalent to the condition of the zero cycle-average conduction current on the electrode, determines the phase ϕ_1. But the requirement of (2.24) does not necessarily provide the boundary arrival back at the electrode at exactly this moment ϕ_1. For this reason, if the sheath expansion is described by the integral curve (2.24) of equation (2.23), its reduction should be described by an integral curve that satisfies the condition $x_s = 0$ at $\phi = \phi_1$

$$x'_s = \frac{d_0}{1 + \xi^2} \left\{ \xi \cos \phi + \sin \phi - (\xi \cos \phi_1 + \sin \phi_1) \exp \left[-\xi(\phi - \phi_1) \right] \right\} \quad (2.25)$$

Function (2.24) should be extrapolated forward in time while function (2.25) must be extrapolated backward until they intercept at point ϕ_2 where the two branches of the solution of $x_s(t)$ should be matched. The corresponding coordinate $x_s(\phi_2) = x'_s(\phi_2)$ actually coincides with $x_{s\,max}$. By substituting equations (2.24) and (2.25) into the integral of the conduction current to the electrode and by equating the integral over the whole cycle to zero, we obtain an equation for finding ϕ_1.[4]

These preliminary operations are a necessary premise to the principal integral relation for the ionization and conduction current self-maintenance in a sheath. This relation will not, in principle, differ from (2.20) if we want to describe ionization with the Townsend coefficient $\alpha(E)$ varying with the local field. But since the electrons go through a thin γ-sheath very fast, the field phase variation over the time of flight can be neglected, so (2.20) is considerably simplified. Eventually, we have

$$\gamma \int_{-\pi/2}^{\phi_1} E(0, \phi) \left\{ \exp \left(\int_0^{x_s(\phi)} \alpha[E(x, \phi)] dx \right) - 1 \right\} d\phi = \int_{-\pi/2}^{\phi_1} E(0, \phi) d\phi \quad (2.26)$$

We have passed from t to $\phi \equiv \omega t$; the function $x_s(\phi)$ or $x'_s(\phi)$ is taken as the upper integral limit in x depending on ϕ, and the integration over time (ϕ) is over the cathode phase only.[5] The set of equations (2.22)–(2.26) serves for finding maximum sheath voltage $V_{s\,max}$ as a function of n_s or j_a, in other words, for evaluation of the sheath CVC.

2.5.3 The CVC and normal sheath parameters subject to nonlocal effects

We will omit discussion of the results obtained from equation (2.26) [2.3, 2.9]; these results are in reasonable agreement with the numerical simulation and this fact makes the major assumption of the $n_s = $ const model, acceptable with certain reservations. Instead, we will focus on the CVC calculated within the same model

---

[4]This equation was deduced in [2.28]; a simplified model corresponding to the limit $\xi \ll 1$, which is realized at moderate pressures (Figure 2.18), was discussed in [2.3].

[5]An averaged Townsend condition of the kind (2.26) was first formulated in [2.10] for the α-mode in which, in contrast, it should not be fulfilled.

but using a better description of ionization, in which equation (2.26) is replaced by a more general equation. Account will be taken of the influence of nonlocal effects on ionization. This permits a refinement of the sheath CVC and a clear interpretation of the origin of negative glow (NG), of Faraday dark space (FDS), and of some other effects observable in a γ-discharge.

In a strongly nonuniform field, the energy spectrum of electrons at a given point of space x_1 depends not so much on the local field $E(x_1)$ as on the total distribution of $E(x)$. For example, in the absence of inelastic collisions, the energy of an electron is, in general, determined by the potential difference rather than by the local field. The nonlocal character of the electron spectrum radically changes the spatial distribution of the sources of charge generation. In particular, the plasma bordering on the space charge sheath contains, in spite of a very low field, a large number of energetic electrons that have gained their energy in the sheath. So, the gas can be ionized at some distance from the sheath where the field is practically zero. This nonlocal effect is the major source of NG regions and then of the FDS in a glow discharge [2.18]. The electron energy spectrum necessary for finding the ionization rate should be searched by solving a kinetic equation over the whole nonuniform field space. Such a calculation was made in [2.31] by solving a simplified kinetic equation in a 'forward-backward' approximation for a dc glow discharge. As a result, the full distribution pattern of charge densities and field in the cathode fall (CF) layer, of the NG and the FDS as far as the positive column, were obtained.

The same approach was used to obtain similar parameters for the RF γ-discharge. Considering that the time of flight for the electrons in a thin γ-sheath is small compared with the field oscillation cycle, one can assume the electron spectrum and ionization rate $q(x,t)$ [cm^{-3} s^{-1}] to be quasistationary, i.e., corresponding to the instantaneous field distribution $E(x,t)$. The latter is given by (2.22) at $x < x_s$ and $E \approx 0$ at $x > x_s$. The number balance of ions is now described by the equation

$$\int_{-\pi/2}^{\phi_1} \mathrm{d}\phi \int_0^\infty q(x,\phi)\mathrm{d}x = \int_{-\pi/2}^{\phi_1} n_s \mu_+ E(0,\phi)\mathrm{d}\phi$$

The results of the calculation based on the approximate solution of the kinetic equation are given in Figure 2.19 for nitrogen, $p = 15$ Torr and $f = 13.56$ MHz. The CVC has a well-defined minimum corresponding to the normal current density amplitude $j_{na} = 55$ mA cm^{-2}, the ion density in the sheath $n_s = 1.1 \times 10^{11}$ cm^{-3} and the maximum voltage fall on the sheath $V_{n\,max} = 334$ V. The cycle-average normal fall coinciding with the constant plasma potential $\overline{V}_n = \overline{V}_p = 124$ V fits well the measurements made in [2.4]. These values are of the same order of magnitude as the normal cathode falls in a nitrogen glow discharge (180–230 V, depending on the cathode metal). The maximum sheath thickness in the normal mode $x_{n\,max} = 0.058$ cm and $px_{n\,max} = 0.85$ Torr cm. The average value of $p\bar{x}_n = 0.4$ Torr cm is close to the normal values in the glow discharge $(pd)_n \approx 0.3$–0.4 Torr cm. The anode stage lasts for $3\pi/2 - \phi_1 = 0.4\pi$, i.e., 20%

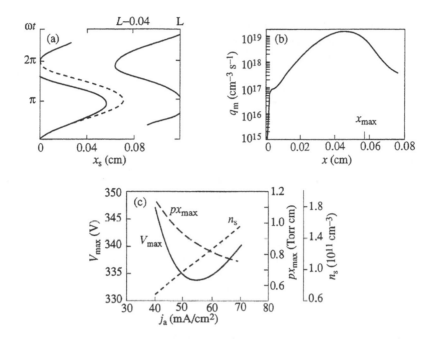

FIGURE 2.19
The 'cathode' sheath at the left-hand electrode of a nitrogen γ-discharge at $p = 15$ Torr, $f = 13.56$ MHz. (a) The sheath thickness variation x_s over a cycle at normal current density $j_{an} = 55$ mA cm^{-2}; the trajectory $x_s(t)$ with no 'anode' stage is given by the dashed curve for comparison. (b) Ionization rate distribution $[q(x)]_m$ for maximum sheath thickness and field (also at j_{an}). (c) The sheath parameters versus the current density amplitude j_a: maximum sheath voltage $V_{s\,max}$, i.e., the sheath CVC, maximum sheath thickness, or $px_{s\,max}$, and the sheath ion density.

of the cycle. Maximum ionization rate is observed at the very edge of the space charge sheath. A considerable portion of the ions is produced in the plasma (in the NG regions) where the field is very weak [Figure 2.19(b)], which is associated with nonlocal effects, as in a glow discharge.

The maximum ionic current on the electrode is $0.25j_{na}$, i.e., the RF current from the plasma is closed on the electrode primarily by displacement current—a situation typical of moderate pressures. The contribution of ionic current to the total current grows with pressure, which is clear from Figure 2.18 and follows from the proportions $j_{ni} \sim p^2$ and $j_{n\,dis} \sim p$. The physical meaning of this correlation may be interpreted as follows. The ratio of the maximum ionic current to the displacement current amplitude at the electrode is

$$\frac{j_{i\,max}}{j_{a\,dis}} \approx \frac{en_+\mu_+ E_{0\,max}}{\omega(E_{0\,max}/4\pi)} = \frac{4\pi en_+}{E_{0\,max}}\frac{\mu_+ E_{0\,max}}{\omega} = \frac{A_+}{x_{n\,max}} \qquad n_+ = n_s \qquad (2.27)$$

where $A_+ = \mu_+ E_{0\,max}/\omega$ is a value approximately equal to the ion displacement over a cycle. Like the electron oscillation amplitude, it weakly depends on pressure, whereas in the normal γ-mode, $x_{n\,max} = E_{0\,max}/4\pi e n_+ \sim p^{-1}$. At low pressures, $A_+ \ll x_{n\,max}$. Because of the ion escape into the electrode during the cathode phase, the sheath part of thickness A_+ becomes depleted of ions, and this deficit is not compensated during the cycle by ions arriving from the remote sheath edge where they are produced. The ionic current becomes weak and is compensated by the displacement current. At elevated pressures, when $A_+ \approx x_{n\,max}$, the ion escape is immediately, within a cycle, compensated by arriving ions, so the ionic current is high. At fairly high pressures, when $A_+ > x_{n\,max}$, the ionic current prevails. However, the replacement of $j_i < j_{dis}$ by $j_i > j_{dis}$ at higher pressures is not accompanied by abrupt changes or transitions. This has been demonstrated experimentally and evidences against the opinion, that is sometimes expressed, that the α- and γ-modes differ in the proportions of the conduction and displacement currents.

The calculation made at $p = 5$ Torr has yielded $V_{n\,max} = 340$ V, $j_{na} = 20$ mA cm^{-2}, $x_{n\,max} = 0.14$ cm, $j_{i\,max} \approx 0.14 j_{na}$ and $3\pi/2 - \phi_1 = 0.3\pi$, which nearly satisfies an approximate similarity principle with respect to p. Calculation of the previous version with $p = 15$ Torr but without account of nonlocal effects yields $V_{n\,max} = 370$ V, $j_{na} = 60$ mA cm^{-2}, $n_s = 1.9 \times 10^{11}$ cm^{-3}, $x_{n\,max} = 0.045$ cm and $j_i/j_{dis} = 0.28$, which does not differ much from the calculation with the kinetic equation. This is a favorable result, since it opens up the possibility of calculating normal parameters in a simpler way, using the Townsend coefficient $\alpha(E)$ and equation (2.26).

2.5.4 Negative glow and the Faraday dark space

Bright regions of negative glow (NG) on the side of each electrode followed by dark regions of the Faraday dark space (FDS) separating a NG region from a rather bright PC are clearly seen in a γ-discharge shown in Figures 2.11 and 2.12. They arise from the same reasons as in the glow discharge. Intense NG is due to the excitation of atoms by energetic electrons arriving from the sheath, being accelerated there by a strong field. The field in the plasma adjacent to the sheath reduces in amplitude and becomes so weak that it cannot give off its energy to the electrons anymore. Where the flux of energetic electrons is exhausted, the processes of ionization and excitation cease [Figure 2.19(b)] and the gas stops emitting light. This is the FDS (Figures 1.11 and 1.12). Since the charge losses are not compensated anymore, the plasma density n decreases with distance from the electrode. Decreasing with x in the same direction, the field passes 'by inertia' the zero point and reverses its direction. This effect of field reversal is well-known for the glow discharge. The electrons drift in the weak field very slowly, and in the reverse field region they drift even in the opposite direction. The current in the FDS is transported by free electron diffusion towards decreasing n. The electron drift in the reverse field partly compensates for the diffusion flux, reducing it to a value of discharge current. When n becomes so low that the diffusion is incapable to maintain the current, the field recovers and the diffusion current is again replaced by the drift current. If the gap is large enough, the field is recovered to a value at which the recovered ionization (this time, by electrons accelerated by the recovered field) compensates for the losses. In this way, the FDS transforms to the PC.

The distribution of n and E in the NG region, FDS and PC are described by the same equations of the charge number and current conservation as for the dc plasma [2.28], but the charge sources should be averaged over a cycle (plasma density fluctuations are weak)

$$-\frac{d}{dx}\left(D_a \frac{dn}{dx}\right) = \bar{q} + \bar{\nu}_i n - \nu_d n - \beta n^2 \qquad (2.28)$$

$$j_a \cos \omega t = e\mu_e n E + eD_e \frac{\partial n}{\partial x} + \frac{1}{4\pi}\frac{\partial E}{\partial t} \qquad (2.29)$$

Here $\bar{q}(x)$ is the cycle-average rate of gas ionization by fast electrons coming from the sheath; it is found by solving the previous problem (Section 2.5.3). In addition to recombination losses, these equations make allowance for the diffusion plasma losses with the frequency of ambipolar diffusion in the lateral direction $\nu_d = D_a/\Lambda^2$ (Section 1.4.3). The latter losses are often dominant in atomic gases. For simplicity, the frequency of gas ionization by plasma electrons is assumed to be the function of the local field and is found from equation (1.36). (When this assumption is not made, a nonmonotonic FDS-to-PC transition appears. For this effect, see [2.31].) The electron temperature on which the coefficients of free and ambipolar diffusion, $D_e = \mu_e T_e$ and $D_a = \mu_+ T_e$, depend may be taken to be constant, $T_e \approx 1$ eV. Equation (2.29) defines the field on which the ionization frequency $\bar{\nu}_i$ depends

$$E = \frac{j_a}{e\mu_e n}\cos(\omega t + \psi) - \frac{T_e}{n}\frac{\partial n}{\partial x} \qquad \psi = \arctan\left(\frac{\omega}{4\pi e n \mu_e}\right) \qquad (2.30)$$

The cycle-average (steady) field

$$\bar{E} = -\frac{T_e}{n}\frac{\partial n}{\partial x} \qquad (2.31)$$

is related only to the plasma polarization due to electron diffusion. In a uniform PC, $\bar{E} = 0$. A recovered field is actually harmonic, and formula (1.36) is valid. The boundary conditions for (2.28) are: at the plasma–sheath boundary $n = n_s$; far into the PC, $dn/dx = 0$ at $x \to \infty$.

Figure 2.20 presents the calculations for nitrogen at $p = 15$ Torr, $f = 13.56$ MHz, and normal γ-current density $j_{na} = 55$ mA cm^{-2}, $n_s = 10^{11}$ cm^{-3}. At currents $I \sim 1$ A, typical of moderate-pressure γ-discharges in molecular gases, recombination losses are dominant, $\beta n \gg \nu_d$. The plasma density grows rapidly from the boundary with the sheath, reaches a maximum, and drops down to the PC value corresponding to ionization–recombination equilibrium at a given

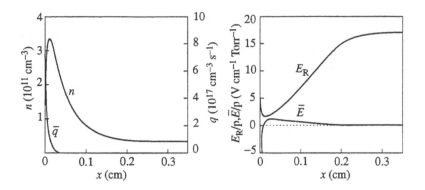

FIGURE 2.20
Distribution of plasma density n, of cycle-average rate of gas ionization by energetic electrons $\bar{q}(x)$, and of rms and average fields E_R/p and $\bar{E}/p$ in the NG, FDS, and in the PC at the left-hand electrode in nitrogen; $p = 15$ Torr, $f = 13.56$ MHz, and normal current density $j_{an} = 55$ mA cm^{-2}.

current density. Maximum n value lies in the NG-to-FDS transition region, where $\bar{q}$ almost vanishes. The rms field recovers up to a PC value calculated neglecting the gas heating, so it is overestimated. The FDS length $L_F \approx 0.25$ cm. The mean constant field at the plasma–sheath boundary is still directed to the left-hand electrode, as in the sheath. The rapidly decreasing field $\bar{E}$ passes the zero point and changes its sign. In the FDS, $\bar{E}$ is reversed and directed to the gap center; it vanishes only in the PC.

The dependence of the FDS length on discharge conditions is quite clear from an approximate expression that is easy to derive by integrating (2.28) and omitting the terms for the ionization, which is really absent in most of the FDS [2.28]

$$L_F \approx \frac{R}{2.4} \ln\left[\frac{(Z_c + 1)(Z_m - 1)}{(Z_c - 1)(Z_m + 1)}\right]$$

$$Z_m = \left(1 + \frac{2\beta n_{max}}{3\nu_d}\right)^{1/2} \qquad Z_c = \left(1 + \frac{2\beta n_c}{3\nu_d}\right)^{1/2} \qquad (2.32)$$

where R is the discharge column radius, $\nu_d = D_a(2.4/R)^2$, n_{max} is the peak plasma density in the NG region, and n_c is the PC density. In the limiting cases of dominant recombination losses, $\beta n_c/\nu_d > 1$, or diffusion losses, $\beta n_{max}/\nu_d < 1$, assuming $n_{max} \gg n_c$, one can get very simple expressions

$$L_{F\,rec} \approx \left(\frac{6D_a}{\beta n_c}\right)^{1/2} \qquad L_{F\,dif} \approx \frac{R}{2.4} \ln\left(\frac{n_{max}}{n_c}\right) \qquad (2.33)$$

In the first case, characteristic of molecular gases at moderate pressures, $L_{F\,rec} \sim 1/p$ and is independent of current $I_a = j_{na}\pi R^2$ ($D_a \sim 1/p$, $n_c \sim j_{na} \sim p$). The second case is for atomic gasses and more low pressures: the logarithmic factor weakly depends on p and I, and $L_{F\,dif} \sim R \sim (I_a/p)^{1/2}$. [Formulas (2.32), (2.33) and their implications are equally valid for a dc glow discharge.] The NG region length L_{NG} is determined by the relaxation length of an energetic electron flux, $L_{NG} \sim 1/p$; $L_{NG} \approx 0.025$ cm in nitrogen at $p = 15$ Torr.

Although E_a and T_e of α- and γ-PCs do not differ much at the PC middle, they differ strongly at its edges. For the γ-mode, E_a and T_e are low here, since the PC contacts the FDS regions. But for the α-mode these values are higher at the PC edges than at the middle, because the PC contacts the sheaths. Let us recall (Section 2.3.4) that the ionization rate maxima in the α-mode are just at the boundaries between the PC and the sheaths.

2.5.5   PC-free γ-discharge: PC in the α- and γ-modes

It is clear from the foregoing that the PC is absent and the FDSs from both electrodes come into contact at the gap center if L is small and the pressure is low. At still lower pressures, when $L_{NG} > L/2$, the FDS may also disappear. The gap plasma is then the plasma of two NG regions, and energetic electron fluxes go through the whole gap. What we have then is something like a hollow cathode. Figure 2.21 illustrates the distribution of n and E for the first case when $L \lesssim 2L_F$, $p = 5$ Torr, and $L = 1.2$ cm. The calculated rms field has a maximum and the plasma density has a minimum at the gap center. The mean field distribution qualitatively coincides with that observed in neon at $f = 13.56$ MHz, $L = 6$ cm, and $p = 0.1-1$ Torr [2.32]. Near the electrodes the mean field is directed to them, far from the electrodes the mean field is weak and directed to the gap center.

The plasma of a short low-pressure γ-discharge exhibits all properties of the FDS or NG region but not of the PC. The field amplitude in it and, hence, the electron temperature are low. This does not, however, contradict the condition for

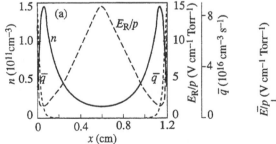
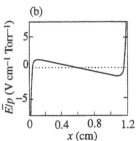

FIGURE 2.21
Distributions of plasma density n, of cycle-average ionization rate $\bar{q}$, of rms and average fields E_R/p and $\bar{E}/p$ when two FDSs come into contact in absence of PC, for nitrogen, $p = 5$ Torr, $L = 1.2$ cm, $f = 13.56$ MHz, normal current density, $j_{an} = 20$ mA cm^{-2}.

plasma maintenance, because there is no need to maintain it via local ionization. Charges are generated only in the vicinity of the electrodes (in the sheaths or NG region); but in the gap bulk where the plasma degenerates, the losses are compensated by ambipolar charge diffusion from the two NG regions at the electrodes. One may say that the plasmas of low-pressure γ-discharges or of high-pressure short γ-discharges are non-self-sustained, contrary to the classical uniform PC which is self-sustained and autonomous because charge losses in it are compensated by local ionization. It follows from the above estimations for nitrogen that the PC vanishes at $p = 5$ Torr for $L \approx 1$ cm and at $p = 1$ Torr for $L \approx 5$ cm. For example, nitrogen γ-plasma is equivalent to two FDSs in a gap of $L = 5$ cm at $p \lesssim 1$ Torr.

Such was the situation in the experiments [2.12] and in the calculations [2.13], which led the authors of the former work to draw the conclusion that γ-discharge plasmas are characterized by low field amplitudes and electron temperatures. We believe that this conclusion is justified only with a reservation concerning the discharge length. If under these conditions the electrode spacing is made much larger, a PC is formed at its center with a much greater field amplitude and electron temperature.

The field amplitude E_a and the electron temperature T_e in a long (relative to the sheath thickness) and approximately uniform PC are limited by the necessity of providing the charge number balance in the plasma, i.e., by the condition of compensation of all processes of charge production and loss. In this respect, the PC of the α- and γ-modes do not differ from one another. On the other hand, the field and T_e necessary for the maintenance of a certain plasma density n do not depend strongly on n. For instance, when the major role is played by electron impact ionization of unexcited atoms and ambipolar diffusion charge losses, E_a and T_e do not depend on n at all, $\nu_{iRF}(E_a) = \nu_d n$. For this reason, E_a and T_e do not differ much in sufficiently long and uniform PC of the α- and γ-discharges. The difference primarily lies in the ionization levels that serve as a background for the local processes of charge production and loss in the field and for the local electron oscillations inducing oscillating current. In the uniform PC, these ionization levels are determined by the processes occurring in the sheaths. In a γ-discharge, the high plasma density is associated with intense multiplication of charges in the strong field of a self-sustained sheath. This maintains normal parameters in the sheath and a fairly high normal current density. In an α-discharge, ionization occurs in the plasma where the field is substantially weaker than in the sheaths, and the current density is much lower than in the γ-discharge.

2.5.6   Discharge contraction and its absence

If the current between two large plane electrodes of a normal γ-discharge is increased, the discharge space expands and the current column diameter D becomes larger. Experiments [2.33] have shown, however, that the current and the column diameter may be increased infinitely (if the electrode area permitted), provided

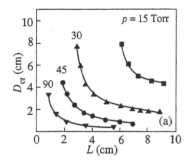
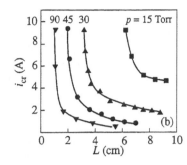

FIGURE 2.22

Contraction-critical (a) discharge column diameter and (b) current strength versus the gap size at various pressures in air, $f = 13.56$ MHz. On the right of and above the curves, the uniform γ-discharge column loses its stability and contracts.

the interelectrode distance L is smaller than a certain critical size L'_{cr}, which is inversely proportional to pressure (Figure 2.22). If $L > L'_{cr}$, the column uniformity is retained only up to a maximum current value I_{max} corresponding to a maximum column diameter D_{max}. If the current is increased further, the discharge column contracts to a bright filament. The contraction occurs more readily when the pressure and the difference between L and L'_{cr} are increased. For instance, in air $L'_{cr} = 2.5$ cm at $f = 13.56$ MHz and $p = 30$ Torr. At $L = 4$ cm, the current and the diameter may be increased only up to $I_{max} = 3.5$ A and $D_{max} = 3$ cm with a subsequent contraction of the discharge column. At $L = 8$ cm, the contraction occurs earlier at the same f and p, i.e., at $I_{max} = 2$ A and $D_{max} = 2$ cm.

The fact that the contraction is stimulated by growing pressure is well known for the dc glow discharge [2.18]. The reasons are the same as above. The discharge column contracts due to the ionization overheating instability when the heat release into the discharge $\langle jE \rangle$ exceeds a threshold value, which is independent of pressure in case of the discharge cooling through heat conduction. Since $E \sim p$ in the PC plasma, the threshold current $j \sim p^{-1}$ decreases with increasing pressure. Without going into the details of RF discharge stability theory [2.33], we will comment briefly on the phenomenological nature of this effect in an attempt to explain why there is a critical interelectrode distance L'_{cr}, below which no contraction occurs.

Stabilization of a uniform state and contraction suppression are stimulated by an outward heat transport, whose rate is characterized by the transport 'frequency' $\nu_\chi = \chi/\Lambda^2$, where χ is thermometric conductivity and Λ is a value close to a characteristic diffusion length. For a cylindrical column, $\Lambda^{-2} = (\pi/L)^2 + (4.8/D)^2$. The relaxation rate of hazardous longwave perturbations at a given spacing L decreases with increasing cylinder diameter, so that starting with a certain maximum diameter D_{max}, the heat transport cannot any longer cope with the growing gas temperature, causing the column contraction. However, at sufficiently small L, the heat transport frequency ν_χ is large at any, even 'infinite,' diameter. Instability is suppressed due to the heat transport nearly electrodes. Indeed, the heat release rate at a given pressure $\langle jE \rangle$, with which the heat transport competes, is limited by the respective normal current density, so at $L < L'_{cr}$ the transport rate exceeds any possible heat release rate. This effect is important for discharge practice. If one sets oneself the task of filling a maximum volume with a uniform RF plasma at a maximum pressure, the chances are that one will be able to achieve this without special tricks, merely by making the gap between the large plane electrodes sufficiently narrow. This is also evidenced by the theoretical and experimental data of [2.34] on RF discharge stability to contraction.

2.6   The $\alpha-\gamma$ transition at moderate and low pressures

The subject of this section is an explanation why the $\alpha-\gamma$ transition occurs abruptly at moderate pressures but smoothly at low pressures. A clear, up-to-date concept of this subject has been shaped in [2.36, 2.37, 2.4, 2.12, 2.28] on the basis of experimental studies and theoretical analyses.

The reader has already been introduced to normal α- and γ-discharges, to the electrode sheath structure, and to the $\alpha-\gamma$ transition mechanism. So, the reply to the above question should be quite clear.

Figure 2.23 schematically shows the Paschen curve and the plots for the sheath thicknesses in the α- and γ-modes as a function of pressure: d_α is weakly dependent on pressure, and $d_\gamma \approx d_m \sim p^{-1}$ nearly fits the Paschen curve. Figure 2.24 illustrates experimental plots for d_α and d_γ versus p [2.35]. Let us first revise briefly what we already know about the $\alpha-\gamma$ transition at moderate pressure. Denote with p_c the pressure at which the sheath thicknesses in both modes are identical. It will correspond to the interception point of the curves $d_\alpha(p) \approx$ const and $d_\gamma(p) \approx$ const$'/p$. For example, in nitrogen $d_\alpha \approx 0.35$ cm, $pd_\gamma \approx 0.85$ Torr cm, and $p_c \approx 2.5$ Torr. Suppose $p > p_c$, i.e., pd_α relates to the right-hand branch of the Paschen curve. If the normal α-current amplitude is increased, the current density j_a, the plasma density n and, hence, the average sheath voltage $\overline{V}_p$ will also increase after the electrode coverage by current. When $\overline{V}_p$ has increased to about breakdown voltage V_t, the sheath is broken down and an ionization wave travels from the boundary to the electrode, raising the sheath ion density. The sheath abruptly contracts from d_α to $d_\gamma \approx d_m = d_n$. The average sheath voltage falls from V_t to $V_m = V_n$. Since the right-hand branch of the Paschen curve with $pd > (pd)_m$ is fairly flat in the vicinity of the minimum, the sheath voltage decreases more slowly than the thickness. As a result, the sheath field $E = V/d$ and the current density $j = j_{dis} \approx \omega V/4\pi d$ rise. The total current, being limited by the circuit resistance (reactive or active), grows much slower, so the discharge space contracts to produce a normal γ-discharge.

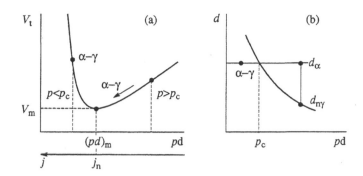

FIGURE 2.23
A diagram illustrating the abrupt and continuous character of the α−γ transition as a
function of pressure based on the same principles as suggested in [2.36, 2.37, 2.12]. The
current density scale has been added for clarity.

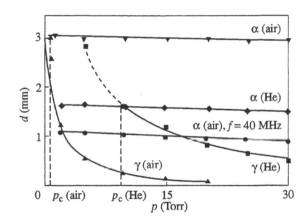

FIGURE 2.24
Experimental plots of the sheath thickness in the α- and γ-discharges in helium and in air.
All curves, except the one marked with $f = 40$ MHz, are for the frequency 13.56 MHz.
One can see the points p_c indicating the interception of d_α and d_γ.

Suppose now that the gas pressure is low, less than p_c, and the value of pd_α
lies in the left-hand Paschen curve. Breakdown of the α-sheath occurs due to
the increased discharge current and voltage. In principle, a 'deionization' wave
might arise after the breakdown, expanding the produced γ-sheath from d_α to
a normal thickness $d_n \approx d_m$ optimal for charge multiplication. In this case,
however, the density of the sheath displacement current and of the discharge
current $j \approx \omega V/4\pi d$ would be inevitably reduced since V falls and d grows.
But the low-pressure α-discharge has already covered the whole electrode prior

to the breakdown (Section 2.3). Further decrease in the current density is quite impossible without a decrease in the current magnitude. On the other hand, a discharge current decrease is prevented by the external circuit resistance. As a result, at $p < p_c$, the α–γ transition is not accompanied by abrupt changes in d and j, but the alteration in the electron multiplication mechanism reveals itself only in the CVC bending. As the voltage in the arising γ-discharge grows, V and d_γ remain in the left-hand Paschen curve, i.e., d_γ decreases while j and the total current grow.

The dynamics of the α–γ transition correlates fairly well with the shape of the CVC, or, to be more exact, with the plot of V versus j. At moderate pressures, the CVC is given by a drop-down curve at the α–γ transition, corresponding to an unsteady state and its abrupt change to a more favorable state. At low pressures, the CVC is given by a rise-up curve corresponding to a steady state, so there is no abrupt change.

In Section 2.1 we discussed the similarity between the α–γ transition at moderate pressures and the formation of a normal glow discharge after gap breakdown. In either case, the ionization wave reduces the sheath to a size optimal for charge multiplication and corresponding to the Paschen curve minimum. The α–γ transition at low pressures, when an optimal sheath size cannot be achieved, also has an analog. In breakdown of a gap d with very low gas pressure, when pd relates to the left-hand Paschen curve, no normal glow discharge can be excited, because the gap is too small for a normal cathode sheath to be formed, $d < d_n$. The electrode voltage then does not fall after the breakdown, and so-called obstructed discharge is excited with $V > V_n$.

2.7 Coexistence of two RF modes in the gap

2.7.1 'Parallel' operation of the α- and γ-discharge modes

Generally, an RF discharge operates either in the α- or γ-mode, depending on the experimental conditions. A mode can be easily identified even from its appearance (Figure 1.11). Situations are possible, however, in which both discharge modes may be observed in the gap simultaneously. We will describe such observations and try to explain why this happens. For example, this effect was observed in helium at $p = 100$ Torr and $f = 13.56$ MHz in the installation described in Section 1.8.1 [2.33]. Two brass discs of 11 cm diameter were used as electrodes, and the spacing between them could be varied up to 10 cm. A typical picture of coexisting α-and γ-modes is shown in Figure 2.25 with radial distributions of the sheath thickness and the electrode current density ($L = 2$ cm). At the center of the discharge column, there is a typical γ-discharge of about 2 cm radius with a thin sheath of $d_\gamma \approx 0.01$ cm and a high current density $j_\gamma \approx 100$ mA cm^{-2}. The γ-column is surrounded by a circular cylinder of the outer radius ~ 5 cm in which a typical α-mode operates. The sheath thickness in it is an order of magnitude larger, ~ 0.1 cm, and the current density an order lower, ~ 10 mA cm^{-2}. As the current rises, the γ-column expands and the α-circle reduces. When the current is decreased, the α-space expands towards the center, forcing out the γ-discharge which eventually disappears, leaving behind only the α-discharge.

The reason for this effect is that in the experimental conditions, the operating voltage of the normal α-mode V_α is lower than that of the normal γ-mode V_γ. If the gap current and voltage are gradually increased starting with small values, there is no discharge at first, and only a capacitive current passes through the gas capacitor formed by the electrodes. Then, at a sufficiently high voltage, a breakdown occurs exciting a normal α-discharge while the voltage drops to V_α. With further current increase, the α-discharge covers the whole electrode and becomes abnormal; the voltage grows, and at the α–γ transition voltage V_{tr} a normal high current density γ-mode is excited. The discharge immediately contracts to a rather thin filament while the electrode voltage drops to V_γ corresponding to a normal γ-discharge. This voltage embraces the whole gap, including the currentless periphery. But voltage V_γ is sufficient to maintain here a normal α-discharge with a weak current since $V_\gamma > V_\alpha$. The latter state is preferable, and the currentless state around the γ-filament is unstable. Breakdown and ignition of an α-mode in the currentless zone are stimulated by the heating and thermal expansion of the gas due to a considerable heat release in the γ-filament, to an outward ambipolar diffusion of the bulk plasma and, perhaps, to the gas ionization by the filament radiation.

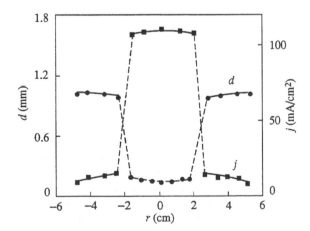

FIGURE 2.25

Coexistence of the α- and γ-modes in the gap. Measured radial distributions of sheath thickness d and current density j for helium, $p = 100$ Torr, $f = 13.56$ MHz, $L = 2$ cm.

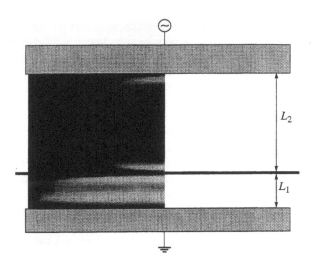

FIGURE 2.26
Two different α-modes in the gap divided into two unequal parts with an insulated plate parallel to the electrodes. Mixture He:air=1:1, $p = 15$ Torr, rms current is 0.35 A, $L_1 = 1.5$ cm, $L_2 = 3.5$ cm.

2.7.2 'Series connection' of two α- or α- and γ-discharges

It is worth discussing another nontrivial experiment, in which the specific features of the α- and γ-modes are clearly manifested [2.21]. A 0.1 cm insulated metallic (or dielectric) plate was introduced between the plasma electrodes parallel to them. Being immersed in an ionized gas, the plate acquires the local plasma potential, or so-called floating potential. The discharge column splits into two successive (at constant pressure) columns with a space charge sheath on each side of the plate. The sheaths are formed by electrons oscillating near the plate surface, alternatively exposing or covering the ionic gas, as in the vicinity of the electrodes.

If the plate is placed at the spacing center, both gaps turn out to be under identical conditions and have an identical discharge. At appropriate voltage and current, these are normal α-modes. But the normal current density of an α-discharge is known to depend on the interelectrode distance (see Section 2.3); so when the plate is shifted closer to one of the electrodes, the current density in the narrower gap becomes lower while the column diameter becomes larger than in the wider gap (Figure 2.26). For example, in a mixture of helium and air in equal proportions at $p = 15$ Torr, $f = 13.56$ MHz and $i = 0.35$ A, in the narrow gap of $L_1 = 1.5$ cm, $j_{n\alpha} = 5$ mA cm^{-2} while in the wide gap of $L_2 = 3.5$ cm, $j_{n\alpha} = 15$ mA cm^{-2}; the respective column diameters differ by a factor of $\sqrt{3}$. Moreover, if L_2 and pressure are so large that pL_2 exceeds the critical value of $(pL)_{cr}$ for the mixture, a γ-discharge is ignited in this larger gap. If the plate is shifted slowly, the γ-mode arises directly from the normal α-discharge that does not cover the whole electrode. This happens because the normal α-mode current density corresponding to a large value of pL_2 turns out to be higher than that of the transition current j_{tr} at which an inevitable breakdown occurs. Meanwhile, the α-mode is still operative in the narrower gap, so that there is a 'series connection' of the two modes. These phenomena have also been observed in other gases (He, Ar, N_2, and CO_2) and at other frequencies (40.7 and 81.3 MHz). At higher pressures, the α- and γ-modes coexist at smaller L_2, in full agreement with the principal role of the product pL_2 in the RF discharge.

2.8 High-pressure RF capacitive discharges

RF capacitive discharges operating at high pressures have not been studied as thoroughly as moderate- and low-pressure discharges. However, there are some practical problems whose solutions require understanding of the physics of the phenomenon. There are three kinds of RF capacitive high-pressure discharges: those occurring in very narrow air gaps, in RF capacitive plasmatrons and torch discharges. A plasmatron discharge is created in a dielectric tube, normally with external annular electrodes. A gas is generally pumped through the tube at atmospheric pressure. When flowing through the discharge space, the gas partly transforms into an arc-type plasma, leaving the tube in the form of a plasma jet. The processes occurring in an RF capacitive plasmatron, the plasma properties and applications differ but little from those of inductive RF plasmatrons [2.18]. So we will not discuss them in this book but refer the reader to respective publications [2.38–2.40]. Here we will briefly discuss discharges in narrow air gaps and torch discharges.

2.8.1 RF discharges in narrow air gaps

Interest in these discharges is due, among other things, to their parasitic nature, because they normally occur between any closely spaced metallic objects near high-power RF transmitters. The discharge possesses nonlinear characteristics. It creates noises when emitting basic frequency harmonics. Air gap discharges have been studied in several works [2.41–2.43]. It is shown experimentally that either a glow-type nonequilibrium plasma or an arc-type equilibrium plasma is produced in the narrow high-pressure air gap. The electrodes in the two situations behave, respectively, as a glow cold cathode or as a cathode in an arc with a cathode spot. In either case, the electrode transforms from the cathode to the anode and back in each halfcycle.

Figure 2.27 presents voltage and current oscillograms in a symmetric glow-type discharge in an air gap of $L = 0.25$ mm at the frequency 1 MHz [2.43]. The

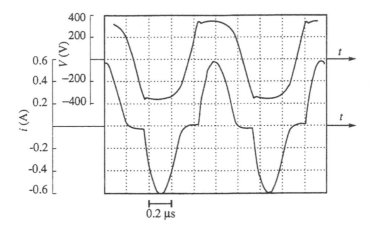

FIGURE 2.27
Oscillograms of the electrode voltage (upper curve) and discharge current (lower curve) in atmospheric air between plane electrodes separated by a gap of $L = 0.25$ mm; $f = 1$ MHz [2.43]. (©1971 IEEE)

plane copper electrodes were smoothly polished and cooled with water to avoid local heating and arc formation. Resistance of 500 Ω was connected in series with the external circuit to stabilize the current. At the beginning of each halfcycle, the electrode voltage V grows up to a certain value $V_i \approx 350$ V, at which a discharge is ignited. Prior to the ignition, the discharge current is low, although it does exist. Then it grows at a nearly constant voltage V_g, which varies little as compared to the ignition voltage. When V decreases at the end of the halfcycle, the current drops, remaining low until the moment of ignition in the next halfcycle but with the reverse polarity. In the time interval between the ignition and the decay, the discharge behaves as a dc glow discharge, with all inevitable implications. The current density is normal, the voltage is close to the normal value $V_n \approx 300$ V, and there is a blue cathode glow at the 'cathode' electrode. In each cycle, the cathode sheath at this electrode is formed anew and decays at sign reversal. These features are independent of frequency in the range studied, $f \approx 1$–30 MHz, and do not change, according to preliminary data, even at 80 MHz. At the frequency of 1 MHz, the 'dark' current prior to ignition is $0.03i_{\max}$ and at 50 MHz it is $0.5\ i_{\max}$. It is induced by the charge movement while the near-electrode region transforms from the 'anode' to the 'cathode' type (the displacement current is very small even in a strong field of the cathode sheath). At low frequencies and small L, the reignition voltage V_i increases since the charges contributing to breakdown have the time to leave the gap. An RF discharge similar to the abnormal glow discharge can sometimes be produced in the case of even a single small electrode (of 1 mm in size). This is, however, very hard to achieve, because the discharge prefers to transform into the arc-type as the current rises, filling the electrode.

A steady RF arc-type discharge commonly arises between graphite or rough copper electrodes. Arc formation is stimulated by increasing electrode spacing. An arc is also ignited when the electrode voltage becomes high, $V_i \approx 300\text{--}350$ V, especially with preceding week current flowing through the gap for some time. Following ignition, the voltage drops to $V \approx 20\text{--}40$ V and the current rises to $i \approx 1$ A. The current density and the glow brightness become much greater than in a glow-type discharge. The arc rapidly moves across the electrode surface as a conventional arc with a cathode spot. This movement is often followed by transition to a glow-type discharge. Sometimes, the discharge types alternate in each halfcycle or no discharge is ignited in one of the halfcycles. One may attribute this behavior to different properties and surface states of the electrodes.

The glow-type high-pressure RF discharge has much in common with the ac discharge operating at moderate pressures and $f = 10$ kHz. It was studied for application to CO_2 lasers [2.1]. Similarly to an atmospheric pressure RF discharge, a moderate-pressure low-frequency discharge shows the same sequence of events in each halfcycle: ignition, cathode sheath formation, extinction, and the sheath decay with subsequent formation of a new cathode sheath at the other electrode in the next halfcycle. There is a current 'pause' between the extinction in a given halfcycle and the ignition in the other halfcycle. The question arises as to why a high-pressure discharge of the megacycle-per-second range behaves very much as a moderate-pressure discharge of the kilocycle-per-second range but is different from a conventional RF γ-discharge, which also has many common features with the glow-type mode.

When the 'cathode' stage at a given electrode is over and its potential becomes higher than that of the other electrode, the created cathode sheath with a higher ion density than in the PC is filled by electrons that turn it to a high-density plasma. The field at the electrode becomes weak and the plasma starts decaying due to dissociative recombination (in a molecular gas) or to attachment (in air), the latter going on much faster than the recombination. If the near-electrode ion density drops to a value of about the positive column plasma density before the next cycle starts, a new cathode sheath should be formed, and the discharge will then be of the ac glow-type. If, however, the ion density does not drop to that value within a cycle, the elevated ion densities at the electrodes will persist, producing a conventional RF γ-discharge with 'permanent' dense ion sheaths at both electrodes, which will be periodically 'flooded' by electrons. The plasma recombination time from a density n_s in the cathode sheath to a density n_{PC} in the positive column is approximately

$$\tau_r \approx \frac{1}{\beta}\left(\frac{1}{n_{PC}} - \frac{1}{n_s}\right) \lesssim \frac{1}{\beta n_{PC}}$$

where $n_{PC} \approx j/e(\mu_e p)(E_{PC}/p)$ and j is a current density equal to normal $j_n = Cp^2$. The RF discharge behaves as an ac glow discharge if τ_r is less than the halfcycle $1/2f$, or if the frequency is lower than the critical frequency

$$f < f_{cr} = \frac{\beta C p^2}{e(\mu_e p)(E_{PC}/p)} \approx 330p^2 \text{ Hz} \qquad (2.34)$$

The numerical formula (2.34) refers to air and copper electrodes: $C = 2.4 \times 10^{-4}$ A cm^{-2} Torr^{-2}, $\mu_e p = 4.5 \times 10^5$ cm^2 Torr V^{-1} s^{-1}, $E_{PC}/p \approx 10$ V cm^{-1} Torr^{-1}, effective recombination coefficient (with account of attachment) $\beta \approx 10^{-6}$ cm^3 s^{-1}. For moderate pressures, say for $p = 30$ Torr, the discharge behaves as a low-frequency one only at f below 300 kHz; above this value, over the whole RF range, it is a conventional RF discharge. However, at $p = 760$ Torr the frequency limit of equation (2.34) increases up to 190 MHz. So, it is quite natural that in the range $f \sim 1$–30 MHz, and even at 81 MHz [2.43], an RF discharge behaves as an ac discharge. As for the appearance of an arc as gap size increases, it can be clearly interpreted within the model discussed in Section 2.5.6. The arc can be considered as a result of the glow discharge contraction, which is induced by the temperature rise due to the slower heat transfer to the electrodes as the gap becomes larger.

2.8.2 'Single-electrode' torch discharge

An RF torch normally arises in air near pointed metallic objects under high RF voltage. A plasma torch is attached to the object tip [Figure 1.1(e)]. The central bright region is surrounded by a less bright sheath. At low pressures, e.g., in pumped vessels, torch discharges may also arise but with a more diffuse glow. The impression that this process involves only one electrode is delusive. In reality, the other electrode is the earth or the nearby grounded object on which the discharge current is closed by displacement current. Detailed experiments have been carried out in air with a high-voltage tip and a grounded plane located at a distance $L = 5$ cm from the tip [2.44]. Basic parameters of this discharge are as follows. At the frequency $f = 70.6$ MHz and the input power $P = 100$ W, the discharge produces the current $i = 53$ mA. The torch length $l = 1.1$ cm, the bright channel radius $r_1 = 0.1$ cm and the radius of the less bright envelope is 0.24 cm. The temperature along the channel axis $T_m \approx 4000$ K. With decreasing frequency but the same input power, the current and temperature slightly decrease, and the torch becomes longer. Its radius, however, does not change, because it is determined by the input power. In a simplified model [2.44], the torch is treated similarly to a channel arc (see, e.g., [2.18]), assuming the plasma to be in equilibrium. A more detailed theoretical treatment of the torch is given elsewhere [2.45].

The authors of [2.44] did not measure the electrode voltage, but from the above data one can conclude that the torch voltage fall $V_T = P/i \approx 1.9$ kV, the ohmic resistance $\Omega = P/i^2 \approx 40$ kΩ and the conductivity $\sigma \approx 3.5 \times 10^{-3}$ Ω^{-1} cm^{-1} (if one takes the plasma displacement current to be small and the bright core radius to be twice the radius of the conducting channel). Although the estimated conductivity coincides, by order of magnitude, with the conductivity of equilibrium air plasma $\sigma \approx 10^{-2} \ \Omega^{-1} \, \mathrm{cm}^{-1}$ at $T = 4000$ K, $p = 1$ atm, there may be a substantial difference between the plasma electron and gas temperatures in a strong field $E \approx V_T/l \approx 1.8$ kV cm^{-1} [2.18]. The capacitance of the conducting torch above the plane is approximately $C = l[2\ln(l/r_1) - b]^{-1}$ cm ≈ 0.28 pF, where $b = 0.61$ at $L \gg l$. The reactive resistance of the air gap $X = 7.5$ kΩ is much less than the plasma resistance. The electrode voltage $V = i\sqrt{\Omega^2 + X^2} \approx 2.1$ kV. This value is consistent with the generator voltage.

2.9 RF discharge with coated electrodes

A practically valuable advantage of the RF discharge is the possibility to insulate the electrodes from the plasma with dielectric material. The electrodes may as well be placed outside a dielectric discharge chamber, and this makes the RF discharge preferable to dc discharges, in which the contact between metallic surfaces and an ionized gas is inevitable. This feature of RF discharges becomes especially attractive when a discharge is to be excited in a chemically aggressive gas, or when the plasma is to be kept uncontaminated by the electrode material, for example in cathode sputtering, or when high voltage electrodes to be placed in an evacuated volume necessitate changes in the discharge design, etc. The possibility of ion treatment of dielectric material covering the electrode and of dielectric film growth on its surface makes RF discharges indispensable for plasma technologies, for etching semiconducting and dielectric materials, for diamond-like film growth and other applications. Clearly, the physical mechanisms involved in the excitation and operation of a coated-electrode discharge must be well understood. We have pointed out above that the α- and γ-modes can operate regardless of whether the electrodes are naked or coated, and that if the gas, pressure, gap size and gap voltage (but not electrode voltage!) are identical, discharges between coated electrodes do not principally differ in the appearance, structure and characteristics from those for naked electrodes. But still, there are some quantitative as well as qualitative features that distinguish coated-electrode discharges. Some of these will be discussed here primarily with reference to moderate-pressure experiments.

2.9.1 Normal current density decrease in the α-discharge

The α-discharge between coated electrodes is characterized by a decreased normal current density and a wider range of parameters. These features are very important for physical measurements. The former permits taking current-voltage characteristics of the whole discharge as well as locally at current densities unfeasible for naked electrodes, $j < j_{n\alpha}$ (Section 2.3.1). The latter opens up greater options for controlling the discharge parameters in RF-excited lasers and other devices. In order to explain the physical nature of the effects involved, return to the derivation of equations (2.3) and (2.4). In the case of coated electrodes, the total capacitance of nonconducting space charge regions and dielectrics per unit area is defined by $[4\pi(d_\alpha + 2\delta/\varepsilon)]^{-1}$ rather than by $(4\pi d_\alpha)^{-1}$, where δ is the dielectric plate thickness, ε is the dielectric permittivity, and d_α is a value close to maximum sheath thickness. Expression (2.3) for the electrode voltage takes the form [2.22]

$$V = \left[\left(\frac{4\pi(d_\alpha + 2\delta/\varepsilon)j}{\omega} \right)^2 + \left(\frac{CpL_1}{j} \right)^2 \right]^{1/2}$$

(2.35)

Here, for simplicity, m is taken to be unity, $m = 1$; i.e., the CVC of the positive column (PC) is approximated by the function $V_{PC} = CpL_1/j$, which is acceptable in this section but will be specified in Section 5.3. For better accuracy, the gap size L has been replaced by the PC length $L_1 = L - d_\alpha$, where L within this section is the distance between the dielectric plates. The normal current density defined by the minimum $V(j)$

$$j_{n\,\alpha}^\delta = \left(\frac{\omega CpL_1}{4\pi(d_\alpha + 2\delta/\varepsilon)} \right)^{1/2}$$

(2.36)

is now lower than in the case of naked electrodes, $j_{n\,\alpha}$. The minimum electrode voltage at which the α-discharge operates

$$V_{\alpha\,\min}^\delta = \left(\frac{8\pi(d_\alpha + 2\delta/\varepsilon)CpL_1}{\omega} \right)^{1/2}$$

(2.37)

has naturally become larger. At $m = 1$ the voltage amplitudes on the PC and the sheaths plus on the dielectrics at the minimum are identical, irrespective of the dielectric thicknesses.

If the current is increased further after the dielectric area above the electrodes has been covered by the α-discharge (outside it, the field rapidly decreases even if the dielectric extends further), then the current density eventually reaches a value j_{tr} at which the sheaths are broken down, and an α–γ transition occurs. Since the critical value $j_{cr} \equiv j_{tr}$ is unrelated to the dielectric thickness, the current density range of the α-mode $j_{n\,\alpha}^\delta \leq j < j_{tr}$ becomes larger than in the case of naked electrodes. This is especially evident at high pressures (therefore, it is significant for lasers) and at a large distance L between the dielectric plates. Indeed, j_{tr} is slightly dependent on pressure and in experiment even somewhat decreases with growing p (see experimental data in Table 2.2). At the same time, $j_{n\,\alpha}$ grows with pressure and L_1, due to which the α-mode range necessary for lasers becomes narrower. This effect can be removed by using a sufficiently thick dielectric coating on the electrodes. In some instances, when $j_{n\,\alpha}$ grows with frequency more rapidly than j_{tr}, this is also good for high frequencies, say 81 MHz. Dielectric coatings make the α-mode feasible in a wide range of j values without the risk of transition to the γ-mode unfavorable for lasers.

It is clear from equation (2.36) that the effect of dielectric coating on the normal current density can manifest itself only at $\delta/\varepsilon \gtrsim d_\alpha$, while thin sheaths with $\delta/\varepsilon \ll d_\alpha$ have no effect whatsoever. For instance, at 13.56 MHz in air, the α-sheath thickness d_α is about 0.18 cm. It weakly depends on pressure, because in order of magnitude, $d_\alpha \sim v_{\mathrm{d}}/\omega$, where $v_{\mathrm{d}} = \mu_{\mathrm{e}}^0(E_{\mathrm{p}}/p)$ is electron drift velocity in the plasma ($\mu_{\mathrm{e}}^0 \equiv \mu_{\mathrm{e}} p = \text{const}$) and E_{p}/p in the plasma depends but slightly on p. For quartz coatings ($\varepsilon = 7$), the thickness must be larger than $\delta \approx \varepsilon d_\alpha \approx 1.25$ cm in order to exert a noticeable effect on $j_{\mathrm{n}\,\alpha}^\delta$ and the discharge.

2.9.2 Normalization of 'subnormal' current densities in the γ-discharge

When the α–γ transition occurs in a naked electrode system at rather large ('moderate') pressures, a normal current density $j_{\mathrm{n}\,\gamma}$ corresponding to the minimum CVC of the space charge region is established in the γ-mode. The electrode current area reduces to a value S_γ, which corresponds to $j_{\mathrm{n}\,\gamma}$ and the new value of the discharge current i ($S_\gamma = i/j_{\mathrm{n}\,\gamma}$). This happens because the states with smaller current densities $j < j_{\mathrm{n}\,\gamma}$ present in the α-mode before the transition, now become unstable, for they belong to the descending CVC branch of the γ-sheath (Section 2.6, Figure 2.23). The electrode insulation from the gas, however, provides 'subnormal' states with $j < j_{\mathrm{n}\,\gamma}$. We use the inverted commas to mean that states with $j < j_{\mathrm{n}\,\gamma}$ subnormal to naked electrodes may be treated as 'normal' for coated electrodes. Denote the respective current density as $j_{\mathrm{n}\,\gamma}^\delta$, especially as this value does depend on the dielectric thickness, more exactly, on δ/ε.

In order to understand the nature of the stabilizing effect on subnormal modes, turn to the schematic representation of the CVC of the various elements between the electrodes (Figure 2.28). This set of elements includes the sheaths, the positive column and the dielectrics. At moderately high pressures, when the ionic current to the electrode is substantially smaller than the displacement current, the γ-sheath CVC $V_{\mathrm{s}}(j)$ is derived from the Paschen curve $V_{\mathrm{t}}(pd)$ for a breakdown of a gap of size d. Indeed, the voltage oscillation amplitude on a sheath of thickness d coincides in order of magnitude with the average voltage $V_{\mathrm{s}} \approx V_{\mathrm{t}}$ which makes the Townsend condition (2.14) valid in the sheath. The displacement current density is, in order of magnitude, equal to $j \approx j_{\mathrm{dis}} \approx \omega V_{\mathrm{s}}/4\pi d$. This relation, together with equation (2.15) describing the Paschen curve $V_{\mathrm{t}}(pd)$ approximately, defines the sheath CVC $V_{\mathrm{s}}(j)$ but only in a parametric form, the parameter being d. Like the Paschen curve, the sheath CVC has a minimum (cf. Figure 2.23 in which the CVC is shown schematically by arbitrarily scaling j). The CVC depicted in Figure 2.28 is analogous to the cathode sheath CVC in a glow discharge in the Engel–Steenbeck theory [2.18]. Figure 2.19 shows an approximate CVC of the γ-sheath calculated for one set of conditions with allowance for the ionic current and nonlocal effects.

The PC CVC is known to have a descending character. When the gap length is small as compared to the transverse size and the discharge column diameter is nearly the same throughout the column, two-dimensional effects are unessential,

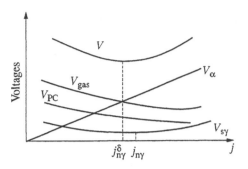

FIGURE 2.28
Schematic γ-CVC of the sheath $V_{s\,\gamma}(j)$, of the plasma $V_{PC}(j)$, of the gap $V_{gas}(j)$, of the dielectrics $V_d(j)$, and of the whole interelectrode system $V(j)$; $j_{n\,\gamma}$, normal current density for naked electrodes and $j_{n\,\gamma}^{\delta}$, for insulated electrodes.

and the PC current density j is the same as in the sheath at the given point of the cross section. The PC CVC $V_{PC}(j)$ is presented in Figure 2.28. Together with the sheath CVC, it forms the CVC of the discharge gap, $V_{gas}(j) = V_s(j) + V_{PC}(j)$, which is a descending curve at $j < j_{n\,\gamma}$ and even at slightly larger j, because the gap CVC minimum is shifted to the right with respect to the sheath CVC minimum $j_{n\,\gamma}$. The CVC for the dielectrics that relates j to the voltage fall V_d on the dielectrics $V_d = (8\pi\delta/\varepsilon\omega)j$ has an ascending character and represents a straight line. The total CVC of the system between the electrodes is given in Figure 2.28 as $V(j)$ with account of the phase shift between the components. It is seen to have a minimum which shifts to smaller current densities with increasing thickness of the dielectric (increasing δ/ε).

If the dielectric completely covers the electrode and if the current density at each point of its surface coincides with that at the given transverse discharge coordinate, this kind of dielectric serves as a distributed reactive ballast resistance to the discharge. It stabilizes the discharge in the state of the total CVC minimum if the minimum lies in the left, descending section of the γ-sheath CVC. The stability is provided by a strong negative feedback between the dielectric and the discharge. For example, suppose the discharge current density is to increase by Δj as compared with $j_{n\,\gamma}^{\delta}$ corresponding to the minimum in the $V(j)$ curve. The dielectric voltage will inevitably increase; and at the constant RF electrode voltage, the voltage across the gap V_{gas} will drop below the value $V_{gas}(j_{n\,\gamma}^{\delta} + \Delta j)$—which provides a stable state in the discharge, because the stable state at the current density $j_{n\,\gamma}^{\delta} + \Delta j$ will require a higher electrode voltage than V_{min}. Due to the shortage of voltage across the discharge, its current density will decrease and return to the steady value $j_{n\,\gamma}^{\delta}$ at the CVC minimum.

Thus, dielectric-coated electrodes can induce new 'normal' states with $j_{n\,\gamma}^{\delta} < j_{n\,\gamma}$; and by varying the dielectric thickness, one can obtain the whole spectrum of

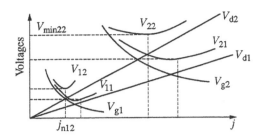

FIGURE 2.29
Schematic CVC of the gap V_{g1} and V_{g2} at pressures p_1 and $p_2 > p_1$, of the dielectric V_{d1} V_{d2} with thickness δ_1 and $\delta_2 > \delta_1$ as well as the resulting CVC of the interelectrode system V_{11}, etc. for the combinations $p_1, \delta_1; p_1, \delta_2; p_2, \delta_1; p_2, \delta_2$. Dashed lines correspond to the normal current densities j_{n11}, etc. and to the minimum operating voltages of the respective γ-discharges V_{m11}, etc.

states with current densities from $j_{n\gamma}$ to j_{tr}.[6] This effect can be clearly seen in experiment [2.36, 2.46] (it was at first observed experimentally and then interpreted as shown above [2.47]). It is very hard to separate in experiment the γ-sheath voltage from the $\pi/2$-shifted voltage of PC present at fairly large pL values. It is easier to measure the electrode voltage V, the dielectric voltage V_d, and the voltage across the whole gas space V_{gas}. The experimental CVC $V_{gas}(j)$ in the γ-mode is commonly a descending curve. The γ-discharge data for various gases and a wide ·range of p, from a few Torr to several hundred Torr, permit the various schematic CVCs to be plotted: $V(j)$, $V_{gas}(j)$, and $V_d(j)$. These are presented in Figure 2.29 for two values of the dielectric thickness and for two pressure values [2.47]. These curves clearly demonstrate how one can produce γ-discharges in a wide range of 'subnormal' current densities by varying the discharge parameters.

One should note a certain subtlety in the consideration above concerning the mechanism of 'subnormal' γ-sheath states with $j_{n\gamma}^{\delta} < j_{n\gamma}$. One can easily imagine a situation with the total CVC minimum lying at $j_{n\gamma}^{\delta} > j_{n\gamma}$, which occurs when the PC voltage is not high and the dielectric is thin, so that the straight $V_d(j)$ line crosses $V_{gas}(j)$ to the right of the γ-sheath CVC minimum. Following the above reasoning, one could assume that, as before, the state at the total CVC minimum, $V(j)$, is stabilized. But this would mean that an abnormal γ-mode is realized. It is very likely that such a situation might become unstable due to the instability of the side boundary of the current spot in the γ-sheath, similarly to the unstable side boundary in an abnormal dc glow discharge if it does not cover the whole cathode and there is a currentless zone on it [2.18]. This problem is interesting in principle and requires a special treatment. Of primary

---

[6]Stabilization of subnormal states by a dielectric in the γ-discharge is quite similar to that in a moderate pressure glow discharge sustained by an electron flux. The rising CVC of the discharge PC, like that of the dielectric, serves as a stabilizer for the descending branch of the cathode sheath CVC [2.18].

importance here are two-dimensional effects that remain outside the scope of a one-dimensional model which deals with CVC like $V(j)$ and assumes the current density to be independent of transverse coordinates and the discharge cross section to be independent of the x-axis along the current.

Another important point is that the dielectric over the electrodes represents a distributed ballast capacitive resistance only with respect to a possible instability in the sheath. The distributed ballast stabilizes the 'normal' state at the minimum of the one-dimensional CVC of the gas space (i.e., of the sheath plus plasma CVC). In the γ-discharge, however, the same dielectric behaves as a lumped capacitive resistance with respect to the PC contraction, which is possible because of the descending character of the PC CVC. Indeed, the negative glow region has a higher plasma density than the PC. This region acts as a good conductor and can be considered as one of the 'capacitor' plates, the electrode being its other plate. Together with the space charge sheath, the dielectric fills up this 'capacitor' and behaves as lumped ballast capacitance with respect to the PC, contributing to its stabilization.

No doubt, many additional details arise in real conditions, making one introduce appropriate changes in the principal schemes presented in Figures 2.28 and 2.29. A major detail is the distortion of the 'unperturbed' CVC of the sheath when the electrode is coated with dielectric material. An evident effect is a change in the secondary γ-emission coefficient: though the dependence of the results on γ is weak, it is logarithmic in formula (2.15). (Recall that the secondary emission in the case of a dielectric represents liberation by ions of electrons deposited from the gas and attached to the surface, while in the case of a metal this is a knockout of electrons that belong to the metal [Sections 1.6.3, 2.1.2]). The effect of the gas heating in the sheath is much stronger and manifests itself clearly in experiment. Insulation of a metallic electrode with a dielectric plate or even with a thin film abruptly impairs heat removal from the gas to the water-cooled electrodes. As a result, the gas in the sheath is heated (in the similarity laws $j_{ni} \sim p^2$, $j_{n\gamma} \sim p$, and $d_{n\gamma} \sim p^{-1}$, the pressure actually represents the gas density), and its density becomes smaller. For this reason, the sheath swells, the 'true' normal (for naked electrodes) current density decreases, and so on. Incidentally, the thermal effect of the electrode insulation slightly influences the value of current density in the α–γ transition. The sheath gas heating reduces only the upper limit of current densities in 'subnormal' γ-discharges, $j_{tr} < j_{n\gamma}^{\delta} < j_{n\gamma}$, decreasing $j_{n\gamma}$ but leaving j_{tr} unchanged. Experimental data on heat balance disturbance in discharges with dielectric-coated electrodes are still scarce, though they would be of much use for the operation of RF discharges and RF-excited lasers.

It has been found experimentally that one and the same dielectric may influence the normal current density in a γ-discharge and may not do so in an α-discharge. For instance, at the frequencies 13.56 and 27 MHz, a quartz layer of thickness $\delta = 0.5$ cm did not practically affect the α-discharge, while in the γ-mode, $j_{n\gamma}^{\delta}$ decreased a few times as compared to $j_{n\gamma}$ at $\delta = 0.05$ cm. This is easy to explain. The dielectric well stabilizes the discharge only if $\delta/\varepsilon \gtrsim d_s$, where d_s is the sheath

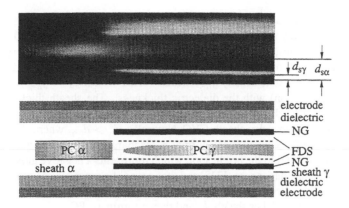

FIGURE 2.30
Coexistence of a 'subnormal' γ-mode and an α-mode in a gap with insulated electrodes.
The photograph shows the discharge half symmetrical to the right vertical boundary due
to the cylindrical symmetry of the discharge. The focal point is on the lower electrode, so
the sheaths at the top are slightly shifted and fuse together. Discharge in air,
$f = 13.56$ MHz, $p = 7.5$ Torr, $L = 0.7$ cm. Below is a schematic 'negative film': bright
spots are made dark while dark areas are left white.

thickness. Otherwise, the reactive resistance of the dielectric is much smaller than
that of the sheath, so it has no effect. But in the α-mode the sheath is thick while
in the γ-mode it is thin, hence the values of δ/ε sufficient for the γ-mode are
inadequate to have an influence on the α-mode.

2.9.3 Coexistence of two discharge modes

We mentioned in Section 2.7.1 a possible coexistence of α- and γ-modes in the
same gap when, after the transition, the voltage across the γ-discharge that has
just emerged becomes higher than the minimum voltage necessary for sustaining
the α-mode. Insulation of the electrodes with dielectric material stimulates such
situations. Indeed, the dielectric stabilizes new normal current densities in the
emerging γ-discharge, which would correspond to subnormal states in the case of
naked electrodes and to higher voltages in the γ-sheath CVC (Figure 2.28). After
the γ-mode is excited, the voltage across the gap becomes higher than it would
be in a discharge with naked electrodes, corresponding to the Paschen minimum.
This new voltage is generally sufficient to sustain both discharge modes.

In experiment, this situation looks as follows [2.47]. As the power applied to the
anomalous α-mode and its current density increase—the latter up to a critical value
j_{tr}—there appears a brighter flat spot near each of the column ends (or near one of
them, if the dielectric coatings are different), while the α-mode along the dielectric
surfaces still remains uniform. The spot creeps randomly across the surface. With
increasing current, new spots appear, first randomly but then becoming arranged in

a certain order until they cover up the whole end surface of the discharge column. This is the way the discharge behaves in molecular gases at $p \sim 20$–150 Torr. At lower pressures, and in helium even at larger p, a single spot generally emerges, and its area growth is proportional to the current i. This is a γ-discharge with a 'normal' current density $j_{n\gamma}^\delta$, which varies within $j_{tr} < j_{n\gamma}^\delta < j_{n\gamma}$ (see the previous section) with the dielectric thickness. The sheath thickness under the spot is smaller than d_α, its value d_γ^δ corresponding to the current density $j = j_{n\gamma}^\delta$ 'subnormal' for the γ-sheath CVC. Naturally, $d_\gamma^\delta > d_{n\gamma}$, where $d_{n\gamma}$ is the sheath thickness of a conventional normal γ-discharge corresponding to the Paschen minimum. Next to the bright spot, which represents a negative glow (NG) region, is the Faraday dark space (FDS) and then a weakly luminous positive column (PC). Outside the γ-region, on the discharge column periphery, there is the α-mode with a respectively thicker sheath d_α and then a luminous PC.

All this is clearly seen in the photograph of Figure 2.30 illustrated with a schematic diagram. The photograph shows only the half of the discharge possessing a cylindrical symmetry. The bright NG has a spectral composition different from that of the α-PC. The spot spectrum is similar to the NG spectrum of a dc glow discharge. It has been checked experimentally that the current density and the sheath conductance are much larger within the bright spot radius than outside the spot. One can clearly see the dark space charge sheaths for both the γ- and α-modes, because these have substantially different thicknesses. What is curious is that the α-mode is extinguished if the electrodes are brought close together so that the α-mode PC on the periphery (the light stripe in the middle of the photograph, Figure 2.30) considerably contracts, and the two α-sheaths come nearly in contact. This is easy to explain, because the energy release and gas ionization sustaining the α-mode are primarily localized outside the sheaths but very close to them (see the two maxima in Figure 2.12).

2.9.4 Dielectric-stabilized subnormal γ-discharge

Figure 2.31 demonstrates the variation in the γ-sheath thickness and conductance over the whole range of dielectric-stabilized 'subnormal' states from the α- to the 'truly' normal γ-discharge. The horizontal straight lines for d_α in Figure 2.31(a) represent the anomalous α-mode: at higher current density $j \sim V_s/d_\alpha$ the sheath voltage rises while the sheath thickness remains practically unchanged. The last lower points of the descending curves correspond to a γ-mode with naked electrodes, i.e., $j_{n\gamma}$, $d_{n\gamma}$. The whole curve from j_{tr}, d_α to $j_{n\gamma}$, $d_{n\gamma}$ describes the discharge states after the α–γ transition for dielectric coatings of varying thickness. At the given dielectric thickness δ/ε, i.e., at the given point in the $d_s(j)$ curve, all states above it are unfeasible, but below it they can be produced by increasing the current after the transition of the γ-mode to the anomalous regime. The foregoing also refers to Figure 2.31(b), which shows dc probe measurements [2.5, 2.48] of the active sheath conductances normalized to the sheath thickness σ/d_s Ω^{-1} cm^{-2}. They can be calculated from the ratio of the

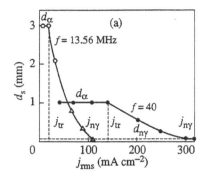

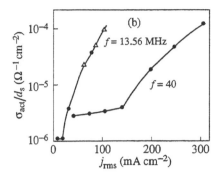

FIGURE 2.31

Measured (a) space charge sheath thickness d_s and (b), related to the thickness, mean active sheath conductance σ/d_s in the α-mode as well as in the normal and 'subnormal' γ-modes as a function of current density. Air, $p = 15$ Torr, $L = 1$ cm, $f = 13.56$ MHz, and 40 MHz, quartz-insulated electrodes ($\varepsilon = 7$). Circles, for quartz thickness $\delta = 1$ cm; triangles, for $\delta = 0.3$ cm.

active current density toward the electrode to the average sheath voltage, because $j_{act} = \sigma V_s/d_s$. One can see that the value σ/d_s grows by 2 orders of magnitude from the α- to the normal γ-mode, while in 'subnormal' γ-discharges it takes all intermediate values, indicating an increasing contribution of the active current toward the electrode to the α–γ transition.

Thus, the insulation of electrodes in an RF discharge provides a means of controlling the parameters of the electrode sheaths and the plasma in a wide range of values by changing the current density from $j_{n\,\alpha}$ to j_{tr} and further continuously to $j_{n\,\gamma}$. This can be done by choosing the appropriate dielectric thickness and material (dielectric permittivity). Dielectric coatings provide discharge regimes experimentally unfeasible in γ-discharges with naked electrodes and glow discharges. This permits a more effective use of RF discharges in laboratory experiments and applications.

3

Low-Pressure RF Discharges and Asymmetry Effects

3.1 Self-bias in an asymmetric capacitively coupled discharge

We have pointed out above that low-pressure RF discharges are widely used for materials treatment. Treatment technologies are based on bombardment of a target surface by positive ions produced in a discharge and accelerated by a constant sheath field up to high energies of several dozens and hundreds of electron volts. The cycle average constant voltage fall in the target sheath $\overline{V}_1$, which determines the ion energy, primarily depends on the applied RF voltage amplitude V_a. The values of V_a commonly applied in practice vary between 10^2 and 10^3 V. One can, however, nearly double $\overline{V}_1$ and ion energy without raising the electrode RF voltage but by employing an asymmetric discharge excited between electrodes of substantially different areas. Higher ion energy contributes to the efficiency of many technologies, and this explains the increasing interest in studies of asymmetric discharges.

The majority of effects inherent to asymmetric RF discharges can be observed at any pressure. In low-pressure discharges, asymmetry arises naturally and is sometimes very hard to remove, while at moderate pressures discharge asymmetry is created intentionally. Owing to the normal current density effect that was discussed in Chapter 2, a moderate-pressure discharge is often symmetric even with different electrode areas. In order to excite a well-defined asymmetric discharge, one often uses a cylindrical design, in which one electrode is a thin tube and the other a coaxial cylinder. For this reason, we think it appropriate to discuss all asymmetry effects, including those occurring at moderate pressures, in this chapter.

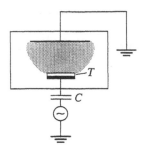

FIGURE 3.1
Technological planar reactor. RF voltage is applied to the smaller electrode through blocking capacitor C. Target T on the electrode is bombarded by ions; the shaded region is plasma.

3.1.1 Experimental data

Discharge chambers with plane electrodes, or planar reactors, are most commonly used in practice (Figure 3.1). A plane target to be treated is mounted on the smaller electrode, and RF voltage is applied to it through a blocking capacitor C connected in the circuit between an RF generator and the smaller electrode. The larger electrode and the reactor walls are grounded. At low pressures, the discharge configuration may, in fact, be asymmetric even with identical electrodes, because the RF current is closed on the ground not only through the grounded electrode but also through the chamber walls—which actually increase the electrode area. For this reason, a low-pressure discharge may externally differ very much from a moderate-pressure discharge. Due to the normal current density effect, the latter looks, at moderately high currents, as a glowing column with a diameter varying but little along the column length and independent of the electrode size. At low pressures, the glow fills up the reactor if, of course, its size is not too large relative to the electrodes. When the difference in the area between the smaller high-voltage (loaded) electrode A_1 and the larger grounded electrode A_2 is considerable and when a blocking capacitor C is connected, the average voltage fall $\overline{V}_1$ in the smaller electrode sheath turns out to be much greater than $\overline{V}_2$ in the grounded electrode sheath.[1] The value of $\overline{V}_2$ coincides with constant plasma potential $\overline{V}_p$ relative to the ground. Within the limit $\eta = A_1/A_2 \ll 1$ and at low blocking capacitor resistance (large C), the smaller electrode gains a large negative (relative to the ground) potential $V_{dc} = -\overline{V}_1 + \overline{V}_2$, known as self-bias, which is close in value to the RF voltage amplitude V_a.

Self-bias V_{dc} and the ratio of constant sheath voltage falls $\overline{V}_1$ and $\overline{V}_2$ vary with the electrode area ratio. In principle, ion energy can be controlled by varying the degree of discharge asymmetry. But the area ratio cannot always be ade-

---

[1] A loaded electrode is also termed as excited, active or living. Below, it will always be marked with subscript 1 and the grounded electrode with subscript 2, irrespective of their size ratio.

quately controlled and directly measured, because the grounded electrode may be affected to some extent by other, uncontrollable grounded surfaces. Therefore, it is important to find an experimental relationship between the $\overline{V}_1/\overline{V}_2$ ratio and the electrode area ratio when both areas are known exactly.

One of the first carefully designed experiments of this kind [3.1] used a planar reactor of the type shown in Figure 3.1. RF voltage of the frequency 13.56 MHz was applied through a blocking capacitor to the upper, small electrode A_1 of a fixed area. The larger electrode A_2 separated from it by a gap of size $L = 1.88$ cm was grounded. The active area of A_2 was regulated by enclosing the discharge in a short Pyrex (dielectric) cylinder of height L placed on the lower electrode. A set of such cylinders of varying diameter was used to obtain a fixed A_1/A_2 area ratio in each experiment. It is relatively easy to take a voltage oscillogram from the loaded electrode $V(t)$ and to find RF amplitude V_a and self-bias V_{dc} equal to the difference between the sought-for values of $\overline{V}_1$ and $\overline{V}_2$. Measurement of constant plasma potential $\overline{V}_p$ in order to find $\overline{V}_2 = \overline{V}_p$ and $\overline{V}_1 = -V_{dc} + \overline{V}_p$ presents a greater problem. To this end, a small window was made at the larger electrode center for ions accelerated in the sheath. The ions entered an energy analyzer and a mass-spectrometer to yield the energy spectra of certain ion species. The working gas was argon and the operating pressure $p = 5 \times 10^{-2}$ Torr. The energy spectrum for $\overline{V}_p = \overline{V}_2$ represented a fairly narrow peak. Figure 3.2 shows a series of experimental plots of $\overline{V}_2$ versus $\overline{V}_1$ for different area ratios $\eta = A_1/A_2$. All curves are close to a straight line, each being obtained by varying V_a at a fixed η.

Description of the measurements by a power formula

$$\frac{\overline{V}_1}{\overline{V}_2} = \left(\frac{A_2}{A_1}\right)^q \tag{3.1}$$

renders this interpolation admissible. Though the power index q depends on η, being as variable as it is, it varies within a narrow range $q \approx 1$–1.4. This makes the convenient relation (3.1) applicable in experimental data analysis and in constructing theoretical approximations. The matter is that the power equation (3.1) has been derived from simplified models, and the value of q varies with the assumption as to the nature of ion motion through a sheath [3.2–3.4] (see Section 3.2.1).

The above approach was used in a series of plasma potential measurements made under variable conditions [3.5]. The smaller (upper) loaded electrode was an aluminum disk of 13 cm diameter, the larger (lower) grounded electrode was 35 cm and the separation $L = 6.3$ cm. The chamber diameter was over 40 cm. Ions entered an energy analyzer through a 0.8 mm window in the large electrode. The discharge was excited in argon at $f = 13.56$ MHz and $p = 2 \times 10^{-2}$ Torr. Some experiments registered Ar_2^+ ions, others ArH^+ ions. Self-bias was found to be proportional to the RF electrode voltage amplitude in a wide amplitude range $V_a \approx 200$–1000 V and only slightly smaller than the amplitude: $V_{dc} \approx -0.934\,V_a$. The constant plasma potential was low, $V_p \approx 20$–23 V, and close to the floating potential (Section 3.5). As the degree of asymmetry was decreased by reducing the lower electrode area and discharge space with a grounded cylinder of 15.9 cm

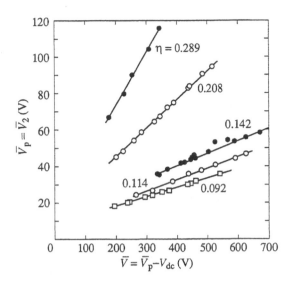

FIGURE 3.2
Constant plasma potential $\overline{V}_p$ or average voltage fall across the larger (grounded) electrode sheath versus voltage fall across the smaller (loaded) electrode sheath; argon, $p = 50$ mTorr, $f = 13.56$ MHz, $L = 1.88$ cm. Digits at the curves indicate area ratio $\eta = A_1/A_2$ [3.1].

diameter and height L, the self-bias reduced, too: $V_{dc} \approx -0.813\,V_a$. The plasma acquired a higher constant potential $\overline{V}_p \approx 40$–50 V but was still much lower than the RF voltage amplitude. Quite a different thing happened when self-bias was removed. For this purpose, the loaded electrode was grounded through a large inductance which actually passed no RF current. The current took its usual path through the discharge, but no matter whether a blocking capacitor was used or not, the loaded electrode gains a cycle-average potential equal to the earth's potential ($V_{dc} = 0$). The constant plasma potential then rose sharply and became close to the applied voltage amplitude $\overline{V}_p \approx 0.85\,V_a$; these measurements were made over the range $V_a \approx 30$–300 V. At moderate RF voltages, $\overline{V}_p$ weakly depends on V_a and is close to the floating potential (Figure 3.3).

It has been shown experimentally and theoretically that moderate-pressure discharges excited between naked and coated electrodes do not practically differ, all other things being equal [3.6]. The effect of the dielectric coating is, in fact, the same as that of an equivalent capacitor connected in the external circuit. Therefore, if a target is made of a poorly conducting material and if it really insulates the active electrode from the plasma, the mere presence of such a target is equivalent to a blocking capacitance equal to the target capacitance. In this case self-bias arises in an asymmetric system even in absence of a blocking capacitor. As for symmetric designs, there is no self-bias in them in any case, since the electrodes have, on

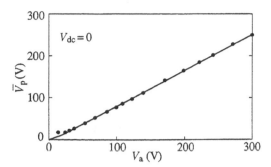

FIGURE 3.3
Constant plasma potential plotted against the RF voltage amplitude in experiment with the loaded electrode grounded though a large inductance; argon, $p = 200$ mTorr, $f = 13.56$ MHz, $L = 6.3$ cm [3.5].

the average, identical potentials (zero potential if one electrode is grounded).

3.1.2 Physical mechanism: A linear 'electrical engineering' model

The primary reason for the appearance of self-bias in a loaded electrode when it is connected to an RF generator through a blocking capacitor is the irreversible escape of some negative electric charge into the electrodes, making the gap gas positively charged. In a symmetric system, both electrodes operate in identical regimes, so they gain equal charges and, therefore, possess equal constant potentials whether or not there is a blocking capacitor. In an asymmetric design, different amounts of charge are gained by the electrodes. If the external circuit is broken for direct current, as is the case with a capacitor, the different amounts of charge accumulated in the circuit from the side of each electrode persist, causing a difference in the constant electrode potentials, or self-bias. Strictly speaking, if the loaded electrode is not grounded through inductance, the external circuit is always disconnected for dc due to the design of the RF generator and matching element themselves (see Section 4.7.2).

This interpretation fits well with a simplified 'electrical engineering' model of electrical processes occurring in an RF discharge circuit. In this model, the sheaths possess a negligible active conductivity and behave as constant capacitors C_1 and C_2. The plasma between the sheaths is considered to be an ideal conductor, and the floating potentials that really exist between the plasma and a solid body (see Section 3.7) are neglected. This kind of conception has been employed by several investigators for theoretical treatment of electrical processes in RF capacitive discharges [3.2, 3.5, 3.7–3.9]. The most detailed and self-consistent electrical engineering model has been described in [3.5], and we will largely follow it in this book.

Due to the constancy of the three capacitors, C_1, C_2 and C, connected in series, the electric circuit is linear (Figure 3.4). With a sine-shaped generator emf,

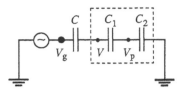

FIGURE 3.4
Electric circuit of a simplified electrical engineering discharge model.

any potential in it may be only sine-shaped against the background of constant magnitudes. If the loaded electrode has a potential

$$V(t) = V_{dc} + V_a \sin \omega t \qquad (3.2)$$

relative to the ground, the plasma potential relative to the grounded electrode may be written as

$$V_p(t) = \overline{V}_p + \Delta V_p \sin \omega t \qquad (3.3)$$

Since equal currents pass through C_1 and C_2, $\omega C_1(V_a - \Delta V_p) = \omega C_2 \Delta V_p$, hence,

$$\Delta V_p = \left(\frac{C_1}{(C_1 + C_2)} \right) V_a \qquad (3.4)$$

Return temporarily to the behavior of a real plasma. Neglecting the floating potential, i.e., the electron flux from the plasma into the electrode when the plasma potential is higher than the electrode potential, we must assume that at a certain moment the plasma will necessarily come in contact with the electrode. Indeed, when the plasma has moved away from the electrode, there is only an ion flux into the electrode. Since there is no direct current through the broken circuit, the ion flux must be compensated by an electron flux as soon as the plasma touches the electrode. However, the plasma potential can never drop below the electrode potential, because there are no 'electronic' sheaths. When the plasma is separated from the electrode by a positive space charge sheath, its potential is higher. When it touches the electrode, its potential becomes equal to the electrode potential, because both are assumed to be ideal conductors. Therefore, the minimum plasma potential $V_{p\,min} = \overline{V}_p - \Delta V_p$ at the moment $\omega t = 3\pi/2$ when the plasma touches the grounded electrode is zero. Hence, $\Delta V_p = \overline{V}_p$. At $\omega t = \pi/2$ half-cycle earlier, when the plasma touched the loaded electrode, its potential $V_{p\,max} = \overline{V}_p + \Delta V_p = 2\overline{V}_p$ coincided with the electrode potential which was also maximal at that moment, $V_{max} = V_{dc} + V_a$. This yields the well-known formula

$$\overline{V}_p = \frac{V_a + V_{dc}}{2} = \Delta V_p \qquad (3.5)$$

Expressions (3.4) and (3.5) define self-bias, constant plasma potential and average

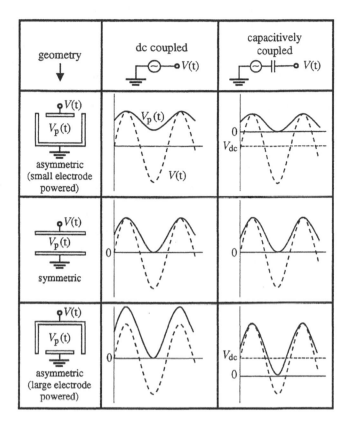

FIGURE 3.5
Time behavior of the plasma (solid curves) and loaded electrode (dashed curves) potentials under varied conditions. In contrast to Figure 3.1, the upper electrode is loaded [3.5].

voltage fall in a loaded electrode sheath through the RF voltage amplitude

$$V_{dc} = \frac{C_1 - C_2}{C_1 + C_2} V_a \qquad \overline{V}_p = \frac{C_1}{C_1 + C_2} V_a \qquad \overline{V}_1 = \frac{C_2}{C_1 + C_2} V_a \qquad (3.6)$$

With a natural assumption that sheath capacitance grows with sheath area, we conclude that self-bias is negative when the smaller electrode is loaded, $C_1 < C_2$, and that $\overline{V}_p < V_a/2$. In a symmetric system ($C_1 = C_2$), $V_{dc} = 0$ and $\overline{V}_p = V_a/2$. If the larger electrode is loaded, self-bias is positive and $\overline{V}_p$ is closer to V_a. These regularities are clearly illustrated by the right-hand column of Figure 3.5.

Consider now an electric circuit capable of passing direct current with no other sources of constant voltage in it. In a symmetric design, everything is the same as described above. The plasma must alternatively touch both electrodes so that the electron flux would compensate the ion flux, since direct current is absent simply because of the symmetry. In an asymmetric system, the plasma must periodically touch at least one of the electrodes, otherwise ions would continuously leave it, making the discharge process unsteady. On the other hand, if the plasma touched both electrodes, we would have the previous situation with self-bias, which is not the case because of the 'zero' resistance to direct current in the external circuit. Hence, the oscillating plasma contacts only one electrode, and some of the positive ion charge cannot be compensated during electron gas oscillations within a cycle. In terms of electrical engineering, one of the capacitors, C_1 or C_2, can never be completely discharged.

On the above assumption that the larger electrode has a larger sheath capacitance, this uncompensated charge should be present in the sheath of the larger electrode, and the plasma should then contact the smaller electrode. Indeed, the positive charge fractions of both ion sheaths, which are flooded by the oscillating electron gas, are identical due to the conservation of the total gap charge or to the equality of displacement currents in both electrodes. Therefore, only the smaller capacitor can be discharged. It follows from this statement that if the smaller electrode is loaded, equality (3.4), which is still valid, is supplemented with the relation $V_a = V_{p\,max} = \overline{V}_p + \Delta V_p$ from which

$$\overline{V}_p = \frac{C_2}{C_1 + C_2} V_a > \Delta V_p = \frac{C_1}{C_1 + C_2} V_a \qquad \text{at} \quad C_1 < C_2 \qquad (3.7)$$

If, however, the larger electrode is loaded, equation (3.4) is supplemented with $V_{p\,min} = \overline{V}_p - \Delta V_p = 0$, hence,

$$\overline{V}_p = \frac{C_1}{C_1 + C_2} V_a = \Delta V_p \qquad \text{at} \quad C_1 > C_2 \qquad (3.8)$$

These correlations are illustrated by the left-hand column in Figure 3.5.

When applied to the experiments with a blocking capacitor and unlimited discharge volume [3.5], the first formula in (3.6) yields $C_2/C_1 = 29.4$ from the measured slope $|V_{dc}|/V_a = 0.934$. When the larger electrode area is limited by a cylinder, probably reducing the effect of the grounded chamber walls, the same formula gives $C_2/C_1 = 9.68$ from the measured slope $|V_{dc}|/V_a = 0.813$. These data describe to some extent the degree of symmetry in both cases. For the experiment with an inductively grounded loaded electrode and unlimited discharge volume, the respective formula (3.7) yields $C_2/C_1 = 5.9$ from the measured slope $\overline{V}_p/V_a = 0.85$. One can see a substantial difference between the two C_2/C_1 ratio values: 29.4 and 5.9. The matter is that the conditions in these experiments differ in spite of the identical surface geometry. In the first case the loaded electrode and the chamber walls are under essentially different constant potentials, while in the second case under the same, zero potential. For details and discussion of the effects of floating potential, sheath conductivity, etc., the reader is referred to [3.5]. Discharge asymmetry has been studied by measuring self-bias and other parameters elsewhere [3.2, 3.10–3.13].

3.2 Correlations between plasma and sheath parameters in an ambipolar diffusion-controlled discharge

In order to understand what factors determine such parameters as thickness, field, voltage fall and ion density in a sheath as well as ion density at the gap center (in the positive column), one should have a clear idea of the discharge mechanisms. It is reasonable to point out at first some differences in the respective parameters of moderate- and low-pressure α-discharges. In both cases a steady discharge is provided by a balance between the production and the loss of charges, ions in particular. At sufficiently high pressure, when diffusion processes are retarded, charges are largely lost through bulk recombination. Ion current into the electrode is negligible as compared to discharge current (Section 2.1), so it cannot noticeably affect the sheath parameters. Assuming $n_+(x) = $ const, we can entirely ignore the ion current but still have a self-consistent and reasonably realistic discharge pattern of the α-discharge (Section 1.5). Average sheath thickness $\bar{d}$ is determined by the plasma electron oscillation amplitude in the field E_a required for the maintenance of ionization-recombination equilibrium $\bar{d} = \mu_e E_a / \omega$. This result has been deduced from the plasma boundary oscillation pattern and from the condition of current continuity, i.e., the equality of the plasma conduction current and the electrode displacement current. The existence of a real ion flux into the electrode leads to a lower ion density at the electrode, as compared to the plasma density $n_+ = n$ determined by the discharge current density. Consequently, $\bar{d}$ and average voltage fall $\bar{V}$ in a sheath grow but not very fast, approximately twice the estimate from the $n_+(x) = $ const approximation (Section 2.1).[2]

Low-pressure discharges are quite another matter. Charges are lost not in the plasma bulk but on the electrodes and walls, being carried to them by an ambipolar flux from the plasma. In the sheaths the ambipolar ion flux becomes a purely ionic flux (which is compensated by an electron flux at the short moment of plasma contact with the electrode). The sheath is formed in such a way that, first, the discharge current would close on the electrode by the displacement current, as in the case of moderate pressures; and, second (this is a novel factor), the incoming ambipolar flux would be carried to the electrode by an ion flux.

---

[2]The ratio $\bar{V}/\bar{d}$ is a measure of field at the electrode which varies much slower than $\bar{V}$ and $\bar{d}$, since the negligible effect of ion current into the electrode cannot change the displacement current which is still equal to the discharge current. We would like to explain here a contradiction of the model [3.14], also described in [3.15] (see Section 1.5.5). One should not ignore equation (1.45) closing the set and obligatory for a model with $n_+(x) = n$. If one replaces (1.45) with another equation for the ion flux into the electrode, as in [3.14], one should introduce a new unknown, e.g., sheath ion density. Otherwise the set of equations becomes overdetermined, or we ignore the violation of equality (1.45).

3.2.1 Ratios of average voltage fall and sheath thickness

The ratio $\overline{V}_1/\overline{V}_2$ is the most revealing and practically important characteristic of an asymmetric capacitively-coupled discharge. For this reason, all theoretical and experimental studies of this kind of discharge are aimed at finding a relationship between $\overline{V}_1/\overline{V}_2$ and A_1/A_2 and the power index q in a similarity law approximation (3.1). The electrical engineering conception is, of course, suggestive of approximation (3.1), because $C_1 = A_1/4\pi\tilde{d}_1$, where $\tilde{d}_1$ is a characteristic sheath thickness. But the capacitance ratio in the relation

$$\frac{\overline{V}_1}{\overline{V}_2} = \frac{C_2}{C_1} = \frac{A_2}{A_1}\frac{\tilde{d}_1}{\tilde{d}_2} \tag{3.9}$$

also contains the sheath thickness ratio dependent, in an unknown way, on A_1 and A_2. The general relation (3.9) follows from the electrostatic and charge conservation laws (or from the equality of displacement currents in the sheaths). Taking sheath ion densities n_1 and n_2 to be constant, we find from (1.39) the instantaneous field at the electrode $E_1(t) = 4\pi e n_1 d_1(t)$, where $d_1(t)$ is the instantaneous sheath thickness, and the average voltage fall $\overline{V}_1 = 2\pi e n_1 \overline{d_1^2}$. When the plasma alternatively touches the electrodes, all the positive gas charge is pumped from one electrode to the other, so

$$n_1 d_{1\,\mathrm{max}} A_1 = n_2 d_{2\,\mathrm{max}} A_2 \tag{3.10}$$

By eliminating n_1 and n_2 and keeping in mind that $\tilde{d}_1 \sim (\overline{d_1^2})^{1/2} \sim d_{1\,\mathrm{max}}$, we obtain equation (3.9).

To advance further, we should consider the relationship between the plasma and sheath processes and to understand how large are the ambipolar ion fluxes entering the sheaths. The first diffusion model of an RF discharge (plane, symmetric) was suggested in [3.16] (also see [3.15]). It considered the balance between ambipolar transport of charges from the bulk and their production at constant ionization frequency ν_i, i.e., at constant electron temperature T_e. At the plasma-sheath boundary, actually near the electrode, because the sheath is thin, ions enter the sheath at a Bohm velocity $u_B = \sqrt{kT_e/M}$, where M is the mass of an ion (Section 3.7). This relates the density of an ambipolar flux to the ion density n_p at the effective plasma-sheath boundary

$$J = -eD_a \left(\frac{\partial n}{\partial x}\right)_{n=n_p} = en_p u_B \tag{3.11}$$

where D_a is the coefficient of ambipolar diffusion. These conditions represent the boundary conditions for the diffusion equation with sources. Detailed theory was developed in [3.3], principally on the same grounds as in [3.16, 3.15], for an asymmetric discharge in one-dimensional spherical geometry. The author analyzed the influence of ion motion through a sheath (collisionless motion, motion at constant collision frequency, and that with a constant path length corresponding to the case of resonance charge exchange dominant over elastic collisions) on the power index q in equation (3.1) and discussed the ionization effects. The problem was solved approximately for two-dimensional geometry in [3.4].

We are attempting to present here in a simple and clear way only the basic aspect of the problem. Like in [3.3], we will consider a spherical geometry but simplify the theory with some additional assumptions which are unable to distort the general picture of the process. Suppose a steady discharge has been excited in a gap between two spherical electrodes: a smaller, loaded electrode of radius R_1 and a larger electrode of radius R_2. As in the above references [3.16, 3.15, 3.3, 3.4], we take the electron temperature T_e and the frequency of electron impact ionization ν_i to be constant, but we consider the ambipolar diffusion coefficient $D_a = \mu_+ T_e$ to be constant, too. Assume the so-called 'extrapolated diffusion length' equal to D_a / u_B in this case, as well as the sheath thickness, to be small as compared with R_1, $R_2 - R_1$. When calculating the plasma density distribution $n(r)$ in the gap, we can replace (3.11) by $n = 0$ at the boundaries and then find plasma fluxes into the electrodes J_1 and J_2. These values can be used as the densities of ion fluxes through a sheath (which is thin, 'plane,' with no ionization). The plasma density n_p at the boundary can be evaluated as $n_p = J / e u_B$.

With the accepted assumptions, the diffusion equation for plasma density

$$\operatorname{div}\left(D_a \operatorname{grad} n\right) + \nu_i n = 0 \tag{3.12}$$

with $n = 0$ at the boundaries has a simple solution for all one-dimensional geometries. In a plane gap of size L

$$n(x) = n_{\max} \cos\left(\frac{\pi x}{L}\right) \qquad \nu_i = D_a \left(\frac{\pi}{L}\right)^2 \tag{3.13}$$

if x is counted off from the gap middle. Formulas for spherical geometry are more cumbersome, and the position of $n_{\max}$ can be found by solving an algebraic transcendental equation. But the relation (3.13) for ionization frequency providing a steady state remains valid; now $L = R_2 - R_1$. The distribution of $n(r)$ is illustrated in Figure 3.6.

The ratio of ion fluxes into spherical electrodes, which can also be regarded as the plasma density ratio at the boundary, is

$$\frac{J_1}{J_2} = \frac{n_{p1}}{n_{p2}} = \frac{R_2}{R_1} = \left(\frac{A_2}{A_1}\right)^{1/2} \equiv \eta^{-1/2} \tag{3.14}$$

In [3.15, 3.3], the diffusion was described by more complex equations than equation (3.12). The authors assumed the ion path length l to be constant but ion mobility μ_+ not; the ion drift velocity in a polarization field was taken to be higher than thermal velocity. As a result, the ambipolar diffusion coefficient turned out to be variable. In this case the power law of the type (3.14) is valid only in the limit $\eta \ll 1$ with the power of η equal to $-7/24$ instead of $-1/2$.

Ion fluxes depend on both radii, but in the limit $R_2 \gg R_1$ the plasma density maximum lies at a distance $r_m - R_1 \sim R_1$ from the smaller electrode, so by order

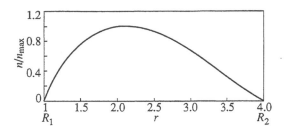

FIGURE 3.6
Plasma density distribution in the gap between two 'absorbing' spheres calculated from
the diffusion equation (3.12); $R_1 = 1$ cm, $R_2 = 4$ cm, $(\mathrm{d}n/\mathrm{d}r)_1 = 2.34\, n_{\max}/R_1$,
$(\mathrm{d}n/\mathrm{d}r)_2 = -2.41\, n_{\max}/R_2$ [cf. equation (3.15)].

of magnitude (Figure 3.6)

$$J_1 \sim \frac{eD_a n_{\max}}{R_1} \qquad J_2 \sim \frac{eD_a n_{\max}}{R_2} \qquad (3.15)$$

The value of $n_{\max}$ is determined by discharge current, to be more exact, by its
density at the radius r_{m}. If the plasma electron current (i.e., conduction and polarization currents) dominates—as it usually does—over the vacuum displacement
current (Sections 1.3.2 and 1.3.5), the amplitude $j_a \sim nE_a$, where the field amplitude E_a must provide adequate T_e and ionization frequency in equation (3.13).
Such is the sequence of cause-effect relationships for plasma. As for sheaths, they
must be formed so as to transport fluxes J_1 and J_2 to the electrodes as well as to
transport discharge current via plasma boundary oscillations.

Sheath dynamics responsible for current transport will be discussed in Section 3.3. Following the above references, we will deal here with average sheath
parameters and their ratios, like in [3.3], without evaluation of the absolute sheath
thicknesses. Denote the cycle-average velocities of ions in the average sheath
field as u_1 and u_2. If there is no ionization in the sheath, the density of an ion
flux through a 'plane' sheath remains constant and is $J_1 \approx n_1 u_1$. The sheath
ion densities n_1 and n_2 are taken to be constant and do not coincide with the ion
densities near the plasma boundary: $J_1 \approx n_1 u_1 = n_{\mathrm{p}1} u_{\mathrm{B}}$. By employing an
electrostatic relation, $n_1 \approx \overline{V}_1/2\pi e \tilde{d}^2$, it is convenient to eliminate the sheath ion
densities from expressions for the fluxes to the electrodes.

Velocity u_1 varies with the kind of ion motion. If ions fly through a sheath
without collisions, $u_1 \approx (e\overline{V}_1/M)^{1/2}$, and we have the Child–Langmuir law
for the flux: $J_1 \sim M^{-1/2}\overline{V}^{3/2}\tilde{d}_1^{-2}$. If they move at constant mobility μ_+,
$u_1 \approx \mu_+ E_1 \approx \mu_+ 2\overline{V}_1/\tilde{d}_1$. The obtained relation $J_1 \sim \mu_+ \overline{V}^2/\tilde{d}_1^3$ naturally
coincides with the formula derived by Engel and Steenbeck from the same concepts
for an ion flux to the glow discharge cathode layer ($\overline{V}_1$ is equivalent to the cathode
fall, $\tilde{d}_1$ is the cathode layer thickness). For monatomic gases operating within

the pressure range $p \approx 5 \times 10^{-3}$–0.3 Torr, the charge-exchange ion motion with path length $l = $ const and drift velocity far exceeding random velocity is the closest to reality [3.15, 3.3] (e.g., in argon $l \approx 5 \times 10^{-3}/p$ cm). In this case $u_1 \approx (eE_1 l/M)^{1/2}$ [cf. formula (1.10a)] and

$$J_1 = en_1u_1 \approx \frac{1}{2\pi}\left(\frac{2el}{M}\right)^{1/2}\overline{V}_1^{3/2}\tilde{d}_1^{-5/2} \tag{3.16}$$

Exact calculation with account of the self-consistent distribution of $n(x)$ and $E(x)$ within a sheath [3.3] yields a value by a factor of 2.5 smaller

$$J_1 = \frac{1}{4\pi}\left(\frac{500}{243\pi}\right)^{1/2}\left(\frac{2el}{M}\right)^{1/2}\frac{\overline{V}_1^{3/2}}{\tilde{d}_1^{5/2}} \tag{3.17}$$

By combining the ratios of capacitor voltages (3.9) of ion fluxes from the plasma (3.14) (or the analogue of this formula from [3.3]) and of ion fluxes into the electrodes (3.17) (or any of the above laws), one can obtain the similarity law (3.1) with the respective power index q. Possible ratio variants were given in a table [3.3]. For instance, in the first theoretical treatment of equation (3.1) [3.2] it was suggested that $J_1 = J_2$ [instead of equation (3.14)] and the Child–Langmuir law was used instead of (3.17), which yielded an overestimated power index $q = 4$.

It should be noted that the uncertainty in choosing the right variant is due to the imperfection of the diffusion problem solution for the plasma, because there are no serious objections to equation (3.17) and equation (3.9) is just taken for granted. There seems to be too much speculation in the treatment of plasma diffusion: in general, the real discharge and electrode geometries are poorly fitted into idealized patterns, like the spherical or cylindrical design; the ambipolar diffusion coefficient may be calculated in different ways; the assumption of $\nu_i = $ const is too idealized since the real RF field in a nonuniform plasma is also nonuniform; there may be other ionization sources; and so on. Therefore, it appears reasonable to tentatively represent the final result of the diffusion problem solution as

$$\frac{J_1}{J_2} = \left(\frac{A_2}{A_1}\right)^\beta \tag{3.18}$$

and to find the effective power index β from the experimental value for q, as was done in the treatment of sheath dynamics [3.17] (Section 3.3). With (3.1), (3.9), (3.17) and (3.18), $\beta = 5/2 - q$. For $q = 1.5$, which is close to the experimental value, $\beta = 1$, a result different from both (3.14) and [3.3].

To turn back to the eliminated average sheath ion density $n_1 \approx \overline{n(x)}$, it is scaled by the near-electrode density values n_{el} rather than by the plasma boundary values n_p. In the self-consistent sheath model with $l = $ const, when the local ion velocity is related to the local average field by $u_i(x) = [e\overline{E}(x)l/M]^{1/2}$ and $J = en(x)u_i(x) = $ const, we have $n(x) \sim n_{el}(\overline{d}/x)^{1/3}$, where the x-coordinate starts at the effective plasma boundary. If we assume that ions enter a sheath at 'zero' velocity (this assumption is possible in treatments of the sheath structure and was used, for instance, in deducing the Child-Langmuir law), the sheath density at the plasma boundary turns out to be infinite and $n(x) = (3/2) n_{e1}$. But even at a finite velocity, the average density ratio differs from the boundary density ratio; and this difference increases with increasing difference in the electrode areas. For example, for identical (Bohm) entrance velocities it follows from the first equality in equation (3.16), from electrostatics ($E_1 \sim n_1 \tilde{d}_1$) and the charge conservation law (3.10) that

$$\frac{n_1}{n_2} = \frac{J_1}{J_2} \left(\frac{A_1}{A_2} \right)^{1/2} = \frac{n_{p1}}{n_{p2}} \left(\frac{A_1}{A_2} \right)^{1/2}$$

Thus, according to these equations, as well as to (3.18) and (3.10), we have

$$\frac{n_1}{n_2} = \left(\frac{A_2}{A_1} \right)^{\beta - 1/2} \qquad \frac{\tilde{d}_1}{\tilde{d}_2} = \left(\frac{A_2}{A_1} \right)^{3/2 - \beta} \qquad \frac{\overline{V}_1}{\overline{V}_2} = \left(\frac{A_2}{A_1} \right)^{q} \qquad q = \frac{5}{2} - \beta \quad (3.19)$$

If we choose $\beta = 1$, taking into account the experimentally close value of $q = 3/2$, we arrive at the proportionality laws

$$\frac{n_1}{n_2} = \frac{\tilde{d}_1}{\tilde{d}_2} = \left(\frac{A_2}{A_1} \right)^{1/2} \qquad \frac{\overline{V}_1}{\overline{V}_2} = \left(\frac{A_2}{A_1} \right)^{3/2}$$

The smaller electrode has a thicker sheath and a higher ion density in it, and this causes a substantially greater voltage fall than in the larger electrode.

3.2.2 Evaluation of sheath and plasma parameters and of discharge current-voltage characteristic at moderately low pressures

The analysis of ion transport from bulk plasma to the electrodes (Section 3.2.1) has yielded only the ratios of asymmetric sheath parameters but not their values. Physically, this is quite clear because these ratios represent cycle-average ratios defined by ion motion in an average constant field. The respective values, however, depend on the oscillatory dynamics of sheath boundaries within a cycle. This is clear if one bears in mind that at low pressures much of the applied voltage falls on the sheaths but not on the plasma, where $E_a/p \sim 10$ V cm^{-1} Torr^{-1} and the field is relatively weak (see a numerical illustration below). Since discharge current is closed through a sheath by displacement current, the current density is, by order of magnitude, $j \approx \omega \overline{V}/2\pi d$. In a symmetric discharge it is nearly straightforwardly related to sheath thickness $\tilde{d}$, as the constant sheath voltage fall is then equal to about half of the applied voltage amplitude, $\overline{V} \approx V_a/2$.

To better understand the causal relationships, let us consider a plane symmetric discharge of length L. A characteristic field value in the plasma—say, its amplitude in the gap middle E_a—is defined through electron temperature T_e by the compensation condition for the production and diffusion losses of charges, i.e., by equation (3.13)

$$\nu_i[T_e(E_a)] = D_a\left(\frac{\pi}{L}\right)^2 \equiv \nu_d \tag{3.20}$$

Due to a linear dependence of the charge production and loss rates on n, the plasma field only slightly depends on the plasma and discharge current densities. Ion transport to an electrode is described by the equality of flux J through a sheath (3.17) and an ambipolar flux from the plasma to the sheath found from (3.13)

$$J = a\left(\frac{2el}{M}\right)^{1/2}\frac{\overline{V}^{3/2}}{\tilde{d}^{5/2}} = \frac{\pi e D_a n_{\max}}{L} \qquad a = \frac{1}{4\pi}\left(\frac{500}{243\pi}\right)^{1/2} = 0.064 \tag{3.21}$$

Electric current transport through a sheath is described by the equality of the electrode displacement current and the current in the gap middle. Keeping in mind that vacuum displacement current in the plasma is generally smaller than conduction and polarization currents, which are proportional to n (Section 1.3.2), we can roughly write

$$j_a \approx \frac{\omega\overline{V}}{2\pi\tilde{d}} \approx \frac{e^2 n_{\max} E_a}{m(\omega^2 + \nu_m^2)^{1/2}} \tag{3.22}$$

where ν_m is the frequency of electron-atom collisions (stochastic heating is neglected). In the plane case $j_a(x) = $ const, therefore, the field E_a at the gap periphery, where the plasma density is smaller than $n_{\max}$, is stronger. Electron temperature at low pressures is likely to be spatially more uniform than the field; this fact somewhat justified the approximation with $\nu_i = $ const often used for solving the diffusion equation (3.12).

In the frame of this approximation, equations (3.20)–(3.22) define the CVC and other discharge parameters. By eliminating $n_{\max}$ from equations (3.21) and (3.22) and putting $\overline{V} = V_a/2$, we find the sheath thickness

$$\tilde{d} = \left[2a\left(\frac{2e\overline{V}}{M}\right)^{1/2}\left(\frac{eE_a}{m\nu_m}\right)\frac{Ll^{1/2}}{D_a\omega(1+\omega^2/\nu_m^2)^{1/2}}\right]^{2/3} \tag{3.23}$$

Substitution of (3.23) into (3.22) yields j_a and $n_{\max}$ as a function of V_a, i.e., the discharge CVC $V_a(j_a)$. These relations are applicable only to moderately low pressures, when ions fly through a sheath with collisions, and stochastic electron heating is unimportant. In this case one can also assume that $\nu_m^2 \gg \omega^2$, which renders the CVC a fairly clear frequency dependence. Thus we have

$$\tilde{d} \sim V_a^{1/3}\omega^{-2/3} \qquad j_a \sim V_a^{2/3}\omega^{5/3} \qquad V_a \sim j_a^{3/2}\omega^{-5/2} \tag{3.24}$$

The CVC is a rising curve. To maintain definite current and plasma densities at higher frequency, it is sufficient to apply lower voltage while at the definite voltage the current is higher.

It is hardly worth comparing this oversimplified theory with experimental findings. Our aim was to demonstrate the principal causal relationships in a discharge

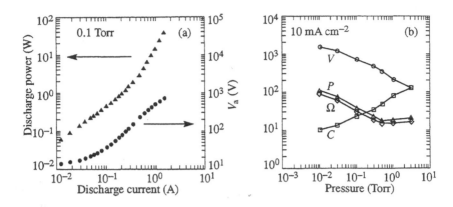

FIGURE 3.7
Experimental data on a plane symmetric discharge in argon, $f = 13.56$ MHz,
$L = 6.7$ cm, $A = 160$ cm^2. (a) CVC $V_a(i_a)$ (amplitudes) and power $P(i_a)$ at
$p = 0.1$ Torr. (b) Voltage V_a (volt), power P (watt), resistance Ω (ohm), and capacitance
C (pF), corresponding to the reactive impedance $X = 1/\omega C$, versus pressure at the given
current amplitude $j_a = 10$ mA cm^{-2} [3.18]. (©1991 IEEE)

and to get a self-consistent model of this phenomenon. We would like to offer a numerical illustration with reference to a comprehensive and thoroughly
designed experiment [3.18]. The subject of the study was a plane symmetric
discharge in argon operating at the frequency 13.56 MHz in the pressure range
$p = 3 \times 10^{-3}$–3 Torr. A glass cylinder of a 14.3 cm inner diameter and height
$L = 6.7$ cm was to limit the discharge space between two plane aluminum
electrodes separated by the same distance L. The area of the electrodes was
$A = 160$ cm^2. Measurements of the CVC, power, active and reactive resistance
of the discharge, etc. provided valuable experimental data on the sheaths and
plasma. Some of the measurements are presented in Figure 3.7.

Let us make evaluations for $p = 0.1$ Torr. Suppose $l = (200\,p)^{-1} = 5 \times
10^{-2}$ cm; $\nu_m = 5.5 \times 10^9\,p = 5.5 \times 10^8$ s$^{-1} \gg \omega = 0.86 \times 10^8$ s^{-1}; $\mu_+ =
10^3/p = 10^4$ cm^2 V^{-1} s^{-1}; $T_e = 3$ eV; $D_a = \mu_+ T_e = 3 \times 10^4$ cm^2 s^{-1}. From
equation (3.20), the frequencies of diffusion losses and ionization are $\nu_i = \nu_d =
670$ s^{-1}, and the relaxation frequencies of electron spectra and temperature $\nu_u =
\nu_m \delta = 1.5 \times 10^4$ s$^{-1} \ll \omega$, $(\delta = 2m/M)$. Under these conditions the ionization
frequency in an RF field of amplitude E_a is close to the ionization frequency in a
dc field $E = E_a/\sqrt{2}$. This gives an opportunity to make use of available data on
the positive column of a dc glow discharge excited in a tube [3.19]. If a discharge
is diffusion-controlled, the plane geometry corresponds to a tube discharge of
radius $R = (2.4/\pi)L$. For the respective $pR = 0.51$ cm in argon, $E/p \approx 7$, i.e.,
$E_a/p \approx 10$ V cm^{-1} Torr^{-1}.

Suppose a voltage $V_a = 300$ V is applied to the electrodes. Putting $\overline{V} \approx
V_a/2 = 150$ V, we find from equation (3.23) $\tilde{d} = 0.74$ cm. In this case, $\tilde{d}/l \approx 15$;

in other words, we really deal with a collisional sheath. The value of $p\tilde{d} = 0.074$ Torr cm lies far left of the Paschen curve minimum, which means that there is no breakdown in the sheath and we have here an α-discharge (Section 2.6). From equation (3.22), $j_a \approx 3.1$ mA cm^{-2}, $n_{max} \approx 6.1 \times 10^9$ cm^{-3}, the average sheath ion density $n_1 \approx \overline{V}/2\pi e\tilde{d}^2 = 3.0 \times 10^8$ cm^{-3}, i.e., n_{max} is 20 times this value. Discharge current $i_a = j_a A = 0.5$ A agrees well with the experimental value of $i_a = 0.45$ A corresponding to $V_a = 300$ V in Figure 3.5. The value of n_{max} fits the average plasma density $n \approx 4 \times 10^9$ cm^{-3}, which can be evaluated from the measured active discharge resistance $\Omega = 15$ Ω for these conditions [3.18] ($\Omega = [L - 2\tilde{d}]/\sigma A$, for argon the conductivity $\sigma \approx 5.3 \times 10^{-14} n_e/p$ Ω^{-1} cm^{-1}). It should be noted, however, that the experimental CVC is better described by the dependence $V_a \sim j_a$ than by $V_a \sim j_a^{3/2}$ from equation (3.24). As for the pressure dependence of V_a, it slightly falls with growing pressure. In order to explain this fall within the frame of the discharge model discussed, one should assume that $E_a/p \sim 1/p$. This is a reasonable assumption if one considers experimental data on the dc glow discharge at low pressures. But here, as in the case of $V_a(j_a)$, one may expect to find effects of some factors that have not been taken into account.

3.3   Sheath dynamics and current anharmonicity in an asymmetric discharge

Available models and theories of the asymmetric discharge [3.3, 3.4] (Sections 3.1, 3.2) deal with some average capacitance and sheath thickness without focusing on their time evolution. An electric circuit with constant capacitances is linear, which means that at a harmonic electrode voltage the discharge current is also harmonic. In reality, approximate harmonicity of potentials (against the background of possible constant components) and current is to some extent a property of symmetric discharges only. In an asymmetric process, the dynamics of the sheath boundaries, voltage fall and current is anharmonic even at applied harmonic voltage. In the simplified discharge model, in which plane sheaths with constant ion densities n_1 and n_2 are separated by an ideally conducting plasma, the conservation law for the total sheath charge is written as

$$en_1 d_1 A_1 + en_2 d_2 A_2 = Q_g = \text{const} \tag{3.25}$$

where d_1 and d_2 are instantaneous sheath thicknesses. For the case of different electrode areas, the equivalent capacitance of both sheaths

$$C_{eq} = \frac{1}{(C_1^{-1} + C_2^{-1})} = \frac{1}{(4\pi d_1/A_1 + 4\pi d_2/A_2)} \tag{3.26}$$

varies over a cycle with varying d_1 and d_2, making the electric circuit nonlinear. Only a symmetric system ($A_1 = A_2$) with constant n_1 and n_2 (in this case $n_1 = n_2$)

would be linear due to a constant total sheath thickness $d_1 + d_2$ and to the constant equivalent capacitance. Within the same elementary model, we will now analyze the sheath dynamics, charge transport and current behavior in an asymmetric capacitively coupled discharge (Figure 3.1). This model is useful not only for the understanding of these processes but for interpretation of experimental data. It also helps to gain more information on the sheath and plasma parameters from electrical measurements.

Instantaneous voltage falls for two sheaths (from the plasma towards the electrodes) are

$$V_1 = 2\pi e n_1 d_1^2 \qquad V_2 = 2\pi e n_2 d_2^2 \qquad (3.27)$$

Since the plasma contacts the electrodes during oscillations, the constant in equation (3.25) is

$$Q_g = n_1 d_1{}_{\max} A_1 = n_2 d_2{}_{\max} A_2 \qquad (3.28)$$

Denote with Q_0 the absolute value of negative charge, which goes forever into the first electrode. Its fraction $e n_1 d_1 A_1$ is concentrated on the electrode surface; the other portion, $Q_0 - e n_1 d_1 A_1$, is concentrated on the plate of capacitor C, which is connected with the electrode (Figure 3.8). The instantaneous generator voltage $V_\sim$ and the potential of the plate C connected to it are identical and equal to

$$V_\sim = V_g \sin \omega t = \frac{Q_0 - e n_1 d_1 A_1}{C} + 2\pi e (n_2 d_2^2 - n_1 d_1^2) \qquad (3.29)$$

The second term in equation (3.29) is the instantaneous potential of the loaded electrode denoted as $V(t)$ in formula (3.2) of the linear model. When the potential $V_\sim$ has a positive maximum, all positive charge Q_g is 'shifted' towards the farther electrode: $d_1 = 0$. When $V_\sim$ has a negative maximum, the positive charge is 'attracted' to the loaded electrode: $d_2 = 0$. Hence

$$V_g = \frac{Q_0}{C} + 2\pi e n_2 d_2^2{}_{\max} \qquad (3.30)$$

$$-V_g = \frac{Q_0 - e n_1 d_1{}_{\max}}{C} - 2\pi e n_1 d_1^2{}_{\max} \qquad (3.31)$$

Equations (3.28), (3.30), and (3.31) can be easily solved relative to $d_1{}_{\max}$, $d_2{}_{\max}$ and Q_0 by expressing these magnitudes through the sheath densities n_1 and n_2. As was pointed out in Section 3.2, the latter are determined by ambipolar charge fluxes from the plasma. Let the densities n_1 and n_2 satisfy the semi-empirical relation (3.19) with $\beta = 5/2 - q$, which eventually follows from the approximation (3.18). This gives us a possibility to express maximum sheath thicknesses, maximum sheath voltage falls from (3.27) and charge Q_0 through the known loaded electrode area A_1, the area ratio $\eta = A_1/A_2$, the experimentally chosen power index q in the similarity law (3.1) and the ion density n_1. The latter is the only magnitude that cannot be measured directly. The equations for sheath

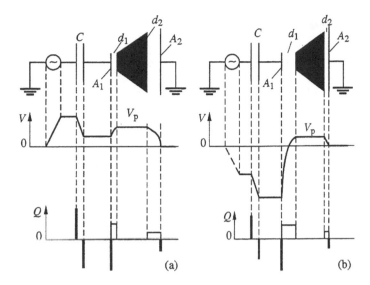

FIGURE 3.8
Distributions of potential V and charge Q along an asymmetric RF discharge with a blocking capacitor. For convenience, the electrodes A_1 and A_2 are arranged vertically in this picture showing the moments of the plasma (shaded region) closest approach to (a) the smaller and (b) the larger electrodes. The charge on the metal is black; in the gas, white.

boundary motion are derived from nonstationary equations (3.29) and (3.25), in which the charge Q_g is defined by equation (3.28). The obtained set of equations is solvable for $d_1(t)$ and $d_2(t)$. The discharge current is equal to

$$i = -en_1 A_1 \dot{d}_1 = en_2 A_2 \dot{d}_2 \qquad (3.32)$$

The anharmonic motion of the boundary and current is associated with the quadratic form of equation (3.29). In the symmetric case with $d_1 + d_2 = d_{1\,max}$, the square terms in (3.29) are canceled out and the equations become linear in d_1 and d_2.

We will not write out the general formulas from [3.17], because some of them are rather cumbersome. Their physical meaning becomes quite clear in the limit of great differences in the electrode area and when blocking capacitance C is much more than the characteristic capacitance of the smaller electrode sheath $\tilde{C}_1 = A_1/2\pi d_{1\,max}$.[3] In the limit of $\eta \to 0$ and $C \to \infty$, with the floating

---

[3] In a symmetric discharge with $d_1 = d_{1\,max}(1+\sin \omega t)/2$, $\tilde{C}_1^{-1}$ exactly coincides with the average reciprocal value of variable sheath capacitance: $\tilde{C}_1^{-1} = \langle C_1^{-1} \rangle = 4\pi \bar{d}_1/A_1$.

potential being neglected, we have $d_2 \rightarrow 0$, $V_2 \rightarrow 0$ and

$$d_1 = d_{1\,max}\left(\frac{1-\sin\omega t}{2}\right)^{1/2} \qquad d_{1\,max} = \left(\frac{V_g}{\pi e n_1}\right)^{1/2}$$

$$V(t) = -V_1 = -V_g + V_g\sin\omega t \qquad V_{1\,max} = 2V_g \qquad \overline{V}_1 = V_g = -V_{dc}$$

$$i = \begin{cases} +\frac{i_{max}}{\sqrt{2}}(1+\sin\omega t)^{1/2} & \text{at } -\frac{\pi}{2} < \omega t < \frac{\pi}{2} \\ -\frac{i_{max}}{\sqrt{2}}(1+\sin\omega t)^{1/2} & \text{at } \frac{\pi}{2} < \omega t < \frac{3\pi}{2} \end{cases}$$

$$i_{max} = \frac{A_1\omega}{2}\left(\frac{V_g e n_1}{\pi}\right)^{1/2} \tag{3.33}$$

The current wave form has a well-defined saw-tooth shape with a gradual rise from $-i_{max}$ to $+i_{max}$ over a cycle and a vertical reverse step to $-i_{max}$ at the end of the cycle. The negative charge $Q_0 = CV_g \rightarrow \infty$ (at $C/\tilde{C}_1 \rightarrow \infty$) enters the electrode connected to the blocking capacitor. Since the positive charge of the gas gap Q_g is limited, a 'slightly smaller' but still 'infinite' positive charge enters the other electrode (the ground). This fact is somewhat unexpected because it has been generally believed that electrons enter the electrodes. Negative charges enter both electrodes only in a symmetric discharge; for this to happen in an asymmetric case, the block capacitance should be small (for exact criterion see [3.17]). It is due to the positive charge escape into the earth that we can reduce the minimum loaded electrode potential below the minimum generator potential, nearly down to $-2V_g$ in the case of a large blocking capacitance. This is what is generally done in practice in order to obtain a large self-bias. The positive charge transported by ions enters the larger electrode as a steady discharge develops after its initiation.

Figure 3.9 illustrates the calculations from general formulas with the following parameters: $q = 1.5$, $\beta = 1$, $A_1 = 100\,\pi$ cm^2 (electrode diameter 20 cm), $\eta = A_1/A_2 = 1/5$, $C = 500$ pF, $f = 13.56$ MHz, $V_g = 300$ V, $n_1 = 5.6 \times 10^8$ cm^{-3}. The calculated value of $d_{1\,max} = 1$ cm corresponds to $\tilde{C}_1 = 28$ pF. Owing to the inequality $C \gg \tilde{C}_1$, $V_{1\,max} = 500$ V is not much smaller than $2V_g$ (e.g., with $C = 50$ pF and other things being equal, $V_{1\,max} = 215$ V, because too much voltage falls on the smaller blocking capacitor). The maximum plasma potential $V_{p\,max} = V_{2\,max} = 50$ V and the constant potential $\overline{V}_p = 15$ V, but we should not attribute much significance to the latter value since it is comparable with the neglected floating potential. Self-bias $V_{dc} = -222$ V. The charge into the smaller electrode $-Q_0 = -1.3 \times 10^{-7}$ C, into the larger one $+10 \times 10^{-7}$ C, the residual charge in the gap $Q_g = +0.3 \times 10^{-7}$ C. The current is closer to a saw-tooth shape rather than to a sine shape. This is because the plasma approaches the smaller electrode and then recedes much faster than it moves at maximum separation. The situation near the larger electrode is quite reverse. Current density on the smaller electrode $j_{1\,max} = 3.8$ mA cm^{-2}, on the larger one $j_{2\,max} = 0.76$ mA cm^{-2}.

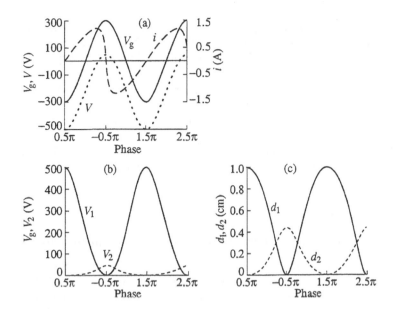

FIGURE 3.9

Evolution of the parameters of an asymmetric discharge with a blocking capacitor calculated in a simplified model; $A_1 = 100\pi$ cm^2, $A_1/A_2 = 1/5$, the larger electrode grounded. (a) Potential at the operating generator terminal (generator voltage) V_g, potential on the smaller electrode (electrode voltage) V and discharge current i. (b) Absolute voltage falls across the sheaths of the smaller (V_1) and the larger (V_2) electrodes; $V = V_2 - V_1$. (c) Sheath thickness of the smaller (d_1) and the larger (d_2) electrodes.

Ion current into the electrodes can be evaluated from equation (3.17) by replacing $\overline{V}_1$ and $\tilde{d}_1$ by instantaneous values and averaging the result over time. For argon at $p = 2 \times 10^{-2}$ Torr ($l = 0.25$ cm), average ion currents $\bar{J}_1 = 0.066$ mA cm^{-2} and $\bar{J}_2 = 0.012$ mA cm^{-2} are two orders of magnitude lower than the discharge current, i.e., than displacement currents. The maximum rate of the sheath boundary motion at the smaller electrode $(\dot{d}_1)_{\max} \approx 4.2 \times 10^7$ cm s^{-1} approximately coincides with the oscillation amplitude of electrons in the boundary RF field. It is clear from the value of $(\dot{d}_1)_{\max}$ that this field is quite strong, $E_a/p \approx 160$ V cm^{-1} Torr^{-1}. The plasma bulk field is not as strong, since the bulk plasma density n is higher than at the boundary: $i(x) \sim nEA \approx$ const (Section 3.2, Figure 3.4). Taking $T_e = 3$ eV, $D_a = \mu_+ T_e = 1.5 \times 10^5$ cm^2 s^{-1}, the gap size $L = 5$ cm, we can find from (3.11) the value of $n_{\max} = 5 \times 10^9$ cm^{-3} in the gap middle ($\bar{J}_1 \sim 2eD_a n_{\max}/L$).

This calculation has been made for conditions close to the experiments of [3.12] for the same gas, p, f, C, A_1, and L. Although the design was symmetric, the chamber was grounded and, judging by the large measured self-bias value, the discharge was essentially asymmetric. At $V_a = V_g = 300$ V, T_e was found to be 2.7 eV and n_{max} was approximately 10^{10} cm^{-3}. This is in agreement with the above estimates, indicating a reasonable choice of the ion density in the sheath n_1.

Like in [3.5], measurement of the potential alone by varying V_a can yield only the ratios of average capacitances of the sheaths. If, however, these data are supplemented by measurements of the discharge current, especially its wave form, one can also get some information on the frequently unknown effective area of the grounded 'electrode,' on the plasma density, ion fluxes to the electrodes, etc. by making simple calculations of the type described above.

3.4 Current anharmonicity in asymmetric and symmetric discharges

The current anharmonicity discussed above for the case of sinusoidal voltage across the electrodes is caused by the discharge asymmetry. As a result, the plasma boundary changes the direction of motion more rapidly on touching the smaller electrode than on touching the larger one. This happens because the boundary motion must provide a higher displacement current density at the smaller electrode, leading to a specific pattern of current variation over a cycle. At the moments of the greatest plasma proximity to this electrode, the current changes fairly rapidly from peak to peak; but then the change between two peaks becomes relatively slow, and the current wave form acquires a saw-tooth profile instead of a symmetric sinusoidal variation (Figure 3.9). A symmetric discharge, however, may also be characterized by current anharmonicity, though its underlying mechanism is quite different. The matter is that the ion density throughout the sheath is nonuniform, decreasing from the plasma towards the electrode. This disturbs the regular sinusoidal current pattern inherent to a discharge model with identical sheaths and a uniform ion density distribution, but it does not disturb the high degree of symmetry of the current 'wave form.' The peaks in it also alternate each half-cycle, and their shape is identical every quarter cycle and even close to the sinusoidal shape.

This can be easily illustrated with a simple example. Suppose $n_+(x)$ within the sheath drops to zero in the power law $n_+ = n(x/d_m)^k$, being equal to the plasma density n at the maximum distance of the plasma boundary from the electrodes $x = d_m$ (Figure 3.10). Assume also the plasma boundaries to be sharp and the plasma to possess an ideal conductivity. Voltage $V_a \sin \omega t$ is applied to the left electrode while the right electrode is grounded. By introducing dimensionless sheath thicknesses $z_1 = d_1/d_m$ and $z_2 = d_2/d_m$, we can write expressions for the charge conservation for the two sheaths and for the instantaneous voltage fall across the gap. We will get the equations describing the plasma boundary motion

$$z_1^{k+1} + z_2^{k+1} = 1 \qquad \sin \omega t = z_2^{k+2} - z_1^{k+2}$$

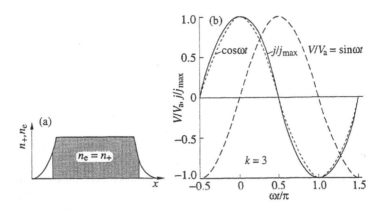

FIGURE 3.10
On anharmonism in a symmetric discharge. (a) Schematic density distribution of ions and electrons in the gap. (b) The respective current wave form at sinusoidal electrode voltage; $k = 3$.

$$V_a = \frac{4\pi e n d_m^2}{(k+2)} \qquad V_1 = V_a z_1^{k+2} \qquad V_2 = V_a z_2^{k+2}$$

At $k = 0$, when $n_+(x) = \text{const} = n$, they transform to equations (3.25), (3.27)–(3.30) with $n_1 = n_2 = n$, $A_1 = A_2$, $C \to \infty$. In general, discharge current is

$$j = -e n_+(d_1)\dot{d}_1 = \frac{e n d_m \omega}{k+2} \frac{\cos \omega t}{z_1 + z_2}$$

The time-dependent factor $z_1 + z_2$ stands for the deviation of $j(t)$ from the cosine law, i.e., anharmonicity. The function $z_1 + z_2$ varies over a cycle from 1 to $2^{k/k+1} \approx 2$ at large k, and this variation characterizes the degree of anharmonicity (Figure 3.10).

Since there are two factors causing anharmonicity of an asymmetric discharge (asymmetry and sheath nonuniformity), the current wave form has a complicated character. The same is true of detailed calculations. For instance, the structure of a single collisionless sheath with an oscillating potential and self-bias was examined rigorously in [3.20]. A constant zero potential was ascribed to the plasma, which is equivalent to extreme asymmetry, i.e., to the zero voltage fall across the other electrode sheath. The current wave form has a well-defined, though very complicated, saw-tooth shape, which qualitatively agrees with the simplified model presented in Figure 3.9. To compare the curve from [3.20] with Figure 3.9, one of the plots should be inverted, because Figure 3.9 refers to the left-hand loaded electrode, while the one in [3.20] refers to the right-hand electrode. Results close to [3.20] were obtained in [3.21] by using simpler equations derived by averaging over high electron motions [3.22]. If a harmonic current is given in the calculations of the sheaths and of the discharge as a whole, the anharmonicity of the sheath and electrode voltages arises for the same reasons [3.23–3.25]. The calculation of a plane symmetric argon discharge with harmonic current at $p = 5 \times 10^{-2}$–0.5 Torr and $f = 13.56$ MHz yielded a very weak voltage anharmonicity, while in an asymmetric discharge the anharmonicity was very pronounced [3.24].

The appearance of the higher harmonics in current oscillograms at sinusoidal generator voltage was observed long ago in experiment [3.15]. This phenomenon has recently been studied experimentally in [3.26]. The effects of the two major factors inducing anharmonicity are very hard to separate practically, because low-pressure discharges are commonly asymmetric, unless special measures have been taken against it. A moderate-pressure plane discharge is usually symmetric, but in this case the anharmonicity is weak because of the small variation in the ion density of the sheaths from the plasma towards the electrodes. This is evidenced by both experiment and calculations (Section 2.1).

Still, experimental data are available that permit observation of changes in anharmonicity when one of its causes is removed. Such data were obtained in magnetron discharge studies [3.13] (Section 3.14). In that work, an axial magnetic field transverse to the current was introduced in an asymmetric argon discharge, which was operating in a gap between coaxial cylinders of 10 and 25 cm in diameter at $p = 3$ mTorr and $f = 13.56$ MHz with a harmonic voltage applied to the smaller electrode. The obtained current oscillograms are reproduced in Figure 3.11. At zero magnetic field, the oscillogram has a complicated shape with a dominant saw-tooth component, as both anharmonicity factors make their contribution. In field $B = 200$ G, the Larmour frequency of electrons $\omega_H = 3.5 \times 10^9$ s$^{-1} \gg \omega$. Therefore, the electron oscillation amplitude, the sheath thickness and, hence, the sheath ion density variations decrease sharply as compared to the zero field values. The Larmour frequency plays the role of collision frequency, and the discharge behaves in this respect as a moderate-pressure discharge. This actually eliminates the effect of the other factor associated with the variable n_+ value in the sheaths, leaving only the asymmetry effect. Accordingly, the current oscillogram acquires a smooth saw-tooth shape quite similar to that obtained from a simplified calculation at constant sheath ion densities n_1 and n_2 (Figure 3.9). The effect is not very pronounced, since the asymmetry is weak: the area ratio $A_1/A_2 = 1/2.5$ [3.13].

3.5 Battery effect in an asymmetric discharge

3.5.1 Sheath dynamics and direct current at low pressures

Let us close the external circuit for dc by grounding the loaded electrode through large inductance (which does not pass RF current). Assume that the voltage applied

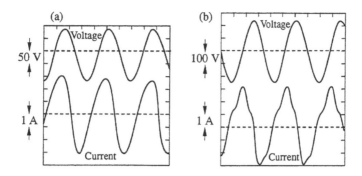

FIGURE 3.11
Experimental (sinusoidal) voltage and current oscillograms from a discharge between coaxial cylinders of 5 and 12.5 cm radii; argon, $p = 3\,\text{mTorr}$, $f = 13.56\,\text{MHz}$ [3.13]; (a) with a longitudinal magnetic field $B = 200$ G and (b) at zero magnetic field.

to the electrodes is harmonic. Then the average potentials of the electrodes are identical, and if one of them is grounded, both potentials are equal to zero. In this case $\overline{V}_1 = \overline{V}_2 = \overline{V}_p$ and no self-bias is observed. Equations (3.27) and (3.25) with the constant of (3.28) prove to be incompatible. For example, at $n_1 = n_2$ equation (3.27) yields $d_1^2 = d_2^2$, whereas from (3.28) the sheath of the larger electrode must be thinner. This means that the oscillating plasma boundary does not actually reach the larger electrode, exposing and covering the ionic sheath only partly. In this case the gas charge in (3.25) is given by other equalities

$$Q_g = e n_2 d_{2\,max} A_2 = e(n_1 d_{1\,max} A_1 + n_2 d_{2\,min} A_2) \qquad (3.34)$$

The term describing the voltage fall across the capacitor C in (3.29) vanishes. At the maximum voltages on the generator and electrodes, equations (3.30) and (3.31) are replaced by

$$V_g = 2\pi e n_2 d_{2\,max}^2 \qquad -V_g = 2\pi e(n_2 d_{2\,min}^2 - n_1 d_{1\,max}^2) \qquad (3.35)$$

Figure 3.12 shows the dynamics of the discharge parameters calculated for the same conditions as in Figure 3.9 but with no blocking capacitor. The current wave form has a shape looking more like a saw-tooth than a sine-shaped pattern. The average plasma potential $\overline{V}_p = 193$ V is much higher than in the previous case. The positive gas charge is also larger $Q_g = e n_2 d_{2\,max} A_2 = 0.73 \times 10^{-7}$ C, most of it $e n_2 \bar{d}_2 A_2 = 0.6 \times 10^{-7}$ C being in average accumulated at the larger electrode, where the ionic charge is never completely neutralized by the electronic charge. In the limit $\eta \ll 1$ at $\beta = 1$, we have

$$\frac{(d_{2\,max} - d_{2\,min})}{\bar{d}_2} = \sqrt{2}\eta^{3/4 - \beta/2} = \sqrt{2}\eta^{1/4}$$

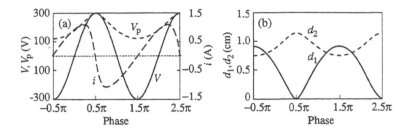

FIGURE 3.12
Evolution of the parameters of an asymmetric discharge without a blocking capacitor.
(a) Potentials at the generator terminal (on the smaller electrode) V, on the plasma V_p, and current i. (b) Sheath thickness at the smaller (d_1) and the larger (d_2) electrodes.

$$\bar{d}_2 = \frac{(d_{2\,\text{max}} + d_{2\,\text{min}})}{2} = \frac{1}{\sqrt{2}} d_{1\,\text{max}} \eta^{1/4-\beta/2} = \frac{1}{\sqrt{2}} d_{1\,\text{max}} \eta^{-1/4} \quad (3.36)$$

which means that the plasma oscillation amplitude at the larger electrode is relatively small, and the average sheath thickness is larger than at the smaller electrode. The constant plasma potential

$$\overline{V}_p = \overline{V}_2 = 2\pi e n_2 \overline{d_2^2} \approx 2\pi e n_2 d_{2\,\text{max}}^2 = V_g \quad (3.37)$$

is close to the applied voltage amplitude.

Since the plasma of an asymmetric closed-circuit discharge never touches the large electrode, the ionic flux enters it but the electronic flux does not. There is certainly an ionic flux towards the smaller electrode, but it is weaker here, because from (3.18) with $\beta = 1$, $J_1 A_1 / J_2 A_2 = (A_1/A_2)^{1/2} < 1$, and is overcompensated by the electronic flux at the moments of the plasma contact with the electrode. As a result, direct current flows through the circuit. In the external circuit it flows from the larger to the smaller electrode. In the larger electrode sheath, the direct current is ionic; but in the smaller sheath and in the plasma, it is electronic. For the above illustration we obtain $\bar{J}_2 \approx 2.6 \times 10^{-2}$ mA cm^{-2} by averaging the formula of (3.17) with the data of Figure 3.12 (as in Section 3.3). The direct current in the circuit $i_{\text{dc}} = \bar{J}_2 A_2 \approx 40$ mA makes up 4% of the maximum RF current.

Thus, the asymmetric RF discharge is a source of direct current, a kind of a 'battery,' which has long been known. It was observed experimentally at low pressures [3.27] with the fixed asymmetry achieved by using cylindrical geometry: the discharge filled up the gap between coaxial cylindrical electrodes with strongly different radii.

There is a direct relationship between the battery effect and the self-bias effect. If the external circuit is disconnected for a direct current, a constant voltage equal to the self-bias V_{dc} on the small electrode arises at the 'battery' terminals, or electrodes. The self-bias acts as the emf of the 'battery,' i.e., of the asymmetric RF discharge. Indeed, the self-bias calculated in Section 3.3 $V_{\text{dc}} = -222$ V nearly

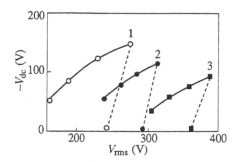

FIGURE 3.13
Measured constant potential difference in the gap between two coaxial electrodes plotted against rms RF voltage V_{rms} applied to the inner electrode of 1 cm radius (outer electrode radius 4 cm); $f = 13.56$ MHz. Air, $p = 7.5$ Torr (curve 1). Air, $p = 15$ Torr (curve 2). Helium, $p = 100$ Torr (curve 3).

coincides in the absolute value with the plasma constant potential $\overline{V}_p = 193$ V just calculated for the 'short-circuited' external circuit. The latter serves as a driving force 'pushing' the direct current through the battery. In the limiting case of a very large asymmetry ($\eta \to 0$), when the larger electrode sheath is the only source of direct current, both values $|V_{dc}|$ and $\overline{V}_p$ are, according to (3.33) and (3.37), identical and equal to V_g.

The battery emf $\overline{V}_p$ in a closed zero-resistance external circuit is totally expended on overcoming the inner resistance of the source. The latter resistance approximately coincides with the electric resistance of the larger electrode sheath, $\Omega_2 = \overline{V}_p/i_{dc}$. Indeed, $\Omega_2 = \overline{d}_2/\sigma_i A_2$, where $\sigma_i = en_2(v_i/E)_2$ is ionic conductivity, which is here nonlinear since the ion drift velocity was taken to be $v_i \sim \sqrt{E} \approx \sqrt{\overline{V}_p/\overline{d}_2}$. We return to the initial expression (3.17) for ionic current, which is now treated in terms of Ohm's nonlinear law $\overline{J}_2 \approx \sigma_i \overline{V}_p/\overline{d}_2$. Simultaneous measurements of the circuit direct current and of the plasma constant potential should be made to extract more experimental information about the ion density n_2 of the larger electrode effective area A_2 and other parameters.

3.5.2 Self-bias and direct current in moderate-pressure discharges

The battery and self-bias effects are inherent to RF discharges of any pressure. It is not the pressure or the charge loss mechanism in the plasma that matters here but the discharge asymmetry. At elevated pressures, one is not always able to provide asymmetry, because the normal current density effect interferes with it. Self-bias at moderate pressures was registered in experiments with anomalous α-discharges operating in a system of coaxial cylindrical electrodes [3.28]. The voltage from an RF generator was applied through a blocking capacitor to the inner, smaller electrode which acquired a negative potential relative to the outer, grounded electrode (Figure 3.11). The large self-bias $V_{dc} = -\overline{V}_1 + \overline{V}_2$ is a direct experimental evidence for high constant potentials of moderate-pressure discharge plasmas. The point is that in some earlier publications, including the pioneering work [3.29], measurements indicated an abrupt decrease in constant plasma potentials down to a few volts at pressures above 1–10 Torr. It was later shown by one of the authors of this book that those results were due to the imperfection of the experimental technique used, and that fairly high constant potentials of a few tens and a hundred volts were also characteristic of RF moderate-pressure discharges (Section 4.2.2).

It is quite revealing that the self-bias abruptly disappears at RF voltages above certain threshold values (Figure 3.11). This results from breakdown of the α-sheaths and appearance of a γ-discharge with a normal current density, when the discharge becomes symmetric in spite of the asymmetry of the electrode system. It is curious that a small self-bias remained only when the electrodes were made from different metals, resulting in different 'cathode' potential falls $\overline{V}_1 \neq \overline{V}_2$ in the γ-discharge.

The effects discussed in Sections 3.3 and 3.5 are clearly illustrated by numerical simulation of the moderate-pressure discharge. The calculation was made of a discharge between coaxial cylindrical electrodes in nitrogen at $p = 25$ Torr and $f = 81$ MHz, a frequency widely used at present for exciting CO_2 lasers (as well as the coaxial cylindrical geometry). The radius of the inner loaded electrode was $R_1 = 0.25$ cm and that of the outer grounded one $R_2 = 0.5$ cm, so the area ratio $\eta = A_1/A_2 = 1/2$. The calculation was made using a set of equations of the type (2.1) taken in the cylindrical coordinates. It was supplemented with an equation of the type (2.5), also in the cylindrical coordinates to allow for the gas heating, assuming the oscillation relaxation to occur rapidly (Section 2.3). For the given parameters, the inequalities $\nu_m \gg \omega \gtrsim \nu_m \delta$ hold, owing to which the mean ionization frequency and the Townsend coefficient can be defined by the rms field E or, more exactly, by the value of $(E/p)(T/T_0)$ with allowance for the gas heating, where $T_0 = 300$ K (Section 2.3). The calculation was performed for the blocking capacitance $C = 50$ pF cm^{-1} (per 1 cm cylinder length) at the generator voltage amplitude $V_g = 200$ V, starting from an initial plasma concentration in the gap. A periodic solution, with an accuracy of 0.01%, was reached in about 10^3 periods.

Figure 3.14 relates to a blocking capacitance discharge. The voltage drop across the capacitor and the phase shift $V(t)$ and $V_\sim(t)$ are insignificant. One can see that the plasma comes in contact with both electrodes and that the ion density at the larger electrode is lower. The self-bias $V_{dc} = -40$ V and the constant plasma potential $\overline{V}_p \approx 38$ V. The current is slightly anharmonic, nearly sinusoidal, with an amplitude $i_a = 200$ mA cm^{-1} (per 1 cm cylinder length) and 0.66π phase-shifted relative to $V(t)$. The average ionic current on the smaller electrode is $i_i \approx 0.7$ mA cm^{-1} and on the larger one 0.35 mA cm^{-1}. The duration of the electron current that compensates for the ionic current is 15% of a cycle, and the maximum temperature near the gap center is $T_{max} \approx 700$ K. Figure 3.15 refers

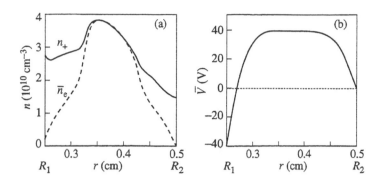

FIGURE 3.14
Calculation of a discharge between two coaxial cylinders of radii $R_1 = 0.25$ cm and $R_2 = 0.5$ cm with a large blocking capacitor; nitrogen, $p = 25$ Torr, $f = 81$ MHz. The amplitude of the voltage applied to the smaller electrode $V_a = 200$ V. (a) Density distributions of ions n_+ and of electrons (cycle-averaged) $\bar{n}_e$. (b) Distribution of constant potential $\overline{V}$.

to a discharge with a closed external circuit. One can see that the plasma does not reach the larger electrode. In this case $\overline{V}_p \approx 64$ V at the gap center. The RF current amplitude $i_a \approx 180$ mA cm^{-1} and heat release have become a little smaller. The current through the larger electrode is ionic alternating current with a dc component $i_{dc} \approx 0.28$ mA cm^{-1}. The average electron current through the smaller electrode exceeds the average ionic current by the same value.

Numerical results are useful for checking on the simplified theory of discharges. To do this, one should slightly modify it, because in the 'numerical experiment' the discharge was assumed to be controlled by bulk recombination rather than by ambipolar diffusion, and $\mu_+ = $ const. So the 'empirical' power index β must be replaced using (3.19) while (3.17) should be replaced by $J \sim \mu_+ V^2/d^3$. The resultant agreement can be considered as an argument in favor of the use of the simplified model in experimental data analysis of low-pressure discharges.

3.6 Plasma 'nontransparency' and fast electron response to RF field and 'oscillationless' sheath ions

The assumptions of the plasma nontransparency to the RF field and of a fast response of electrons to the field variation, together with the concept of 'oscillationless' ions in the sheaths, underlie most of the available models of RF discharge sheaths, including that of asymmetric low-pressure discharge discussed above. (The numerical simulation described in Sections 2.1 and 3.5.2 is free from these assumptions.) Let us establish the quantitative criteria of validity for these

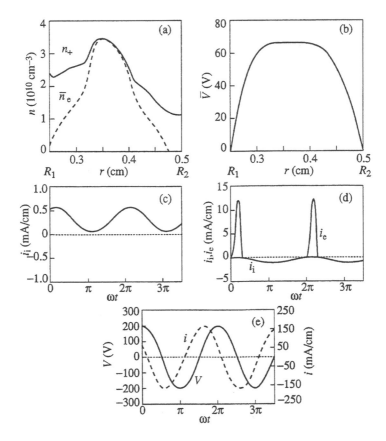

FIGURE 3.15
Calculation of a discharge between two coaxial cylinders with the external circuit closed
for dc; nitrogen, $p = 25$ Torr, $f = 81$ MHz, $V_a = 200$ V, $R_1 = 0.25$ cm, $R_2 = 0.5$ cm.
(a) Distributions of ion density and average electron density, (b) constant potential
distribution, (c) ionic current across the larger electrode (with no electron current),
(d) ionic and electron currents across the smaller electrode, (e) electrode voltages and
discharge current.

assumptions, starting with the first one.

The space charge sheaths around the electrodes or their dielectric coatings are
formed due to the oscillatory motion of the electron gas relative to the slow ions
under an oscillating plasma field $E_p = E_a \sin \omega t$. When one says the plasma is
'impenetrable' to the RF field, one means that the field, which does exist in the
plasma, is much lower than the sheath field. In other words, when the gaps are
quite small, nearly all applied RF voltage falls on the sheaths, because the plasma
acts as a kind of screen to the electrode fields, not letting them in. From (1.2),

the oscillation amplitude of electrons in the plasma is $A = eE_a/m\omega(\omega^2 + \nu_m^2)^{1/2}$. It determines the order of magnitude, and [with the assumption of $n_+(x) =$ const even exactly] the average sheath thickness d ($d \sim A$). Hence, the sheath field, as well as its constant and variable components, are equal in the order of magnitude to $E_s \approx 4\pi e n_+ A$, where n_+ is a characteristic ion density in the sheath, which is comparable to the plasma density n in the vicinity of the sheath. By combining the two expressions and introducing the electron plasma frequency (1.21) corresponding to n, we get the criterion we were seeking:

$$\frac{E_s}{E_a} \approx \frac{\omega_p^2}{\omega(\omega^2 + \nu_m^2)^{1/2}} \gg 1 \tag{3.38}$$

In a 'collisionless' plasma with $\nu_m^2 \ll \omega^2$, the condition of impenetrability (3.38) turns to a common condition of electromagnetic wave reflection from the plasma $\omega^2 \ll \omega_p^2$ (Section 1.3.7). If $\nu_m^2 \gg \omega^2$, the RF field screening by the plasma is described by the inequality $\omega\nu_m \ll \omega_p^2$. It is clear from the list of typical RF discharge parameters given in Section 1.3.5 that inequality (3.38) holds for the majority of cases, except for very high frequencies and high pressures.

Under the condition $\omega^2 \ll \omega_p^2$, the electron gas responds very fast to the field variation. The spatial distribution of electron density and electron fluxes at any given moment of time corresponds to the instantaneous field distribution, i.e., it is quasi-stationary. Indeed, turn to the hydrodynamic equation of electron gas motion

$$mn_e\frac{\mathrm{d}v}{\mathrm{d}t} = -n_e eE - \nabla p_e - n_e m v \nu_m \qquad p_e = n_e kT_e \tag{3.39}$$

Equation (3.39) for the mean electron gas velocity differs from (1.1) for the mean velocity of a single electron by the presence of pressure force $-\nabla p_e$ inherent to the gas particles. Let us compare the orders of magnitude of the various terms in (3.39) as applied to the electron gas motion in an oscillating sheath field and drop the insignificant term, which is proportional to ∇T_e and describes thermal diffusion.

The field in a sheath is equal, in the order of magnitude, to $E \sim 4\pi e n_+ d$; the velocity of electrons periodically flooding the sheath greatly varies over a cycle; and the electron oscillation amplitude is nearly as large as the sheath thickness d. Therefore, $\mathrm{d}v/\mathrm{d}t \sim \omega v$ and $v \sim \omega d$. Divide equation (3.39) by mn_e and write out the orders of magnitude of its terms in the same sequence as they are in this equation:

$$\omega^2 d, \qquad \omega_p^2 d, \qquad \frac{kT_e}{m}\left|\frac{\nabla n_e}{n_e}\right|, \qquad \omega\nu_m d \tag{3.40}$$

Here the plasma frequency ω_p corresponds to the ion density in the sheath. If the sheath is 'collisionless' for the electrons, $\nu_m \ll \omega$; and besides $\omega^2 \ll \omega_p^2$, the inertial (the first) and the collisional (the fourth) terms are small, so the electron density distribution is defined by the other terms in (3.39). The distribution of $n_e(x)$ is quasi-stationary and of the Boltzmann type corresponding to the instantaneous distribution of the potential $\varphi(x)$

$$n_e(x, t) = n \exp\left(\frac{e\varphi(x,t)}{kT_e}\right) \qquad E = -\nabla\varphi \qquad (3.41)$$

where φ is counted off from the plasma. In case $\nu_m \gg \omega$, the inertial term is small, compared to the collisional term, and we obtain a diffusion-drift equation for an electron flux

$$n_e v = -n_e \mu_e E - D_e \nabla n_e \qquad \mu_e = \frac{e}{m\nu_m} \qquad D_e = \frac{kT_e}{m\nu_m} \qquad (3.42)$$

where μ_e and D_e are the mobility and the diffusion coefficient. This electron flux is quasi-stationary, too. In the collisional case the relationship between ω and ω_p is not as significant as in the collisionless case (which we discussed in Chapter 2 with no allowance for diffusion, whose contribution is smaller than that of drift).

Let us now turn to a discussion of ions. Their oscillations in a sheath are insignificant, and ions are affected by a constant average field if their amplitude A_+ is much smaller than the characteristic sheath thickness d. To evaluate A_+, one can make use of the same formula (1.2) but substitute the electron parameters m and ν_m by the ion parameters M and ν_{mi}. Since the variable component of the sheath field has the same order of magnitude as the whole field ($E_a \sim E_s$), we obtain the following criterion

$$\frac{A_+}{d} \approx \frac{\omega_{pi}^2}{\omega(\omega^2 + \nu_{mi}^2)^{1/2}} \ll 1 \qquad \omega_{pi}^2 = \frac{4\pi e^2 n}{M} \qquad (3.43)$$

where ω_{pi} is the plasma ion frequency. At low pressure, when $\nu_{mi}^2 \ll \omega^2$, inequality (3.43) takes the form $\omega^2 \gg \omega_{pi}^2$. Moreover, if $\nu_m^2 \ll \omega^2$ too, the three conditions formulated in the title of this section hold over the frequency range

$$\omega_{pi}^2 \ll \omega^2 \ll \omega_p^2 \qquad (3.44)$$

This is what frequently happens in RF low-pressure discharge practice, so most theoretical publications on these discharges recognize the validity of (3.44).

When inequality (3.43) holds, the ionic current to the electrode is much lower than the displacement current, and the discharge current from the plasma is closed on the electrodes mainly by the displacement currents. Indeed,

$$\frac{j_i}{j_{dis}} \approx \frac{4\pi e n v_i d}{V_s} \approx \frac{v_i}{\omega d} \approx \frac{v_i}{v_e} \qquad (3.45)$$

where v_i is the ion velocity at the electrode and $v_e \approx \omega d$ is the average electron velocity in the plasma, i.e., their drift velocity if $\nu_m^2 \gg \omega^2$. At very low pressures, $p < 10^{-3}$ Torr; when the sheaths are collisionless, $v_i \approx (eV_s/M)^{1/2}$. It follows from (3.45) and (3.43) that $j_i/j_{dis} \approx (A_+/d)^{1/2} \ll 1$. At $p > 5 \times 10^{-2}$ Torr, the ions move from the plasma to the electrodes with collisions, their path length being $l \approx$ const. In this case $v_i \approx (eEl/M)^{1/2}$, and the current ratio in (3.45) is equal to $(\omega_{pi}/\omega)(l/d)^{1/2}$. Besides, the relationship between the current ratio in (3.45) and the length ratio in (3.43) depends on the ratio between ω and ν_{mi}. At moderate pressures with $\nu_{mi}^2 \gg \omega^2$, the ratio of (3.43) coincides exactly with that of (2.27), namely $j_i/j_{dis} \approx A_+/d$. We have made use of the expression for the 'mobility' $\mu_+ = e/M\nu_{mi}$. When $l =$ const, $\nu_{mi} \approx v_i/l \sim \sqrt{E}$ and $\mu_+ \sim 1/\sqrt{E}$. The quantities ν_{mi} and μ_+ should be understood here as corresponding to the characteristic sheath field. The ion oscillations and ionic current are significant $(A_+ \sim d$ and $j_i \sim j_{dis})$ only at rather high pressures used in RF capacitive discharge practice; for instance, $p \gtrsim$ 50–100 Torr at $f =$ 13.56 MHz in nitrogen and air.

3.7 The floating potential

It has been shown experimentally that with decreasing RF voltage amplitude, the constant plasma potential tends to a certain constant positive value (Figure 3.3) rather than to zero, as the simple formulas (3.6)–(3.8) imply. In other words, in a weak RF field the electrodes possess a certain negative potential relative to the plasma. A similar fact is well known from the probe practice. A zero-current probe or an insulated body, introduced in a plasma, acquire a negative, so-called floating potential V_f relative to the plasma. The physical reason for this behavior is that at zero current the electron and ion fluxes, induced by their thermal motion in the plasma, are identical on the body surface. Since the electrons have much greater thermal velocities than the ions, their flux from the plasma is considerably larger; hence, the body becomes negatively charged up to the potential V_f. In a steady state, only electrons with energies larger than $e|V_f|$ can reach the body—the others being retarded on their way and returned to the plasma. For this reason, there is a lack of electrons in the vicinity of a probe or of an insulated body, producing a positively charged ionic layer, in which the potential φ varies from zero to V_f (if φ is counted from the plasma). At low pressures, the density of electron current from the plasma onto the plane body surface is described by the Langmuir formula

$$|j_e| = \frac{en\bar{v}_e}{4}\exp\left(\frac{eV_f}{kT_e}\right) \qquad \bar{v}_e = \left(\frac{8kT_e}{\pi m}\right)^{1/2} \qquad (3.46)$$

which expresses Boltzmann's law.

When finding ionic current density, one should keep in mind that the temperature of the gas and, hence, of the plasma ions is expressed as $T \ll T_e$. The electrical neutrality of the plasma far from the body is violated where the electrons experience an essential retardation by the repulsive field, i.e., where the negative potential relative to the plasma is $|\varphi_B| \sim kT_e/e$ (Figure 3.16). It is this place

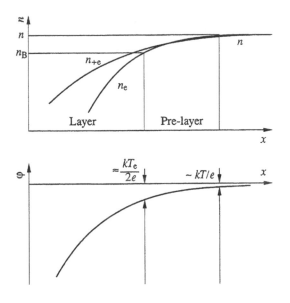

FIGURE 3.16
Schematic density distributions of ions, electrons and potential near a negative probe.
One can see a layer, a pre-layer, and an unperturbed plasma region.

that indicates the edge of the space charge sheath with the density of electrons
$n_B = n \exp\left(-e|\varphi_B|/kT_e\right)$. Outside this boundary, low-energy ions become 'adjusted' to the more energetic electrons so that the electrical neutrality would be preserved, so at the sheath boundary the ion density is also equal to n_B. But in the region between unperturbed plasma and this boundary, which is called a presheath and has $kT \lesssim e|\varphi| \lesssim kT_e$, the ions experience a strong field effect because $e|\varphi_B| \sim kT_e \gg kT$. They thus acquire an energy $e|\varphi_B|$ and enter the sheath at a velocity $(2e|\varphi_B|/M)^{1/2}$. A more rigorous examination [3.30] has yielded the numerical coefficient in φ_B: $|\varphi_B| \approx kT_e/2e$, so the ions enter the sheath at the Bohm velocity $u_B = (kT_e/M)^{1/2}$, while their density, as the plasma density at the sheath boundary, is $n_B \approx n/\sqrt{\bar e}$ ($\bar e$ is the number e). The ion current from the plasma into the sheath and from the sheath onto the body is $j_i \approx e n u_B/\sqrt{\bar e}$. By equating j_i to j_e from (3.46), we can find the well-known expression for the floating potential of a probe

$$V_f = -\frac{kT_e}{2e} \ln\left(\frac{M}{\gamma m}\right) \qquad \gamma \approx \frac{2\pi}{\bar e} = 2.3 \qquad (3.47)$$

Considering that the ions entering the sheath from the pre-sheath possess a certain velocity distribution, one can find $\gamma = 2\pi/C^2$, where $C = 1.247$ [3.31].

Let us analyze the situation when, in addition to a constant component, the potential difference between the body and the plasma has an RF component with no direct current in the body. These conditions usually exist on the electrodes or dielectric coatings in symmetric RF discharges or in asymmetric capacitively coupled discharges. Consider the grounded electrode, the plasma potential being $V_p(t) = \overline{V}_p + \Delta V_p \sin \omega t$. Assume the conditions of (3.44) to hold and the density of electron current to the electrode over a cycle to be described by (3.46) with the instantaneous electrode potential relative to the plasma $-V_p(t)$. The ionic current density is practically constant and coincides with that for the probe (see above). We can now write the condition for the absence of cycle-average current to the electrode

$$\int_0^{2\pi} \frac{e n u_B}{\sqrt{e}} \mathrm{d}(\omega t) = \frac{e n \overline{v}_e}{4} \exp\left(-\frac{e\overline{V}_p}{kT_e}\right) \int_0^{2\pi} \exp\left(-\frac{e\Delta V_p}{kT_e}\sin\omega t\right) \mathrm{d}(\omega t) \tag{3.48}$$

The integral from 0 to 2π in the right-hand part of this equation is twice as large as the integral from 0 to π, the latter being $\pi I_0(x)$, where I_0 is a modified Bessel function of the first kind (the function J_0 is of a purely imaginary argument) and $x = e\Delta V_p/kT_e$. The constant plasma potential relative to the electrode, according to (3.48), is

$$\overline{V}_p = |V_f| + \frac{kT_e}{e}\ln I_0(x) \qquad x = \frac{e\Delta V_p}{kT_e} \tag{3.49}$$

This kind of formula was first derived by Garscadden and Emeleus [3.32], who examined the effect of periodic potential fluctuations on the current-voltage characteristic of the Langmuir probe. Later, similar arguments were applied to the RF discharge [3.33–3.35, 3.5].

In the limit of a very low RF voltage ($x \ll 1$), when $I_0 \approx I_0(0) = 1$, the potential just coincides with the floating potential of the probe. Within this limit, measurements in argon yielded $\overline{V}_p \approx |V_f| = 16.4$ V [3.5]. The electron temperature calculated with (3.47) was found to be $T_e = 3.16$ eV. In the limit of high RF voltages ($x \gg 1$), $I_0 \approx e^x/\sqrt{2\pi x}$, and from (3.49) we have

$$\overline{V}_p = \Delta V_p + |V_f| - \frac{kT_e}{2e}\ln\left(\frac{2\pi e\Delta V_p}{kT_e}\right) \tag{3.50}$$

$$\overline{V}_p = \Delta V_p + \frac{kT_e}{2e}\ln\left(\frac{M}{\gamma m}\frac{kT_e}{2\pi e\Delta V_p}\right) \tag{3.51}$$

These equations give a correction to the formula $\overline{V}_p = \Delta V_p$ in the electrical engineering model (Section 3.1.2, Figure 3.2). The constant plasma potential relative to the grounded electrode (in the symmetric discharge relative to both electrodes) is larger than the variable potential amplitude by the value of the floating potential. The latter is slightly smaller than the probe floating potential because of the effect of alternating voltage on the electron flux to the electrode. Equation (3.50) is consistent with measurements [3.5].

Equations (3.49)–(3.51) can be naturally extended to the case of the loaded electrode. By virtue of the derivation procedure of these equations, we will mean by V_p the potential difference between the plasma and the electrode. With the designations of Section 3.1, it can be expressed as $V_p - V$, where $V = V_{dc} + V_a \sin \omega t$ is the loaded electrode potential and V_p is the plasma potential, both relative to the earth. In equations (3.49)–(3.51), $\overline{V}_p$ should be replaced by $\overline{V}_p - V_{dc}$, and ΔV_p by the amplitude modulus of the variable component $|\Delta V_p - V_a| = V_a - \Delta V_p$ [though $I_0(x)$ is an even function, the argument x should be assumed to be positive in transition from (3.49) to (3.50)]. In addition to equation (3.50), we obtain for high RF voltages

$$\overline{V}_p - \overline{V}_{dc} = V_a - \Delta V_p + |V_f| - \frac{kT_e}{2e} \ln \frac{2\pi(V_a - \Delta V_p)}{kT_e} \qquad (3.52)$$

or, after substituting here ΔV_p, we get with (3.50)

$$\overline{V}_p = \frac{V_a + V_{dc}}{2} + |V_f| - \frac{kT_e}{4e} \left(\ln \frac{2\pi e \Delta V_p}{kT_e} + \ln \frac{2\pi(V_a - \Delta V_p)}{kT_e} \right) \qquad (3.53)$$

This relation specifies the 'electrical engineering' equation (3.5). In calculations, one can substitute ΔV_p under the sign of logarithms in (3.50), (3.52) and (3.53) by the measurable quantity $\overline{V}_p$ equivalent to it in the 'electrical engineering' approximation. An equation of the type (3.52) was also employed in [3.5] for experimental data analysis.

3.8 The α-sheath

There is a large number of publications available on the theory of electrode sheaths in RF discharges. The active interest in this subject has been due to its importance for applications: a detailed knowledge of the sheath parameters and behavior is necessary for the control over the technological processes and characteristics of RF reactors producing materials for microelectronics.

It seems that the idea of a key role of sheaths in determining the RF discharge impedance was first advanced in the work [3.36], which considered sheaths as constant capacitors with an *a priori* given thickness. The author of [3.29] was the first to regard a sheath as having a variable thickness but on the assumption of $n_+(x) = $ const. He pointed out the basic reasons for the appearance of a constant positive plasma potential relative to the electrodes, and he discussed its relation to the electrode bombardment by ions accelerated by this constant voltage. The model of a sheath with $n_+(x) = $ const has been used by many workers since that time, including the authors of this book. This model permits a simple and clear interpretation of many effects observed in RF discharges. Unfortunately, soon after the publication of [3.29], these results were entirely forgotten or remained unnoticed for a long time. During the 1960s–1970s, attempts were made to simulate some processes significant for technological applications, for example, the distribution of ions bombarding the electrodes [3.37, 3.38] and the effect of self-bias in asymmetric discharges [3.2, 3.1]. As in [3.36], sheaths were considered by these workers as structureless constant capacitances.

Modern sheath theories are aimed at getting an insight into the sheath nature, at an understanding of its structure and evolution over a cycle. Most of the theoretical research has been concerned with conditions, commonly encountered in practice, when the inequalities (3.44) hold, thus setting limitations on the field frequency range used and, partly, on the plasma and current densities. In this 'high-frequency' range, usually corresponding to $f \gtrsim 10$ MHz, ions do not respond to the alternating field, and their behavior is affected only by the average dc field. The electron gas periodically covers and exposes the ions. The conduction current at the electrode is very weak compared to the displacement current. Two situations are possible here, depending on the gas pressure: (i) at very low pressures ($p \lesssim 3 \times 10^{-3}$ Torr for $f = 13.56$ MHz) the sheaths are collisionless, while (ii) at higher pressures of the low-pressure range, $p \sim 10^{-2}$–1 Torr (and more) collisions of charged particles with neutrals are essential. These situations are of academic as well as of practical interest. They have been examined in much detail, and the amount of work done on the 'high-frequency' mode is fairly large [3.14, 3.15, 3.34, 3.45–3.47, 3.31, 3.23, 3.25, 3.50, 3.20, 3.21, 3.51–3.53]. It is this mode that we will discuss below.

The 'low-frequency' mode, for which $\omega^2 \ll \omega_{\mathrm{pi}}^2$ and the frequency $f \lesssim 100$ kHz, has been studied in [3.39–3.44]. At these frequencies the sheaths evolve in a quasi-stationary way. At each moment of time, they are similar to a sheath at a plane surface with a constant potential equal to the instantaneous potential. For this reason, in the publications cited above, an RF discharge sheath was assumed to be a plane probe sheath, to which a constant negative potential was applied. The well-elaborated probe theory can be extended to this case. Theoretically, the most difficult case to analyze is that of intermediate frequencies when $\omega \sim \omega_{\mathrm{pi}}$. The Monte-Carlo method was employed in [3.55] to simulate ion kinetics in collisional and collisionless sheaths. Some variants of the calculation just correspond to such an intermediate case, the ion distribution function being sensitive to the phase of ion entry into the sheath. Ion kinetics was also considered in [3.56], using Boltzmann's equation for ions (for the ion distribution function, see Section 3.9).

3.8.1 Analytical theory of an ionization-free sheath

The primary task of sheath theory is to find the ion density and average field (potential) distributions, though the temporal characteristics of oscillations of the plasma boundary, field, etc. are also very important. The sheath problem can be solved analytically for low-pressure discharges in case of sufficiently reasonable assumptions. The formulation of such a problem, its solution and result can

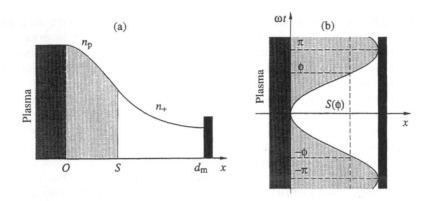

FIGURE 3.17
Schematic ion distribution and plasma boundary motion at the right-hand electrode sheath:
unperturbed plasma (shaded area); sheath occupied by the electron gas (dotted area).

reveal the basic features of the phenomenon and be of use for evaluations. We
will principally follow two publications [3.23, 3.50], though many details of the
analytical solution are also contained in [3.46, 3.51, 3.15, 3.58, 3.57, 3.22, 3.21].

Consider a plane sheath of the right-hand electrode with the axis directed from
the plasma towards the electrode (Figure 3.17). This is convenient because the
field E and the ionic current are then positive. We would like to stress that in our
understanding, a 'sheath' is not only an area devoid of electrons at a given moment
but the whole region around the electrode, in which the electron density varies
over a cycle and there is a time average space charge. Outside the sheath, the
space charge is negligible and is associated only with the plasma density gradient
rather than with electron oscillations. Following this definition, we will place
the reference point $x = 0$ at the boundary between the 'always electroneutral'
plasma and the sheath. The plasma density at this point will be denoted by n_p, as
earlier. At $x > 0$, the ion density $n_+(x)$, equal to n_p at $x = 0$, decreases towards
the electrode, which is located at $x = d_m$, where d_m is the sheath thickness (the
maximum thickness of the space charge region). The reason for decreasing $n_+(x)$
in the sheath is the regular ion flux from the plasma to the electrode.

Suppose inequalities (3.44) hold. This means that the distribution $n_+(x)$ has
a steady-state character, and the ions are affected only by the time-average field
$\bar{E}(x) \equiv \langle E(x,t) \rangle$. The electrons and the plasma front, on the contrary, move
following the field oscillations. Assume the oscillating plasma boundary, or the
plasma front, to be sharp, and denote its coordinate as s [$s = d_m - d(t)$, where
$d(t)$ is a variable space charge region thickness]. The plasma behind the front is
taken to be electrically neutral, i.e., $n_e = n_+(x)$ at $x < s$ and $n_e = 0$ at $x > s$.
This assumption is valid if the Debye radius λ_D is small compared with s, since
it is along s that the plasma density varies appreciably. The relative deviation

from the electrical neutrality is $|n_+ - n_e|/n_+ \sim (\lambda_D/s)^2$, and the electron density decreases from the value $n_+(s)$ right behind the plasma front to zero before the front occurs at a distance of several λ_D. The magnitude $\lambda_D = (kT_e/4\pi e^2 n_p)^{1/2}$ corresponding to the electron temperature and plasma density at the sheath edge is taken as a scale of λ_D. Since the scale of s is the quantity d_m, the validity criterion for the above assumptions is the inequality $\lambda_D \ll d_m$.

The cycle-average field in the sheath $\bar{E}(x)$ satisfies the electrostatic equation

$$\frac{d\bar{E}}{dx} = 4\pi e(n_+ - \bar{n}_e) \tag{3.54}$$

where $\bar{n}_e \equiv \langle n_e(x,t) \rangle$ is the cycle-average density of electrons at the point x. The actual electron density at this point coincides with $n_+(x)$ when $s > x$ and the plasma front is closer to the electrode, while $n_e(x,t) = 0$ when $s < x$. This means that the average value of $\bar{n}_e(x)$ is equal to the product of $n_+(x)$ and the part of the cycle during which the plasma is absent at the point x. The time will be counted from the moment of maximum plasma front distance from the electrode: $t = 0$ at $s = 0$. One can see from Figure 3.17(b) that the plasma is absent at the given point x for $2\phi/2\pi$ of the cycle, where $\phi(x) = \omega t(x)$ is the phase corresponding to the moment of the plasma front crossing the point x. Thus,

$$\bar{n}_e(x) = \left(1 - \frac{2\phi}{2\pi}\right) n_+(x) = \left(1 - \frac{\phi}{\pi}\right) n_+(x) \tag{3.55}$$

and the average field is related to the function $\phi(x)$ by the equation

$$\frac{d\bar{E}}{dx} = 4en_+(x)\phi(x) \tag{3.56}$$

Let us consider the discharge current j to be harmonic. According to the initial time moment chosen, $j = -j_a \sin \omega t$. The conduction current is negligible as compared to the displacement current. Within the assumptions made, the charge conservation law is expressed as

$$-en_+(s)\frac{ds}{dt} = j = -j_a \sin \omega t \tag{3.57}$$

an expression equivalent to the equality of the displacement current in the sheath to the discharge current. Keeping in mind that the function $\phi(x)/\omega$ is a result of inversion of the function $s(t)$ at $s = x$, rewrite (3.57) in the form

$$\frac{d\phi}{dx} = \frac{e\omega n_+(x)}{j_a \sin \phi} \tag{3.58}$$

Dividing (3.56) by (3.58) and integrating the resultant equation

$$\frac{d\bar{E}}{d\phi} = \frac{4j_a}{\omega}\phi \sin \phi$$

with the initial condition $\bar{E} = 0$ at $x = 0$ or $\phi = 0$ (Figure 3.17), we obtain the algebraic expression

$$\bar{E} = \frac{4j_a}{\omega}(\sin\phi - \phi\cos\phi) \tag{3.59}$$

which relates the average field at the point x to the phase of the plasma front interception of this point. This general relation, whose validity is independent of pressure, underlies most analytical and semi-analytical sheath theories. (This relation appears to have been derived independently in the references [3.57, 3.58, 3.23, 3.22] given here in chronological sequence.)

The next step limits the applicability of the theory under consideration to cases of fairly low pressures, when there is no ionization in the sheath (the path length of the electrons with respect to ionization $\bar{v}/\nu_i \gg d_m$). In this case the ion flux entering the sheath from the plasma remains constant throughout the sheath

$$j_i = en_+(x)u_i(x) = \text{const} \tag{3.60}$$

where u_i is the ion velocity at the point x. The assumption (3.60) allows us to replace the function $n_+(x)$ in (3.58) by the velocity u_i, which is, in this way or another, defined by the field

$$\frac{d\phi}{dx} = \frac{j_i}{j_a}\frac{\omega}{u_i\sin\phi} \tag{3.61}$$

The ionic current j_i in this equation represents a parameter external with respect to the sheath. This current enters the sheath from the plasma of the discharge positive column and can, at will, be expressed in terms of the plasma density near the sheath edge and of the incoming ion velocity, taking the latter to be, say, the Bohm velocity $u_B = (kT_e/M)^{1/2}$, as was done in [3.23]. Then $j_i = en_p u_B$, if the sheath is assumed to have a common boundary with the pre-sheath (Section 3.7) rather than with the 'unperturbed' plasma.

The local ion velocity $u_i(x)$ in (3.61) and its relation to the field depend on the way the ions travel through the sheath. There may be three versions of this relation. In a collisionless sheath the ion velocity is defined by the potential difference between the initial and the final points of the ion path. If we start counting the cycle-average potential $\bar{V}(x)$ from the sheath-plasma boundary $\bar{V}(0) = 0$ with $u_i(0) = u_B$, then

$$u_i(x) = \left(u_B^2 - \frac{2e\bar{V}(x)}{M}\right)^{1/2} \qquad \bar{V} < 0 \tag{3.62}$$

$$\bar{E} = -\frac{d\bar{V}}{dx} \tag{3.63}$$

Equations (3.61)–(3.63) and (3.59) form a closed set of equations for this variant. If, however, the ions perform collisions and their mean free path l is constant, which is characteristic of the drift in the charge-exchange regime, then

$$u_i = \left(\frac{2el\bar{E}}{\pi M}\right)^{1/2} \tag{3.64}$$

[the factor $(2/\pi)^{1/2}$ results from averaging with allowance for the probability of the paths being variable around this mean value]. If the collision frequency is constant, i.e., the ions drift at a constant mobility μ_+, then $u_i = \mu_+\bar{E}$. The collisionless variant was analyzed in [3.15, 3.23, 3.21] and the variant of a constant path length in [3.46, 3.15, 3.50]. The variant with a constant ion mobility was examined in [3.57, 3.58] for the case of non-sustained moderate-pressure RF discharge with an external ionization source. Here we will discuss in more detail the first two variants for the case of low-pressure α-discharges.

We would like to stress that the assumption (3.60) of an ionization-free sheath is entirely unacceptable for moderate-pressure α-discharges. For instance, Figure 2.1 in Section 2.1.2 illustrates the calculations for an α-discharge mode, in which the ion flux grows within a sheath by nearly an order of magnitude (about 7-fold). But the cycle-averaging methods and equation (3.59) are applicable to this case, too (Section 2.1.4).

3.8.2 Analytical solution for a collisionless sheath

The complete analytical solution to the set of equations (3.61)–(3.63) was given in [3.23]. Analytical calculations were also made of the high frequency components of the field and potential and of the effective sheath capacitance. The effective resistance is due to the stochastic electron heating considered in the same work. The problem involves a dimensionless similarity factor, which can be used to represent the distribution of all sheath characteristics in the universal dimensionless form. This is the combination

$$H = \frac{4j_a^2}{kT_e\omega^2 n_p} = \frac{1}{\pi}\frac{a^2}{\lambda_D^2} \qquad a = \frac{j_a}{e\omega n_p} \tag{3.65}$$

In its physical sense, the length scale a is the amplitude of free electron oscillations in a collisionless plasma, through which a current of amplitude j_a and frequency $\omega \ll \omega_p$ flows. In such conditions, it is the polarization current (Sections 1.2.1, 1.3.2), and $j_a = en_p u_a = en_p\omega a$, where the amplitudes of the electron velocity u_a and displacement a are expressed by equations (1.3) through the RF field amplitude.

Figure 3.18 presents the plasma front trajectory specifying the schematic plot in Figure 3.17 and the ion density and constant potential distributions calculated in [3.23]. In fact, the theory holds only at high H. In this limit, the sheath thickness and the constant voltage across the sheath $\bar{V}_1 = -\bar{V}(d_m)$ are

$$d_m = \frac{5\pi}{12}Ha = \frac{5\pi^{3/2}}{12}H^{3/2}\lambda_D \qquad \bar{V}_1 = \frac{9\pi^2}{32}H^2\frac{kT_e}{e} \tag{3.66}$$

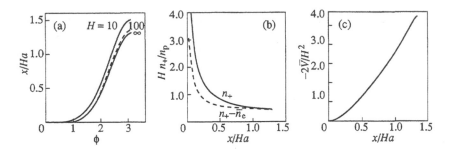

FIGURE 3.18
Results of an analytical solution [3.23]: (a) the plasma boundary trajectory, (b) the
distributions of the ion density n_+ and 'space charge' ($n_+ - \bar{n}_e$), and (c) the distribution
of constant potential in the sheath, all in dimensionless variables in accordance with the
similarity laws. Plots (b) and (c) correspond to the limit $H \to \infty$. (©1988 IEEE)

Only at $H \gg 1$ we have $d_m \gg \lambda_D$, which is one of the basic assumptions of
the theory. For this reason, the distribution $n_+(x)$ and $\bar{V}(x)$ have been plotted in
Figure 3.18 just for the case $H \gg 1$. For the majority of experimental conditions,
this inequality can be considered to be valid: $\bar{V}_1 \sim 100$ V while $T_e \approx 3$ eV.

However, provided that $e\bar{V}_1 \gg kT_e$, the electron temperature has, by itself, a
very little effect. The key parameter for a low-pressure discharge sheath is the ion
flux from the plasma.[4] Indeed, in the limit $e\bar{V}_1/kT_e \gg 1$, in most of the sheath
$u_i \gg u_i(0) = u_B$ and $n_+ \ll n_+(0) = n_p$, so that it is natural to turn to the limit
$T_e \to 0$ and $n_p \to \infty$, retaining the product $j_i = en_p u_B \sim n_p \sqrt{T_e}$. As a result,
we obtain from (3.66)

$$d_m = \frac{5\pi}{3} \frac{e}{M} \left(\frac{j_a}{j_i}\right)^2 \frac{j_a}{\omega^3} \qquad \bar{V}_1 = \frac{9\pi^2}{2} \frac{e}{M} \left(\frac{j_a}{j_i}\right)^2 \frac{j_a^2}{\omega^4} \qquad (3.67)$$

We have intentionally grouped the factors in such a way as to demonstrate the
principal relationships. The matter is that the j_a/j_i ratio varies but slightly with
current j_a. Both quantities are proportional to the same plasma density in the
gap center (Section 3.2.1), but E_a and T_e in the plasma at the given pressure do
not vary greatly, so the sheath thickness is proportional to the discharge current
density while the sheath voltage is proportional to its square value. The thickness
and the voltage decrease rapidly with increasing frequency.

The sheath capacitance and thickness were determined in [3.59] by measuring
the capacitive resistance of a symmetric discharge. Measurement was also made
of the RF voltage amplitude, to which the constant plasma potential $\bar{V}_1$ is close at
large amplitudes [see (3.51)]. Equation (3.67) yields the theoretical dependence

---

[4] As was explained in Section 3.2, a low-pressure discharge differs in this respect from a moderate-
pressure discharge, where the ion flux does not affect the sheath so much.

of d_m on $\overline{V}_1$

$$d_m = \frac{5\sqrt{2}}{9\omega} \left(\frac{e}{M}\right)^{1/2} \frac{j_a}{j_i} \overline{V}_1^{1/2} \tag{3.68}$$

Keeping in mind the above stated approximate proportionality of j_a and j_i at current (and voltage) variation, one can conclude that $d_m \sim \overline{V}_1^{1/2}$. It seems more reliable to compare with experiment the limiting dependence (3.68) rather than the relation $(d_m/\lambda_D) \sim (e\overline{V}_1/kT_e)^{3/4}$, which follows from (3.66). In the latter case one should allow for the variation in T_e and λ_D—i.e., n_p at varying voltage—but this is very hard to do reliably. A comparison of experiment and theory will be made in Section 3.8.4, where we will discuss the work [3.52].

With (3.67) the discharge current can be represented in a form characteristic of the displacement current in the sheath

$$j_a = \frac{40}{27} \frac{\omega \overline{V}_1}{4\pi d_m}$$

The sheath model with $n_+(x) = \text{const} = n_p$ yields a value only $9/5 = 1.8$ times larger

$$d = a(1 + \sin \omega t) \qquad \overline{V}_1 = 2\pi e n_p d^2 = 3\pi e n_p a^2$$

$$d_m = 2a \qquad j_a = e n_p \omega a = \frac{8}{3} \frac{\omega \overline{V}_1}{4\pi d_m}$$

The ionic current density j_i is related to $\overline{V}_1$ and d_m by the Child-Langmuir law $j_i \sim \overline{V}_1^{3/2} d_m^{-2}$ but with the proportionality factor $50/27 = 2$ times larger, which was interpreted in [3.23] as being due to the reduced space charge in the sheath resulting from a non-zero average electron density. The ion density near the electrode, being the true density scale of the sheath ions, is

$$n_1 = n_+(d_m) = \frac{1}{3\pi} \frac{\omega^2 M}{e^2} \left(\frac{j_i}{j_a}\right)^2 \tag{3.69}$$

It is practically independent of the plasma density in the gap center or at the sheath edge, the latter being 'infinite' in this approximation, i.e., actually $n_1 \ll n_p$ (see Section 3.2.1).

To illustrate the theory, we will borrow a numerical example from [3.23], where the parameters used were $\overline{V}_1 = 200$ V, $T_e = 3$ eV, $f = 13.56$ MHz, and $j_i = 0.5$ mA cm^{-2}; the molecular weight of ions was 69, corresponding to CF$_3^+$; the obtained values of u_B and n_p were 2.04×10^5 cm s^{-1} and 1.53×10^{10} cm^{-3}, respectively; $j_a = 8.52$ mA cm^{-2}, the similarity factor $H = 4.9$ being large compared to unity; $\lambda_D = 1.04 \times 10^{-2}$ cm, $a = 4.08 \times 10^{-2}$ cm, and the sheath thickness $d_m = 0.262$ cm. The ionic current releases on the electrode the power $j_i \overline{V}_1 = 0.1$ W cm^{-2}; the power due to the stochastic electron heating (Section 1.7) is smaller, 2.83×10^{-3} W cm^{-2}. The sheath capacitance with respect to the fundamental voltage harmonic $C = 0.414$ pF cm^{-2}. The second harmonic makes up about 12% of the fundamental one, and the third one about 4%. Maximum voltage across the sheath (at $\phi = 0$) is about twice the average $\overline{V}_1$. In the work [3.21], a similar theory for the case of harmonic current in the sheath was extended to a symmetric discharge with two sheaths and harmonic voltage across the electrodes. The calculated anharmonic current is in reasonable agreement with the results of numerical simulation [3.20] for a single sheath, which we will now discuss.

3.8.3 Boltzmann distribution $n_e(x, t)$ for a collisionless sheath

The analytical theory for a collisionless sheath is based not only on the concept of harmonic current but also on the assumption of absolute electrical neutrality of the plasma and its sharp boundary. This assumption corresponds to the limit $T_e \to 0$, or $\lambda_D \to 0$. For this reason, the limiting equations (3.67)–(3.69), probably, should not be refined with equations (3.66) and (3.65) involving T_e and λ_D. With non-zero T_e in the limiting condition at $x = 0$ with the given $u_B \neq 0$ and $n_p \neq 0$, one should simultaneously take into account the corrections for non-absolute plasma neutrality and for its diffuse boundary, which are also characterized by the λ_D/d_m ratio. This ambiguity is not inherent to the formulation of a self-consistent problem of collisionless sheaths at the electrodes, to which harmonic voltage is applied. (For the rigorous formulation, however, one will pay by having to make numerical calculations.)

If the inequalities of (3.44) hold, and the electrons are not subject to collisions, their spatial distribution $n_e(x, t)$ at each moment of time is of the Boltzmann type (Section 3.6). By substituting (3.41) into Poisson's equation for the instantaneous potential φ and using equations (3.60), (3.62) and (3.63), we obtain the equation for a self-consistent field

$$-\frac{\partial^2 \varphi}{\partial x^2} = 4\pi e n_p \left[\left(1 - \frac{2e\bar{\varphi}}{M u_B^2} \right)^{-1/2} - \exp\left(\frac{e\varphi}{kT_e} \right) \right]$$

The left-hand boundary of the integration range may be assumed to be, as before, the sheath-presheath boundary, where $u_i = u_B$ and $n_+ = n_p$. The plasma potential here can be taken to be zero, and the boundary can be shifted to infinity. Then one of the boundary conditions will be $\varphi(-\infty, t) = 0$. The right-hand boundary represents the electrode, to which the coordinate $x = 0$ may be ascribed. Given the harmonic electrode potential with a constant bias $\varphi(0, t) = V(t) = V_{dc} + V_a \sin \omega t$, V_{dc} should be considered to be related to V_a through (3.49), where $\overline{V}_p = -V_{dc}$ and $\Delta V_p = V_a$. The discharge current can be found by calculating the electrode displacement current

$$j(t) = \left(\frac{1}{4\pi} \right) \left(\frac{\partial^2 \varphi}{\partial x \partial t} \right)_{x=0}$$

This problem, but in a more complicated formulation, was solved numerically by the iteration method in [3.20]. The complication was due to the allowance

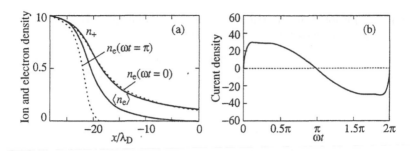

FIGURE 3.19

Numerical simulation of the sheath [3.20]. (a) The density distributions of ions n_+ and electrons n_e at the maximum and minimum distances between the plasma boundary and the right-hand electrode and mean $\langle n_e \rangle$; these densities are related to the plasma density n at infinity. (b) Wave form of dimensionless current density measured in the unit of $\omega k T_e / \lambda_D$ for $V_a = 48.2\, kT_e/e$; λ_D corresponds to T_e and n.

for the velocity distribution of ions on their entering the sheath from the pre-sheath [3.31].[5] Figure 3.19 gives the results of the calculation. Qualitatively, the ion distribution in the sheath $n_+(x)$ looks very similar to the analytical theory distribution (Figure 3.18). The current wave form is quite close to what we get from the limiting formula (3.33) of the simplified theory (see also Figure 3.9). It is just this limiting formula that is compared with the calculations in Figure 3.19. The above problem for a single electrode, zero plasma potential and large voltage $V_a \gg kT/e$, corresponds to an extremely asymmetric discharge, in which the electrode in question is small and loaded while the other is grounded and 'infinitely' large. We also would like to remind the reader that when comparing the anharmonic current in Figure 3.19 with formula (3.33) and Figure 3.9, one should invert the curve in Figure 3.19 because it corresponds to the right-hand electrode, while Figure 3.9 corresponds to the left-hand one.

The results of the numerical calculation are in reasonable agreement with the analytical theory. Really, the value of $e\overline{V}_1/kT_e = 50.9$ from (3.51) corresponds to $eV_a/kT_e = 48.2$ in Figure 3.19. Hence, from the second formula of (3.66) the similarity factor is $H = 4.29$ (we would like to remark that it is close to the numerical example [3.23] given above). In the first formula of (3.66), the voltage corresponds to the maximum sheath thickness $d_m/\lambda_D = 20.6$. In the plot of Figure 3.19 the sheath changes into the pre-sheath asymptotically with the effective maximum thickness $d_m/\lambda_D \approx 23$–25. The diffuse boundary width is $(3$–$4)\lambda_D$.

---

[5]The approximate substitution in [3.31] of the time-average electron density by the exponent of the time-average potential in the Boltzmann equation (3.41) should result in a large quantitative error, although this does not seem to qualitatively distort the results obtained.

3.8.4 Collisional sheath with a constant ion path length: The plasma–sheath boundary condition

A collisional sheath is described by equations (3.61), (3.64) and (3.59), which are also solved analytically [3.50]. The dimensionless similarity factor is expressed as

$$H' = \frac{8lj_a}{\pi \omega k T_e} = \left(\frac{2la}{\pi^2 \lambda_D^2}\right)^{1/2} \qquad a = \frac{j_a}{e\omega n_p} \tag{3.70}$$

Qualitatively, the results do not differ from those for a collisionless sheath, but there is certainly a quantitative difference. The maximum sheath thickness and average voltage fall have a form similar to (3.66), except that for finding the coefficients one has to take two numerical integrals

$$d_m = 1.95\,H'a = \frac{1.95\pi}{\sqrt{2}} H'^2 \left(\frac{a}{l}\right)^{1/2} \lambda_D \qquad \overline{V}_1 = 3.15 \frac{H'}{\pi} \frac{a^2}{\lambda_D^2} \frac{kT_e}{e} \tag{3.71}$$

Like in the previous case, in the limit $e\overline{V}_1 \gg kT_e$, for which the formulas (3.71) hold, it seems expedient to drop the product of two parameters $n_p\sqrt{T_e}$, substituting it by one—the ionic current j_i. The formulas then take a form similar to (3.67) and (3.68)

$$d_m = 2 \cdot 1.95 \left(\frac{2el}{\pi M}\right)^{1/2} \left(\frac{j_a}{j_i}\right) \frac{j_a^{1/2}}{\omega^{3/2}}$$

$$\overline{V}_1 = 8 \cdot 3.15 \left(\frac{2el}{\pi M}\right)^{1/2} \left(\frac{j_a}{j_i}\right) \frac{j_a^{3/2}}{\omega^{5/2}}$$

$$d_m = \frac{1.95}{3.15^{1/3}\omega^{2/3}} \left(\frac{2el}{\pi M}\right)^{5/6} \left(\frac{j_a}{j_i}\right)^{2/3} \overline{V}_1^{1/3} \tag{3.72}$$

The sheath parameters have thus been expressed directly through a quantity (ionic current) that actually defined them, and the dependence of the sheath thickness on the sheath voltage $d_m \sim \overline{V}_1^{1/3}$ is shown here explicitly (if, of course, we take $j_a/j_i \approx$ const, as previously).

According to (3.71) or (3.72), the discharge current can be represented as the electrode displacement current

$$j_a = 1.94 \left(\frac{\omega \overline{V}_1}{4\pi d_m}\right)$$

The ionic current is related to $\overline{V}_1$ and d_m by an equation substituting the Child–Langmuir law

$$j_i = a_1 \left(\frac{2el}{M}\right)^{1/2} \frac{\overline{V}_1^{3/2}}{d_m^{5/2}} \qquad a_1 = \frac{1.95^{5/2}}{4\pi^{1/2}3.15^{2/3}} = 0.133 \tag{3.73}$$

This is the same relation as (3.17) and (3.21), except that the numerical coefficient a_1 has been refined and corresponds to a non-zero average electron density in the sheath. The formulas of (3.24) are also valid.

We will illustrate this situation by a numerical calculation borrowed from [3.50]. The parameters were taken to be $\overline{V}_1 = 400$ V, $p = 4.7 \times 10^{-2}$ Torr, $T_e = 3$ eV, and $f = 13.56$ MHz; the atomic weight of the ions was 40 (argon), $j_i = 0.5$ mA cm^{-2}, $l = 0.07$ cm. The obtained values were $n_p = 1.2 \times 10^{10}$ cm^{-3}, $H' = 4.1$, $j_a = 10.8$ mA cm^{-2}, $a = 6.8 \times 10^{-2}$ cm, $\lambda_D = 1.2 \times 10^{-2}$ cm, and $d_m = 0.54$ cm. The second voltage harmonic is 19% of the fundamental one, the third harmonic is 5.3%, the sheath capacitance relative to the fundamental harmonic $C = 0.25$ pF cm^{-2}, and the ionic current releases on the electrode the power 0.2 W cm^{-2} due to stochastic heating 8.3×10^{-3} W cm^{-2}. It is seen that all the results coincide, in the order of magnitude, with those for a collisionless sheath.

A detailed sheath analysis was made in [3.52]. The ion motion in an average field is described by the hydrodynamic equation [cf. (3.39)]

$$Mn_+u_i\frac{\mathrm{d}u_i}{\mathrm{d}x} = n_+e\bar{E} - kT_i\frac{\mathrm{d}n_+}{\mathrm{d}x} - n_+\frac{\pi}{2}\frac{Mu_i^2}{l} \qquad (3.74)$$

The thermodiffusion term has been dropped, and the ion path length is taken to be constant. As compared with (3.64), equation (3.74) includes the inertial and diffusional terms. Account was taken of the electron current to the electrode in addition to the ionic current. The electron current is determined by the number of electrons leaving the instantaneous plasma boundary and capable to overcome the potential barrier separating this boundary from the electrode, $j_e \sim \exp\left[eV(t)/kT_e\right]$, where $V(t)$ is the instantaneous (negative) electrode potential relative to the plasma. Note that with account of this circumstance, the plasma appears to be separated from the electrode by a distance of about the Debye radius even at the moment of 'contact' with the electrode, never actually touching it. The current in [3.52] was assumed to be harmonic.

Quite a special problem is the condition at the boundary between the sheath and the plasma. In the case of a collisionless sheath, the ions are generally supposed to enter the sheath at a Bohm velocity u_B. This approximation was used to obtain the solution [3.23] we discussed in Section 3.8.2. The same boundary condition was assigned to the collisional case [3.50]. Generally speaking, this contradicts the physics of the process, though in the limit $e\overline{V}_1 \gg kT_e$ the error resulting from this approximation is small; so, from the very beginning, one can put $u_i(0) \to 0$ and ignore the arising singularity in the ion density, $n_+(0) \to \infty$, as we pointed out above. Really, in the transient region between the pre-sheath and the sheath—in which the electrical neutrality is strongly violated—the field is non-zero, since the potential of the order kT_e/e falls along a distance of about the Debye radius. This means that the field here is $E(0) \approx kT_e/e\lambda_D$ [3.60]. But if the ions are involved in numerous collisions and have enough time to acquire a steady velocity corresponding to a drift at a constant path length, their velocity in

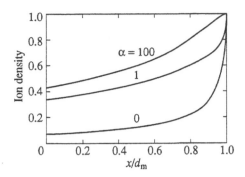

FIGURE 3.20
Distribution of the ion density n_+ in the left-hand sheath related to the density n in the plasma. The parameter $\alpha = \pi\lambda_D/2l$ characterizes the contribution of collisions. The Debye radius λ_D corresponds to T_e and n in the plasma [3.52].

such a field, according to (3.64), $u_i(0) = (kT_e/M)^{1/2}(2l/\pi\lambda_D)^{1/2}$ is $(\pi\lambda_D/2l)^{1/2}$ times lower than the Bohm velocity. It was shown in [3.61] for the general case that the velocity of ions entering the sheath can be described by the expression

$$u_i(0) = \left(\frac{(kT_e + kT_i)/M}{1 + \pi\lambda_D/2l} \right)^{1/2} \tag{3.75}$$

which also takes into account the ion temperature T_i (though $T_i \ll T_e$). In the collisionless case with $l \gg \lambda_D$, the velocity (3.75) practically coincides with the Bohm velocity. When there are many collisions occurring in the Debye sphere and $l \ll \lambda_D$, the result is that the ions enter the sheath at a velocity lower than the Bohm velocity. The plasma density at the boundary for the same ion flux turns out to be larger than the theory has assumed it to be [3.50].

Generally speaking, ions coming into the sheath from the pre-sheath have different velocities. Even in the collisionless case the pre-sheath has a sufficiently large size for a velocity distribution of ions to be established. Such a distribution was found in [3.41] and used in [3.20] for the sheath calculation. It should also be noted in connection with the boundary condition that the classical Bohm criterion and its extensions have been derived for the case of constant potential and current, so that they might seem to be valid only at $\omega \ll \omega_{pi}$. It has been shown in [3.62, 3.63], however, that these results can be extended to RF discharges no matter what is the ratio of ω and ω_{pi}.

Figure 3.20 presents the ion density distributions in the sheath (again for the left-hand electrode) calculated in [3.52]. One can clearly see the way these distributions change from the collisionless to the collisional case, depending on the parameter $\alpha = \pi\lambda_D/2l$, which describes the contribution of collisions (here λ_D corresponds to the plasma density at its boundary with the sheath). Collisions reduce the ion density variation in the sheath because of the smaller ion velocity

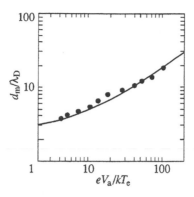

FIGURE 3.21
Calculated dependence of the sheath thickness referred to the Debye radius [3.52] on the RF voltage amplitude. For comparison, experimental data of [3.59].

ratios at the sheath edges. In the collisional case the velocity ratio is determined by the field ratio $\bar{E}(d_m)/\bar{E}(0)$, while in the collisionless case it is determined by the potential ratio $\overline{V}_1/\bar{\varphi}(0) = e\overline{V}_1/kT_e$, which is by a factor of $d_m/\lambda_D \gg 1$ larger than the field ratio (cf. Figure 2.1).

Figure 3.21 gives a calculated plot of the collisionless sheath thickness versus the RF voltage amplitude across the sheath [3.52], the latter being close to constant potential $\overline{V}_1$, as follows from equation (3.51). This calculation was compared with experiment in [3.59]. In the calculation the velocity ratio of ions and electrons at the plasma-sheath boundary was taken to be constant and was actually equal to the ionic-to-discharge current ratio. From equation (3.68), which follows from a simpler theory than that developed in [3.52], we have $d_m \sim \overline{V}_1^{1/2}$. This relationship is consistent with the straight line slope in the logarithmic plot of Figure 3.21, in which the experimental points fit the straight line reasonably well. The deviation of the calculated plot [3.52] from the results of (3.68) is associated with the allowance for details absent from equation (3.68).

3.9 The energy spectrum of ions bombarding the electrode surface

The efficiency of many technologies based on low-pressure RF discharges largely depends on the energy of ions bombarding a target placed on the electrode. The fact of intense bombardment of both electrodes in the RF discharge was established as far back as the 1930s (e.g., see [3.64, 3.65]). An effect similar to cathode sputtering could be attributed only to the action of constant fields directed to the electrode surface and accelerating the ions that escape the plasma. The relationship

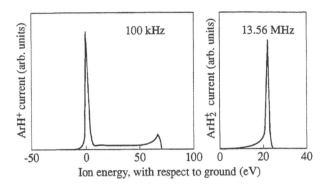

FIGURE 3.22
Typical energy spectra of ions for low and high field frequencies; argon,
$p = 50$ mTorr [3.70].

between the electrode sputtering intensity and the constant plasma potential was
first demonstrated experimentally in [3.29]. Later, the problem of the energy
and energy spectra of ions bombarding the electrode surface in the discharge was
studied by many researchers, both experimentally [3.66, 3.67, 3.1, 3.68–3.74] and
theoretically [3.37, 3.38, 3.40a, 3.55, 3.72, 3.74–3.76].

At sufficiently low energies, when the ions fly through the sheath without
collisions, the energy distribution of ions largely depends on the relationship of
their time of flight through the sheath and the voltage oscillation cycle. If an
ion takes many cycles to go through the sheath, as it happens at relatively high
frequencies $\omega \gg \omega_{pi}$ (Section 3.6), the ions are affected only by the average field,
and most of them gain an energy corresponding to the average sheath voltage, i.e.,
to the plasma constant potential relative to the electrode. At fairly low frequencies
with $\omega \ll \omega_{pi}$, when an ion crosses the sheath over a small cycle fraction, the ions
arriving at the electrode have very different energies. The energy depends on the
external voltage phase in which the ion entered the sheath, because this defines
what potential difference it can cross. The energy spectrum extends as far as the
value corresponding to the maximum sheath voltage, though the number of ions
in the spectrum that have gained maximum energy is not large.

In experiment, one generally registers the spectrum of ions coming from the
plasma to the grounded, usually larger electrode. A small hole is made in the
electrode, through which an ion flux enters a magnetic analyzer and a registration
system, as this was done, for example, in the experiments described in [3.70].
Figure 3.22 illustrates the energy spectra corresponding to extremely low and
extremely high frequencies. Similar spectra were also registered in [3.1, 3.5, 3.69]
and other studies. If the time of flight of ions is comparable with the frequency,
which is the case at $\omega \sim \omega_{pi}$, the energy spectrum has a characteristic saddle
shape that was first registered in [3.66] and later observed and studied in many

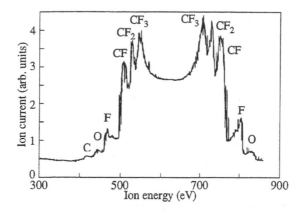

FIGURE 3.23
Energy spectra of ions of different masses for the time of flight through a collisionless sheath comparable with the field oscillation cycle; CF_4, $p = 3$ mTorr [3.74].

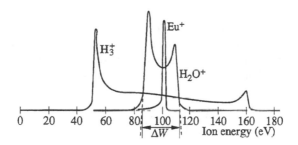

FIGURE 3.24
Illustration of spectral variation with the ion mass at the same frequency $f = 13.56$ MHz (argon, $p = 75$ mTorr). Light ions H_3^+ behave as at low frequency, heavy ions Eu^+ as at high frequency (cf. Figure 3.22). The spectrum of H_2O^+ ions of an average mass has the same saddle shape as in Figure 3.23. Borrowed from [3.1].

experiments [3.67, 3.1, 3.5, 3.73, 3.74, etc.] (Figures 3.23 and 3.24). The reason for this spectral shape is the energy modulation of ions entering the electrode by the variable component of voltage fall across the sheath. This was established theoretically in the framework of simplified analytical theories [3.37, 3.38, 3.74] and by means of sophisticated numerical calculations [3.55, 3.75, 3.40a]. We will explain the origin of this effect in terms of a simple physical model.

Consider a plane collisionless sheath of thickness d constant in time. One of the sheath edges contacts the plasma, the other the electrode. Suppose the plasma possesses the potential $V = \overline{V}_p + V_a \cos \omega t$ relative to the electrode with a small RF voltage modulation, $V_a \ll \overline{V}_p$. Such a situation arises in the large grounded electrode sheath in a strongly asymmetric discharge with the external circuit closed for dc, when the plasma does not come in contact with the electrode during its oscillations. The ratio of the plasma boundary displacement to the maximum sheath thickness, as well as the $V_a/\overline{V}_p$ ratio, are the less, the greater is the difference between the electrode areas (Section 3.5.1). Let us neglect for simplicity the sheath field distortion due to the presence of space charge, the more so as the analytical models ignore the average electron charge and take into account only the ion charge. Then flying through the sheath, ions move in a field spatially uniform but varying in time $E = V/d = (\overline{V}_p/d)(1 + \lambda \cos \omega t)$ ($\lambda \equiv V_a/\overline{V}_p \ll 1$), which imparts energy to them. The initial velocity of the incoming ions can be neglected.

In the zeroth approximation that corresponds to unmodulated voltage, ions move through the sheath at a constant acceleration due to the absence of collisions and arrive at the electrode at velocity u_0, energy W_0 over a period of time τ_0. These quantities are defined as

$$u_0 = \left(\frac{2e\overline{V}_p}{M}\right)^{1/2} \qquad W_0 = e\overline{V}_p \qquad \tau_0 = \frac{2d}{u_0} = d\left(\frac{2M}{e\overline{V}_p}\right)^{1/2}$$

In the next approximation allowing for the voltage modulation, the energy of an ion hitting the electrode at the moment t_1 depends on the moment $t_0 = t_1 - \tau$ at which it entered the sheath, where τ is the time of flight of the ion in a real modulated field. Let us integrate once the equation of motion $M\ddot{x} = eE$ with the initial condition $\dot{x} = 0$ at $t = t_0$. The ion now arrives at the electrode at velocity

$$u = \frac{e\overline{V}_p}{Md}\left(\tau + \frac{\lambda}{\omega}(\sin \omega t_1 - \sin \omega t_0)\right) \tag{3.76}$$

In order to find the time of flight $\tau = t_1 - t_0$, one should integrate once again the equation of motion with the conditions $x = 0$ at $t = t_0$ and $x = d$ at $t = t_1$. However, bearing in mind that $\lambda \ll 1$, we will neglect the difference between τ and τ_0 and substitute $\tau = \tau_0$ into equation (3.76). Up to the value of the second order of smallness in λ, the velocity in equation (3.76) corresponds to the ion energy

$$W = \frac{Mu^2}{2} = e\overline{V}_p\left(1 + \frac{2\lambda}{\omega\tau_0}(\sin \omega t_1 - \sin \omega t_0)\right)$$

$$= e\overline{V}_p\left[1 + \frac{4\lambda}{\omega\tau_0}\cos\left(\omega t_0 + \frac{\omega\tau_0}{2}\right)\sin\frac{\omega\tau_0}{2}\right] \tag{3.77}$$

The cosine value in the last expression varies between -1 and $+1$ with the moment of ion penetration into the sheath. Strictly, the sine value in this expression is quite definite; but if $\omega\tau_0 \gtrsim 1$, the slightest variation in the real time of flight may lead to an appreciable change in the sine. However, the time of flight does depend on the phase, in which the ion penetrates the sheath, i.e., on the value of the accelerating field. One may say for this reason that the whole trigonometric factor in the last expression of (3.77) varies from -1 to 1. Hence the energy range $\Delta W = W_{\text{max}} - W_{\text{min}}$ for the ions being equal to $\Delta W = 8\lambda e\overline{V}_{\text{p}}/\omega\tau_0$.

With account of the ion space charge in a collisionless sheath, which in this model is similar to the Child–Langmuir vacuum diode, the diode field is nonuniform, being distorted by the space charge and $E \sim x^{1/3}$. By integrating the equation of ion motion in the sheath with the respective field $E = (\overline{V}_{\text{p}}/d)(1 + \lambda\cos\omega t)(x/d)^{1/3}$, we obtain a time of flight τ_0 by a factor $3/2$ larger and an energy gap size by a factor of $3/2$ smaller than the values found from the above simple calculation, i.e.,

$$\Delta W = \frac{8\lambda e\overline{V}_{\text{p}}}{\omega\tau_0} = \frac{8}{3}\frac{eV_{\text{a}}}{\omega d}\left(\frac{2e\overline{V}_{\text{p}}}{M}\right)^{1/2} \qquad (3.78)$$

This formula was first derived in [3.37].[6] The energy spectrum of ions is broader for lighter ions. The respective relationship $\Delta W \sim M^{-1/2}$ agrees well with experimental data, including measurements made by the authors of this theory.

Let us now find the energy distribution function of ions bombarding the electrode. Suppose Γdt_0 is the number of ions entering from the plasma per 1 cm² sheath area at the moment t_0 over the time dt_0. Evidently, Γ has a very weak, if any, dependence on time. Assume $\Gamma(t_0) = \text{const}$. The ions arrive at the electrode at the moment t_1 with energies W in the range dW defined by equation (3.77), which can be rewritten as

$$W - e\overline{V}_{\text{p}} = \frac{\Delta W}{2}\cos\left(\omega t_0 + \frac{\omega\tau_0}{2}\right)$$

The number of such ions at the electrode is $F(W)dW = \Gamma dt_0$; hence, the ion energy distribution function is

$$F(W) = \frac{\Gamma}{dW/dt_0} = \frac{2\Gamma}{\omega\Delta W}\left[1 - \left(\frac{W - e\overline{V}_{\text{p}}}{\Delta W/2}\right)^2\right]^{-1/2} \qquad (3.79)$$

This elegant result was also obtained in [3.37]. According to (3.79), the spectral function is symmetric relative to the mean energy $e\overline{V}_{\text{p}}$ and possesses two sharp maxima at $W = e\overline{V}_{\text{p}} \pm \Delta W/2$. Equation (3.79) fits well the experiment (Figure 3.23) and numerical calculations made by the particle-in-cell procedure for the collisionless case [3.75].

The theoretical scheme, whose basic result is equation (3.76) derived in the simplest way, clearly demonstrates the ion behavior in the limiting cases we referred to at the beginning of this section. At high frequencies with $\omega\tau_0 \gg 1$, the energies of all ions, irrespective of the moment of their entering the sheath, are close to the average value $e\overline{V}_{\text{p}}$. The ions respond only to the average field; one

---

[6]Formally, equation (3.78) follows from the equations of [3.37] but with the condition $\omega\tau_0/2 \gg 1$, which allows us to drop one of the two trigonometric terms contained in an intermediate formula— which was not, however, given in [3.37] but yields the final equation of the type (3.77). The complex question of the factor $\sin\omega\tau_0/2$ in the latter equation has been discussed in [3.74].

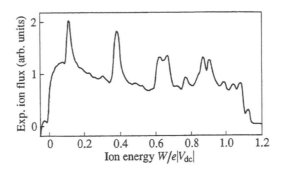

FIGURE 3.25
Ion spectrum. Ion energy is referred to the electrode self-bias of 500 V; oxygen,
$p = 15$ mTorr, $f = 13.56$ MHz [3.72].

can see this from equation (3.77), in which the small correction to the energy is
inversely proportional to $\omega \tau_0$. For low frequencies with $\omega \tau_0 \ll 1$, equation (3.77)
should be refined. In this case the correction to the time of flight of ions $\tau - \tau_0$
has the same order as that to the energy in equation (3.77). If one takes this
circumstance into account when deriving equation (3.77) from (3.76), retaining
the respective term $\tau - \tau_0 \sim \lambda$, the refined equation (3.77) immediately yields
a clear result, too: $W(t_1) \approx eV(t_0)$. The ion energy is determined by the real
potential difference that the ion passes over a given relatively short period of time
τ near the moment t_0—this period being too small for the voltage to change.

When ions collide with neutrals, their energy spectrum becomes more complicated [3.68, 3.72, 3.74]. This situation is illustrated in Figure 3.25. The spectral
pattern presented is characteristic of both atomic and molecular gases. In addition to the voltage modulation across the sheath, as in the collisionless case,
the spectrum is affected by the slow ion production within the sheath due to the
charge exchange of fast ions. A non-self-consistent theoretical model was suggested in [3.72] to obtain an experimentally consistent ion energy spectrum for a
given space charge distribution function, i.e., for a given sheath field distribution.
At present, the Monte-Carlo method [3.55], the particle-in-cell procedure or the
combination of both are being successfully used for calculation of the ion distribution function in collisional sheaths. As was pointed out in [3.77], the charge
exchange effect gives rise to high-energy-neutral species, which significantly contribute to surface modification processes but are difficult to register in laboratory
experiments. Nevertheless, fast neutrals can be examined and evaluated by self-
consistent numerical models, as this was done, for instance, in [3.75].

3.10   RF discharge in electronegative gases

Interest in RF discharges in electronegative gases has largely been due to their applications. Working gases of RF reactors designed for ion bombardment of materials commonly contain electronegative components—halogens and halogen-containing compounds (Cl_2, BCl_3, CF_4), oxygen and others [3.78]. Though these components are often mixed with electropositive ones (He, Ar, N_2, etc.), the RF discharge operating in such mixtures does possess specific features which distinguish it from attachment-free gas discharges (the same is true of dc glow discharges [3.79, 3.19]). There has been much work, both experimental [3.79–3.89] and theoretical [3.84, 3.90–3.105], done on electronegative gases. One of the experiments, for instance [3.83], was aimed at studying discharges in pure BCl_3 and, for comparison, in a mixture of 5%BCl_3 + 95%Ar at a pressure of 0.3 Torr over the frequency range from 50 kHz to 10 MHz. BCl_3 was chosen because, first, the RF discharge in this gas is used for etching Si, Al and other materials and, second, BCl_3 is suitable for field measurements by registering Stark mixed laser fluorescence spectra.

The discharge was excited between flat stainless-steel disks of 7.5 cm in diameter in a gap of 1.6 cm. The gas was slowly pumped through at a rate of 10 cm^3 per minute. Fluorescence was produced by a nitrogen laser, whose radiation excited a tunable dye laser. The radiation frequency of the latter was doubled, and the double-frequency radiation excited fluorescence of BCl_3 molecules present in the discharge at a frequency of 272 nm. The electric field in the sheath and, hence, the sheath thickness variation were measured in short intervals over a cycle. The measurements showed the maximum thickness of the sheath in pure BCl_3 to be 0.2 cm, or 1.5 times less than in the BCl_3 + Ar mixture, 0.29 cm at 50 kHz. At 10 MHz, the maximum thicknesses were 1.5 times those at 50 kHz.

The fact that a sheath in an electronegative gas is thinner than in an electropositive gas was also observed previously [3.79]. The smaller thickness was considered in [3.83] to be due to a smaller Debye radius, which, in the author's opinion, is the measure of the sheath thickness. The Debye radius $\lambda_D \sim (T_e/n)^{1/2}$ in an electronegative gas is smaller because of a higher charge density n owing to the formation of negative ions. Moreover, the ionization potentials of the atoms and molecules of B, Cl, and BCl_3 are lower (8.3, 13, and 11.6 eV, respectively) than that of argon (15.8 eV), which has no low levels. For this reason, the electron temperature in pure BCl_3 is lower than in a mixture with a large amount of argon, which also decreases λ_D.

However, keeping in mind our comments on deducing formulas (3.67), (3.68), and (3.72) in Sections 3.8.2 and 3.8.4, it seems questionable whether the Debye radius, a magnitude primarily related to electron temperature, can be really taken as a measure of sheath thickness. Sheath voltages are fairly high: in experiments [3.83] at 10 MHz, $V_{1\,max} \approx 50$ V; while at 50 kHz, even $V_{1\,max} \approx 400$ V, which is much more than $T_e \approx 1$ eV. Besides, the thermal motion of electrons

must have a smaller effect on a sheath than their regular motion due to the oscillating field. The sheath thickness is determined by the necessity to transport the
ambipolar positive ion flux out of the gap middle to the electrodes and to carry the
discharge displacement current through the sheath. A Debye radius that is small
relative to the sheath thickness determines only the width of the diffuse boundary
between the plasma and the space charge region.

3.10.1   Results of numerical simulation

It is generally difficult to treat experimental data unambiguously, since introducing
new, especially, electronegative components in a discharge gas produces several
simultaneous effects: the charge number balance changes due to attachment;
the energy loss by electrons changes when a molecular gas is introduced in an
atomic gas; the ionization rate also changes; etc. Of interest in this connection
is the numerical simulation [3.92] of an RF discharge in an electropositive gas,
with a simultaneous simulation of a discharge in the same gas under the same
conditions but introducing an artificial attachment, with the other characteristics
and constants being identical. The discharge was described in a hydrodynamic
(drift-diffusion) approximation by a set of equations of the type (2.1) with similar
boundary conditions but with an additional continuity equation for the negative
ion density $n_n \equiv n_-$ ($n_p \equiv n_+$)

$$\frac{\partial n_e}{\partial t} + \frac{\partial \Gamma_e}{\partial x} = (\alpha - \eta)|\Gamma_e| \qquad \Gamma_e = -n_e \mu_e E - D_e \frac{\partial n_e}{\partial x}$$

$$\frac{\partial n_p}{\partial t} + \frac{\partial \Gamma_p}{\partial x} = \alpha|\Gamma_e| - \beta n_p n_n \qquad \Gamma_p = n_p \mu_p E - D_p \frac{\partial n_p}{\partial x}$$

$$\frac{\partial n_n}{\partial t} + \frac{\partial \Gamma_n}{\partial x} = \eta|\Gamma_e| - \beta n_p n_n \qquad \Gamma_n = -n_n \mu_n E - D_n \frac{\partial n_n}{\partial x}$$

$$\frac{\partial E}{\partial x} = 4\pi e(n_p - n_e - n_n) \qquad -\int_0^L E(x, t)\mathrm{d}x = V_a \sin \omega t \qquad (3.80)$$

Here $\eta = 0.2\,p(1 + E/p)^{-1}$ cm^{-1} (p in Torr, E in V cm^{-1}) is an attachment
coefficient, which, in contrast with the Townsend coefficient α, was supposed
to be a decreasing function of E/p (this is the way η behaves in the strongly
electronegative gas, SF$_6$). No account was taken of the detachment and electron-
ion recombination. The calculations were made for pure helium, at $\eta \equiv 0$ and
$n_n \equiv 0$, and for helium 'with attachment' at $p = 1$ Torr, $L = 4$ cm, $V_a = 500$ V,
$f = 10$ MHz and 1 MHz.

In the absence of attachment [Figure 3.26(a)], we get a common picture discussed above and presented in Figures 2.1 and 2.12 (the discharge in [3.92] also
belongs to the α-type). The slight difference in the plasma density distribution in the gap is due to the fact that the equations of [3.92] involved charge
diffusion essential at a pressure of 1 Torr, while bulk recombination of electrons with ions was neglected. In contrast, Figures 2.1 and 2.12 refer to nitro-

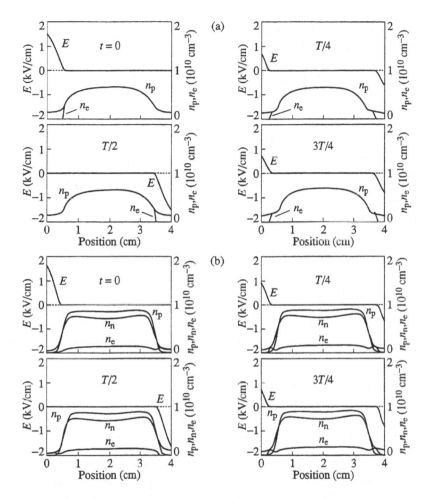

FIGURE 3.26

Spatial density distributions of positive n_p and negative n_n ions and electrons n_e, as well as of field E for each quarter cycle (T); $f = 10$ MHz, $p = 1$ Torr, $V_a = 500$ V, $L = 4$ cm. (a) Electropositive gas (helium), (b) electronegative gas (helium with 'attachment') [3.92].

gen at $p = 15$ Torr, when the recombination is strong while the diffusion is inessential because of high pressure, and can be ignored. The principal effect of attachment [Figure 3.26(b)] is the sharply increased density of both negative and positive ions as compared to the electron density: in the gap middle $n_p \approx n_n \approx 2.4 \times 10^{10}$ cm^{-3}, but $n_e \approx 0.4 \times 10^{10}$ cm^{-3} (all at $f = 10$ MHz). In pure helium, $n_p = n_e \approx 0.6 \times 10^{10}$ cm^{-3}. In the case of attachment, the sheaths are thinner: $d_{max} \approx 0.4$ cm instead of $d_{max} \approx 0.5$ cm, which qualitatively agrees with experiment [3.83]. However, as the frequency decreases from 10 MHz to 1 MHz, the maximum sheath thickness in an electronegative gas increases up to 1.6 cm, in contrast with the tendency registered experimentally [3.83]. The calculated inverse dependence of d_{max} on ω seems to be physically more understandable, and this is supported by equations (3.67), (3.68), and (3.72) for both collisional and collisionless sheaths. Experimental findings may have been affected by some additional factors associated with the complexity of the system under study.

We would like to stress that in an electronegative gas, negative and positive ions have low mobility, and their density varies only slightly in time. It is the electron gas that exhibits oscillations. The ionic current to the electrode is small relative to the displacement current, only positive ions being able to reach the electrode. There are no negative ions near the electrode at all, because they are confined to the plasma by strong dc fields in the sheaths. The rates of ionization and attachment in an α-discharge normally have maxima at the boundary between the plasma and the sheaths. Results similar to [3.92] have been obtained by other investigators [3.84, 3.94, 3.96–3.99, 3.101–3.104] who calculated, in a hydrodynamic approximation, RF discharges in Cl_2 [3.94, 3.95, 3.99, 3.101], SF_6 [3.84, 3.96, 3.100, 3.103], and O_2 [3.104] (see also [3.106]).

The smaller sheath thickness in an electronegative gas consistent with experiment [3.83] is treated in [3.92] without appealing to the Debye radius. Since n_e is smaller due to attachment, the electron current toward the electrode at the anodic stage is lower than in pure helium. Hence, the author of [3.92] concludes that a lower electron multiplication rate in the gap is sufficient to sustain the discharge. Lower multiplication rate is achieved owing to the smaller sheath thickness, with the RF voltage amplitude remaining unchanged.

This interpretation does not seem convincing enough. Reference to the sheath thickness effect on electron multiplication would be appropriate if we dealt with a γ-discharge. But the discharge calculated in [3.92] undoubtedly belonged to the α-type in spite of the introduction of the secondary electron emission coefficient $\gamma = 0.1$. The sheath thickness is regulated by the discharge current to be transported to the electrode, most of which is the displacement current. Unfortunately, the author of [3.92] did not give data on the discharge current in an electronegative gas at the frequency 10 MHz; but it is clear from the results presented that the maximum field at the electrode in such a gas is higher than in helium (1.8 and 1.6 kV cm^{-1}, respectively), i.e., the current in the first case seems to be higher. This is supported by the calculations for 1 MHz, from which it follows explicitly that the current in an electronegative gas is larger. The transport of large current through the sheath is accomplished by the stronger field at the electrode, because $j \approx j_{dis} \sim \omega E_{max}(0)$. It is this, we believe, that leads to a shrinkage of the sheath in an electronegative gas, since the sheath voltage fall $V_{1\,max} \approx E_{max}(0)d_{max}/2$ is limited by the same RF amplitude V_a. (Here again we see that interpretation of results of 'a numerical experiment,' as well as those of a laboratory experiment, permits a certain freedom in choosing the hypotheses, and their check-up sometimes requires more detailed evidence.)

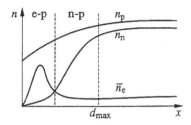

FIGURE 3.27
Sheaths in a strongly electronegative gas.

Experiments reported in [3.81–3.83] have shown that, at low frequencies, there is a local field maximum at the plasma-sheath boundary at the anodic stage, with the sheath filled up by electrons. This field is directed away from the electrode to the gap center. The local field maximum is due to the appearance of a double layer formed by positive and negative ions. This effect was also revealed in the calculations [3.92]. At the frequency 1 MHz, it is quite distinct, while at a high frequency of 10 MHz it is nearly unnoticeable. While in [3.83] this effect is attributed to detachment and to possible nonequilibrium processes, it is asserted in [3.92] that the double layer is formed due to oscillatory displacement of ions leading to their density modulation and to separation of the ion space charges. This idea is favored by the effect attenuation with increasing frequency and by test calculations with account of detachment, which was not found to influence the above effect.

3.10.2 Some effects at $n_n \gg n_e$

The electrode sheath of a discharge operating in a strongly electronegative gas has two characteristic regions (Figure 3.27). Close to the boundary with the plasma, as in the plasma itself, the electron density is extremely low, $n_e \ll n_n$. Nearer to the electrode, negative ions are pushed out by a dc field towards the plasma, so the cycle-average electron density in this region is $\bar{n}_e \gg n_n$. Hence, there are regions of 'electron-ion' and 'ion-ion' composition in the sheath, as they were termed in [3.105]. It has been shown in that work that the positive ion density has a smooth profile (as is the case for attachment-free gases and as is shown in Figure 3.27 for an electronegative gas) when the time τ of the ion flight through the sheath under a mean field is incomparable with the characteristic attachment time ν_a^{-1}, where ν_a is the attachment frequency. This happens no matter whether $\nu_a \tau \gg 1$ or $\nu_a \tau \ll 1$. If the times are comparable, i.e., $\nu_a \tau \sim 1$, a peak or an abrupt increase in the ion density is possible in the ion-ion region, even though the discharge is of the α-type.

In strongly electronegative gases, like pure Cl_2, SF_6 and others, a large number of negative ions is produced due to intensive attachment. The densities n_p and n_n may exceed n_e as much as two orders of magnitude. For instance, according to numerical simulation data [3.94, 3.95], the electron density in the plasma gap middle was found to be $n_e \approx 5 \times 10^8$ cm^{-3} while $n_p \approx n_n \approx 10^{11}$ cm^{-3} for a Cl$_2$ discharge at $f = 13.56$ MHz, $p = 1$ Torr, $L = 1$ cm, and $j_a = 3.3$ mA cm^{-2}. The ionic component of the plasma current was comparable with the electronic one and might even exceed the latter (the Joule heat is spent directly for heating the gas). When $n_p \approx n_n \gg n_e$, there may occur another curious effect revealed in the above simulation and later in [3.100] in the calculation of an SF$_6$ discharge. Common periodic changes in all parameters with the field frequency $f = T^{-1}$ are accompanied by their slow periodic variation with a period $T' \approx 300\,T$. This variation looks like a kind of beats or low-frequency modulation, as the RF voltage amplitude changes, at a given current, by a few dozen percent. The matter is that for many cycles T, electrons—whose number is much less than that of positive ions—cannot reach the electrodes during their oscillations, so that both electrodes act as cathodes for a period of time as large as a long cycle T'. As a result, the sheaths become gradually depleted by positive ions which go on entering the electrodes, while negative ions penetrating into the sheaths via diffusion are accumulated in them. The potential, which locked the electrons out, drops to a value permitting the electrons to 'break through' to the electrodes. The 'anodic' stage occurs, during which the electrons escape the gap, the previous high positive potential is restored in it and everything starts again.

Another exotic effect is the generation of subharmonics at the frequencies $f/3$ and $2f/3$ at such low pressures that ions perform 'collisionless' oscillations and $\omega < \omega_{pi}$ [3.100] (in a plasma of complex ionic composition, ω_{pi}^2 is equal to the sum of squared plasma frequencies of the ionic components). It is clear from the above that the RF discharge in electronegative gases is the seat of many new, perhaps, quite unexpected effects, being a much more complicated phenomenon than the discharge in electropositive gases.

To conclude, the processes of attachment and detachment in electronegative gases may result in a normal current density effect in the α-mode. Such an effect may arise from the descending character of the PC CVC (Section 2.3). It is in this case that a CVC minimum of the whole discharge is formed, as the sheath CVC has an ascending profile. Which mechanism is responsible for the PC CVC decrease does not actually matter. In Section 2.3, we discussed the gas heating and stepwise ionization as possible mechanisms, but under some conditions the CVC of the positive column in an electronegative gas may also have a descending character. This may occur as a result of a competition between attachment and detachment processes [3.19]; therefore, the possible appearance of the normal current density effect should not be entirely disregarded. Briefly, this effect may be produced via any mechanism that provides a descending CVC and, hence, leads to an unsteady state. Attachment–detachment stimulates the so-called attachment instability, though the ranges of pressure, current density and the n_n/n_e ratio for the unsteady state are rather limited [3.19].

3.11   Smooth α–γ transition and the γ-mode

The α–γ transition is accompanied by an abrupt increase in the current density and by contraction of the discharge column at elevated pressures only (Sections 1.8, 2.6). At low pressure, there is no contraction whatsoever, and the discharge plasma still covers the whole electrode after the transition; one observes no abrupt changes in the current and voltage that generally take place at moderate pressures. However, the current-voltage characteristic $i_a(V_a)$ and other voltage-dependent parameters (plasma density n, electron temperature T_e, plasma field amplitude E_a) show a bend at a certain value of V_a ($V_{a\,cr}$). Abrupt changes in the parameters and in their derivatives evidence for some radical alteration in the discharge process. We have pointed out above that these changes are a manifestation of breakdown of the α-sheath and an alternation of the charge multiplication mechanism. In the breakdown and the γ-mode that follows, the Townsend condition for charge self-multiplication in the sheath through secondary electrode emission is fulfilled. More exactly, the process of charge multiplication in the γ-sheath involves not only the sheath itself but also the adjacent plasma region of 'negative glow,' in which the field is weak. However, energetic electrons born and accelerated in the γ-sheath penetrate into this region.

Whether an abrupt change in the current density will occur and the discharge will contract or the α–γ transition will go on smoothly depends on the relation between the sheath thicknesses of the α-mode d_α and of the normal γ-discharge $d_{\gamma n}$. If $d_\alpha > d_{\gamma n}$, the sheath will shrink to $d_{\gamma n}$, and the current density will rise to $j_{\gamma n}$ corresponding to the normal γ-discharge, with the discharge column also contracting. If $d_\alpha < d_{\gamma n}$, nothing of the kind will happen. Because the value of $pd_{\gamma n}$ is close to that of $(pd)_{\min}$ in the Paschen curve (Figures 1.15, 2.23), the inequalities $pd_\alpha \lesssim (pd)_{\min}$ or $p \gtrsim (pd)_{\min}/d_\alpha \approx p_c$ can be taken to be the criteria for abruptness or smoothness of the α–γ transition, as d_α depends but slightly on pressure. In any case, the transition is accompanied by a restructuring of the discharge, since the γ-mode is characterized—regardless of pressure—by quite definite distributions of the plasma density n and the field, which differ from those for the α-mode. We remind the reader of the principal points discussed in Sections 2.6 and 2.5 for convenience, in order to be able to support these important conclusions with experimental evidence [3.107, 3.15] (see also [3.51]). In these experiments, a smooth α–γ transition and a low-pressure γ-discharge were observed.

A helium discharge was excited in a glass tube of a 6 cm inner diameter. Titanium flat disk electrodes were separated by a gap of length $L = 7.8$ cm. The discharge was studied over the frequency range 1–10 MHz and the pressure range 0.4–3 Torr at the voltage amplitude $V_a = 100$–2000 V. Langmuir probes of radius $r_p = 5 \times 10^{-3}$ cm were used to measure the distributions of n and T_e along the axis. Oscillograms of current i were taken with a Rogovsky belt, as well as those of electrode voltage V. Both parameters were harmonic within a 10% accuracy

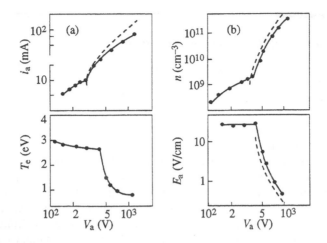

FIGURE 3.28
Measured dependences of current amplitude i_a, plasma density n, electron temperature T_e, and field amplitude in the gap middle on the electrode voltage amplitude indicating the α–γ transition. Helium, $f = 3.2$ MHz, $p = 3$ Torr, $L = 7.8$ cm, tube radius $R = 3$ cm. Dashed line, calculations [3.107]. (©1986 IEEE)

and $\pi/2$-phase shifted, indicating the basic contribution of the sheath capacitances to the discharge impedance. The field amplitudes E_a in the plasma were found from the measured n, i_a and the known area of the current-filled electrodes, using a formula for plasma conductivity.

Figure 3.28 shows the experimental plots for the parameters as a function of the voltage amplitude. Well-defined bends in the curves appear at $V_{a\,cr} \approx 400$ V, signaling an α–γ transition. With further increase of V_a, E_a and T_e drop, though not so sharply as it seems from the pictures, because the scale of V_a is logarithmic. In the gap middle, the electron temperature decreases from $T_e = 4$ eV for the α-mode with $V_a = 100$ V down to $T_e = 0.6$ eV for the γ-mode with $V_a = 1000$ V—the temperature, perhaps, approaching saturation. The decrease in T_e results from a considerable weakening of the RF field E_a in the gap middle, approximately by a factor of 30 from $V_a = V_{a\,cr} \approx 400$ V to $V_a = 1000$ V.

The α- and γ-modes have characteristic measured distributions of T_e and n along the axis (Figure 3.29). In a typical α-discharge ($V_a = 100$ V), the electron temperature is high and nearly uniform along the gap, being slightly higher at the edge near the boundary with the sheath. (Recall that the ionization rate, which is very sensitive to T_e, always has maxima at the plasma-sheath boundary in an α-mode.) As V_a approaches $V_{a\,cr}$, T_e in the middle somewhat decreases while at the boundary falls appreciably. In a well-developed γ-discharge ($V_a = 560$ V), the temperature is low throughout and increases only in the vicinity of the electrode, at the very boundary or, perhaps, in the sheath, too. In a typical

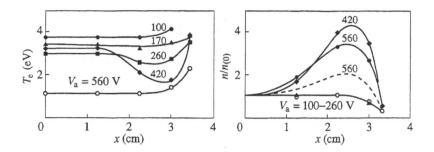

FIGURE 3.29
Probe measurements of electron temperature and plasma density distributions along the gap axis at various electrode voltages for the conditions of Figure 3.28. The x-axis is counted from the gap middle. Dashed line, calculations [3.107]. (©1986 IEEE)

α-mode (100 V$<$ V_a $<$ 260 V) the plasma density is constant nearly in the whole gap and decreases only at its edges, which is supported by numerous calculations discussed above. Under conditions close to the α–γ transition (V_a $\approx$ $V_{a\,cr}$) and in a mature γ-discharge (V_a = 560 V), the plasma density has a maximum at the boundary, which has also been confirmed by calculations (Figures 2.2, 2.3, 2.20, 2.21).

These features can be easily interpreted in terms of the transition theory, in which the transition is attributed to the α-sheath breakdown, and the γ-mode is considered as being analogous to a dc glow discharge (Sections 2.4–2.6). However, unlike the moderate pressure γ-discharge that becomes similar to a normal glow discharge after the transition, the low-pressure γ-mode becomes anomalous right after its excitation. A normal discharge is in this case unfeasible, because the current density should then become lower than the real value; but this is impossible due to the electrodes being already completely covered by current. The low values of E_a and T_e and the plasma density maximum lying near the boundary—these are characteristic features of a negative glow region and the beginning of a Faraday dark space. In this respect, there is no qualitative difference between anomalous and normal discharges and, hence, between low- and moderate-pressure RF discharges.

Let us make calculations to support the idea that in the helium experiment at p = 3 Torr, the α–γ transition is to occur smoothly; in other words, that the α-sheath to be subjected to breakdown really corresponds to the left-hand branch of the Paschen curve. The measurements [3.108] presented in Figure 2.24 for helium at 13.56 MHz show that d_α $\approx$ 0.16 cm and weakly varies with pressure at p $\gtrsim$ 1 Torr. At the frequency 3.2 MHz, at which the measurements were made in [3.107], the sheath is to be thicker. From equation (3.72), d_α $\sim$ $\omega^{-2/3}$, i.e., d_α $\approx$ 0.43 cm, which is consistent with Figure 3.29, considering that the plasma region ends at x_1 $\approx$ 3.5 cm and $x_1 + d_\alpha$ $\approx$ $L/2$ = 3.9 cm. At p = 3 Torr, pd_α $\approx$

1.35 Torr cm and is much less than $(pd)_{min} \approx 5$ Torr cm for helium, corresponding to the left-hand branch of the Paschen curve [3.19]. From the measured normal γ-sheath thickness in helium [3.108], which is independent of frequency and equals $d_\gamma \approx 1.5/p$ [Torr] cm, one may assume that the boundary between abrupt and smooth transitions at 3.2 MHz lies at $p_c = 3.7$ Torr, corresponding to the interception point of the curves $d_\gamma(p)$ and $d_\alpha(p) \approx$ const in Figure 2.24. If the authors of [3.107] had gone a little higher in the pressure, they may have found a slight discharge contraction at the transition indicating the formation of a normal γ-mode. Incidentally, the Paschen curve minimum for helium is gently sloping, and it is hard to predict the excess of p over p_c for the change to become sufficiently remarkable.

The deceleration length for fast electrons with energy 500 eV in helium at $p = 3$ Torr is $l_f \approx 3$ cm. This electron energy corresponds to sheath voltage $V_1 \approx 500$ V and to $V_a = 560$ V in Figure 3.29. The deceleration length of 3 cm makes up about half the gap length L. At lower pressures and higher voltages, it is still greater, but it is this path that determines the negative glow length in a glow-type discharge. Thus, the plasma in the experiments in question represents a plasma of two 'negative glow' layers extending from the sheaths and coming in contact with one another at the gap middle. It is quite likely that at the highest pressure studied (3 Torr), a Faraday dark space may start originating at this site, as $l_f \approx 3$ cm $\approx L/2 \approx 4$ cm. But there is no positive column here with a high electron temperature characteristic for these conditions, and one should not even expect it, because there is no room for it. It can be suggested that if the electrodes could be taken apart far enough for a positive column to appear at 3 Torr, the electron temperature would then be the same, $T_e \approx 4$ eV, as in the α-plasma. Then the field E_a in the gap middle would, respectively, be much higher, as in the pre-bend section of the curve in Figure 3.28. This interpretation of the low-pressure γ-discharge behavior [3.109] has been supported by calculations made for another gas—nitrogen (see Section 2.5).

The structure of a low-pressure discharge was examined theoretically in the same work [3.107]. Numerical simulation in the approximation of slow and fast electrons was performed in [3.110], and the results agreed with the measurements of [3.107]. In the simulation just mentioned, the γ-plasma is maintained via ionization produced by fast electron fluxes that are generated in the sheaths by secondary electron emission from the electrodes. The gap middle is filled up with plasma owing to the ambipolar diffusion from the sources primarily located at the edges. A similar scheme was used in [3.111] for a simplified calculation. In this work, an argon discharge in a commercial plasma reactor with electrodes of 60 cm in diameter was examined. The frequencies used were 380 kHz and 13 MHz, $p \approx 7.5 \times 10^{-3}$–0.5 Torr, $L = 18$ and 36 cm, and $V_a \approx 100$–1000 V. The CVC had an ascending character and showed no bends; the current density rose up to 1 mA cm^{-2}.

The authors of [3.107, 3.110] differentiate between α- and γ-modes by which gas ionization rate is higher: that due to the plasma electrons or to the fast electrons generated by secondary emission and accelerated in the sheath. The criterion for the transition is thought to be the equality of these ionization rates. In principle, this assertion is no doubt correct. However, it provides no means for evaluation of the transition parameters, because the ionization rates due to either mechanism can be found only by solving the general problem, which has made allowance for both ionization mechanisms.

In this respect, the Townsend criterion for the α-sheath breakdown, which was suggested and discussed in Sections 1.8, 2.4, and 2.6, is much more constructive because it better deals with the essential points of the problem. From this criterion, one can say with certainty whether the secondary emission mechanism is involved. If the Townsend criterion is fulfilled, one can be sure that the contribution of fast electrons is dominant. Due to the sharp field dependence of the Townsend ionization coefficient, a small excess of sheath voltage is sufficient for the secondary emission multiplication rate to exceed the ionization rate—which is due to the plasma electrons and for a γ-mode, normal or anomalous, to emerge. Direct calculations presented in Section 2.4 showed the estimated α–γ transition parameters to fit well with experiment and simulation for moderate pressures. Reliable application of the Townsend criterion for low pressures is difficult because of the imperfect theoretical description of the left Paschen curve branch at very low pd, several times smaller than $(pd)_{min}$, and because of the lack of appropriate experimental data; but this is quite another matter. Anyway, if the Paschen curve, even an experimental one only, is available, we are able to make evaluation of the transition parameters for low pressures.

3.11.1 The electron energy distribution function and fast electrons generated by stochastic heating

The electron energy distribution function (EEDF) can be found using Langmuir probes [3.112–3.115]. In the α–γ transition region, the electron spectrum differs from the Maxwellian one, being enriched by energetic electrons. Figure 3.30(a) illustrates some of the spectra, and Figure 3.30(b) gives the respective electron temperatures measured for low-energy sections of the spectra, $\varepsilon \sim 1$–5 eV, as well as the electron densities [3.115]. A helium discharge was studied in a cylindrical chamber with flat disk electrodes of a 10 cm diameter and a gap $L = 3$ cm at the frequency 27 MHz. As the input power and the electrode voltage amplitudes grow, the electron temperature decreases while the plasma density rises. The reasons for this effect are the same as for the situation presented in Figures 3.28 and 3.29. An α–γ transition occurs, and the plasma transforms to a negative glow with an elevated plasma density and a lower electron temperature. The γ-discharge spectra are enriched with high energy electrons, $\varepsilon \sim 5$–20 eV. These are γ-electrons generated by secondary electrode emission or in the space charge region and accelerated by the high sheath field. On the other hand, there is a sharp reduction in the number of electrons with $\varepsilon > 20$ eV. The spectrum becomes depleted by these electrons due to the inelastic energy losses for the excitation and

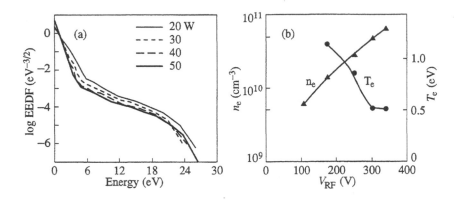

FIGURE 3.30
Probe measurements of (a) electron distribution functions and (b) the respective electron density and temperature. Helium, $p = 2$ Torr, $f = 27$ MHz, $L = 3$ cm. Distribution functions relate to various powers: electrode RF voltage amplitudes are 175, 250, 300, and 330 V, respectively [3.115].

ionization of the atoms.

Registration of high energy electrons by probe techniques is not always effective, because one has to apply a high negative potential to the probe. This gives rise to a strong ionic current towards the probe, which considerably screens the high energy electron current. For this reason, fast electrons and electron fluxes are often studied by indirect methods, for instance, by polarization spectroscopy [3.116, 3.117], which permits investigation of the distribution function anisotropy (the degree of radiation polarization increases when the atoms are excited by an electron beam). Conventional optical spectroscopy is also employed [3.117–3.119]. The distribution of the gas glow intensity with distance from the electrode in γ-RF and dc glow discharges are surprisingly alike under identical experimental conditions (Figure 3.31). In these experiments, a neon γ-discharge was studied at $p = 0.25$ Torr and $f = 4$ MHz, and a dc glow discharge was investigated in the same tube with the same electrode arrangement. Spectral characteristics of the two discharges were also found to be very similar. A study of radiation polarization revealed fast electron beams directed away from the electrode [3.117].

There is still another reason for generation of fast electron beams in low-pressure RF discharges. In addition to common γ-electrons emitted by the electrode and accelerated in the sheath, there are electron beams produced by stochastic heating due to recoil of electrons from the oscillating sheath boundaries. It is these fast electron beams that ionize the gas to sustain the gap plasma. This mode of RF discharge operation was termed in [3.120] as 'electron-sheath collision' mode. This is a new mechanism, different from the conventional α- and γ-modes of operation. High energy electrons due to stochastic heating were observed

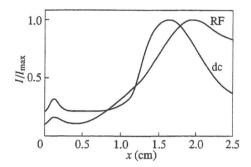

FIGURE 3.31

Distributions of relative glow intensity I/I_{max} with distance from the cathode in RF and dc glow discharges in the same tube under identical conditions: neon, $p = 0.25$ Torr, $f = 4$ MHz, amplitude in the RF discharge $V_a = 500$ V and in the dc discharge $V = 700$ V [3.117].

in experimental spectra of argon discharges [3.121] operating at $p = 0.1$ Torr, $f = 13.56$ MHz, $L = 2$ cm and current density amplitude $j_a = 2.56$ mA cm^{-2}. The electron distribution function was shown by probe measurements to consist of two Maxwellian functions with electron temperatures $T_e = 0.5$ eV and $T_e = 3$ eV. In the workers' opinion [3.121], the low-temperature portion of the spectrum resulted from common local ohmic heating of electrons in a weak RF field, while the high-temperature part was due to stochastic heating. As the pressure was increased, the latter degenerated and vanished, which seems to be due to the decreasing contribution of stochastic heating at higher pressure.

The numerical simulation [3.122] for the experimental conditions of [3.121], made by the super-particle method, provided a good fit with the experimental findings. In particular, the presence in the spectrum of high energy electrons due to stochastic heating at low pressures and the disappearance of this effect at higher pressures were confirmed. Generally, it is not always easy to distinguish experimentally between fast electrons from the two sources (γ-electrons and stochastic heating electrons). Of special value in this case is a combination of numerical simulation and numerical experiment, as was done in [3.123] by a completely self-consistent examination of the discharge process on a kinetic level (helium, $p = 0.1$ Torr, $f = 13.56$ MHz, $L = 4$ cm). Comparison of the two calculation approaches that used identical parameters but different secondary emission coefficients, $\gamma = 0.1$ and $\gamma = 0$, showed that stochastic heating was the principal source of fast electrons under the conditions used. Since the mean electron energy was very low, $\bar{\varepsilon} = 0.5$ eV, one can conclude that the principal sources of ionization are electrons that have gained energy through stochastic heating. Thus, in this case we deal with the third mode of 'electron-sheath collisions' (in [3.124] a similar mode with lower mean electron energy $\bar{\varepsilon} = 0.043$ eV was observed).

3.12   Some aspects of stochastic heating of electrons

The general concepts of the stochastic heating mechanism and of the related collisionless energy dissipation were described above in Section 1.7. The nature of the electron acceleration mechanism is similar to that suggested by E. Fermi in 1949 to explain the origin of fast particles in randomly moving magnetic clouds in the interstellar medium. V. Godyak [3.125] extended Fermi's ideas to stochastic heating of electrons in their collision with the oscillating plasma–sheath boundary, which serves as a potential barrier to the electrons entering the sheath from the plasma. The general case of stochastic heating of plasma by random high-frequency fields was examined in [3.126]. Stochastic heating in RF discharges was later analyzed theoretically in [3.127, 3.23, 3.50] and in many other publications that followed.

3.12.1   Collisionless energy dissipation

The mechanism of stochastic heating of electrons in RF discharges can be clearly observed in low-pressure experiments. The results obtained evidence for a power release that, by one order and more, may exceed the power of common ohmic heating due to electron–atom collisions. The relation between the heating mechanisms involved is described by the relation between the terms in the effective 'collision frequency' proportional to the Joule heat released in the discharge. According to [3.127], this quantity can be written as

$$\nu_{\text{eff}} = \nu_{\text{m}} + \nu_{\text{stoch}} = \nu_{\text{m}} + 2\bar{v}_{\text{e}}/(L - 2\bar{d}) \tag{3.81}$$

where ν_{m} is a common collision frequency of electrons and atoms, ν_{stoch} is a 'collision frequency' of an electron with the sheath boundaries when the electron performs a collisionless flight from one sheath to the other, $\bar{d}$ is the average sheath thickness, and $\bar{v}_{\text{e}}$ is the electron thermal velocity.

Figure 3.32 shows experimental measurements of the total power P_{t} released in a mercury vapor discharge at low pressures of 1.2×10^{-3} and 2×10^{-2} Torr ($f = 40.8$ MHz) and of the power P_{ν} produced in common collisions [3.128]. The latter was calculated from the measured current and known frequency of collisions between electrons and mercury atoms, which, with a good accuracy, is equal to $\nu_{\text{m}} = 2.1 \times 10^{10} p\,[\text{Torr}]\,\text{s}^{-1}$ over a wide electron temperature range, $0.5 < T_{\text{e}} < 5$ eV. It is clear from Figure 3.32 that with increasing RF voltage amplitude both powers grow, but the total power P_{t} noticeably exceeds P_{ν}— the more so as the pressure drops. This is quite understandable: the lower the pressure, the smaller is the relative contribution of electron–atom collisions to the power release. The relative contributions of both mechanisms are illustrated in Figure 3.33, which borrows from the same work [3.128] that compares the effective collision frequency (3.81) and the frequency of collisions with atoms ν_{m} for mercury vapor. One can notice that they differ by an order of magnitude

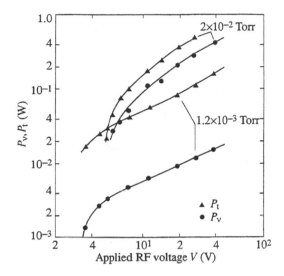

FIGURE 3.32
Measured total power and power associated with electron–atom collisions as a function of RF voltage amplitude. Mercury vapor discharge at two pressures; $f = 40.8$ MHz [3.128].

and more at $p \sim 10^{-4}$–10^{-3} Torr. Only at $p \sim 0.05$ Torr and higher does the stochastic heating become relatively less significant.

3.12.2 Maximum electron energy achievable by stochastic heating

Recent progress in the theory of stochastic heating [3.129–3.131] has primarily been due to the application of stochastic dynamics results and techniques [3.132, 3.133]. Without going into detail and omitting many essential results of this complicated theory, we will explain the physical nature of one of its basic conclusions concerning stochastic heating of electrons in the RF discharge. It has turned out that there are upper limits of velocity v_{max} and energy $\varepsilon_{max} = m v_{max}^2 / 2$ achievable by an electron through a stochastic gain of energy in its repeated reflection from the oscillating plasma–sheath boundaries.

Consider a simple model, a certain modification of so-called Ulam's model. Suppose a sheath boundary oscillates as $d(t) = d(1 + \sin \omega t)$, which is characteristic of a simplified discharge model with $n_+(x) = $ const. Here $d(t)$ is the instantaneous and d the average sheath thickness. We have referred to this model several times above. Note that the particular law of sheath boundary oscillation is of no principal importance for the final conclusion. The boundary velocity $u_s = \dot{d} = \omega d \cos \omega t$ oscillates with an amplitude $u_a = \omega d$. Assume the gap length L to be much larger than the sheath thickness, $L/d \gg 1$, and the pressure to be so low that the electrons reflected by the boundaries (potential barriers) move

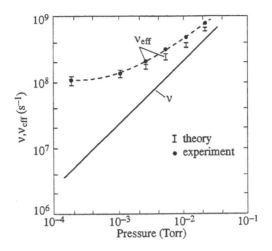

FIGURE 3.33
Effective collision frequency ν_{eff} with account of stochastic heating and electron–atom collision frequency for a mercury vapor discharge [3.128].

to and fro without colliding with atoms. With an accuracy to small values of the order $d/L \ll 1$, the time of flight of an electron between two opposite boundaries is equal to $2L/v$, where v is its velocity normal to the surface at this stage of the process. It will be taken to be quite large compared to the boundary velocity. Assume also for simplicity that the left boundary oscillates, while the other one, being immobile, elastically repels the electron.

We take v_1 to be the absolute electron velocity before a given, say, first collision with the oscillating boundary and v_2 to be the absolute velocity of the electron after the recoil, i.e., before a second collision. For an elastic repulsion from a moving solid wall at moment t_1, we can write

$$v_2 = v_1 + 2u_s = v_1 + 2u_a \cos \omega t_1$$

The average energy gain for the electron is equal to

$$\left\langle \frac{mv_2^2}{2} - \frac{mv_1^2}{2} \right\rangle = mv_1\overline{u_s} + 2m\overline{u_s^2} = mu_a^2$$

It is a positive value if the moment of the electron collision with the boundary is quite accidental, and we have the right to put $\bar{u}_s = 0$. Hence, a necessary condition for stochastic heating of an electron is the absolutely accidental character of collision with the boundaries and, therefore, the absence of correlation between the phases of two subsequent collisions.

The phase difference $\varphi = \omega t$ between the second and the first collisions is approximately equal to

$$\varphi_2 - \varphi_1 = \omega t_2 - \omega t_1 \approx \frac{2\omega L}{v_2} = \frac{2\omega L}{v_1 + 2u_a \cos\varphi_1} \tag{3.82}$$

Suppose the moment of the first collision will slightly change, because of a fluctuation, by δt_1; $\delta\varphi_1 = \omega\delta t_1$. Then the moment and phase of the second collision will also change: $\delta\varphi_2 = \omega\delta t_2$. By differentiating (3.82) with respect to φ_1, we will find the relation between these phase fluctuations

$$\delta\varphi_2 \approx \delta\varphi_1 \left(1 + \frac{2\omega L\, 2u_a \sin\varphi_1}{v_1^2} \right) \tag{3.83}$$

where we have neglected in the first approximation the velocity of the boundary u_s, which is small relative to the particle velocity v_1. In order to violate a correlation between the phase fluctuations $\delta\varphi_2$ and $\delta\varphi_1$ and to be able to consider the new phase φ_2 as arbitrary, the second term in brackets should be markedly larger than unity. Otherwise, the correlation between φ_2 and φ_1 will be quite rigid, because we will then have $\delta\varphi_2 \approx \delta\varphi_1$. Thus, with account of unity as the measure of $\sin\varphi_1$, the condition for stochastic heating of an electron can be written as $2\omega L 2u_a/v_1^2 > 1$. This inequality sets the following restrictions on the velocity and energy that the electron can acquire from stochastic heating at a given frequency ω of the oscillating boundary

$$v < v_{\max} \approx 2\omega d\sqrt{\frac{L}{d}} \qquad \varepsilon < \varepsilon_{\max} \approx 2m\omega^2 dL \tag{3.84}$$

For instance, at $f = 40$ MHz, $L = 10$ cm and $d = 0.025$ cm, we have $v_{\max} \approx 2.5 \times 10^8$ cm s^{-1} and $\varepsilon_{\max} \approx 20$ eV ($u_a = 6.3 \times 10^6$ cm s$^{-1} \ll v_{\max}$).

Inequality (3.84), dealing with frequency but not with velocity, refines the speculative qualitative condition for the violation of the phase correlation $\omega > 2v/L$ discussed in Section 1.7. The latter just expresses the idea that the boundary oscillation frequency must be markedly larger than the frequency of collisionless flights of the electron from one boundary to the other and back. Only in this case, the electron will 'forget' about the previous repulsion phase by the moment it returns to the initial boundary. The refined condition for the frequency (3.84) states that $\omega > v/\sqrt{Ld}$. Since $d \ll L$, it sets a more rigid limitation on the frequency. Stochastic heating of electrons up to the given velocity and energy requires a higher frequency than it follows from the qualitative condition $\omega > 2v/L$. In reality, this process is, of course, much more complicated. An essential factor is the finite time of the electron stay in the sheath during the reflection process. Some electrons are not reflected at all but go through the sheath into the electrode. These and many other aspects of this problem have been considered in [3.130] and [3.131] developing further the ideas of [3.130].

A real discharge plasma is always nonhomogeneous, as is nonuniform the average field representing a polarization (ambipolar) field; we mean just the average field $\bar{E}$ rather than the rms field. As a result, steady-state potential barriers are formed near the electrodes. Slow electrons turn out to be trapped in this potential well located in the gap middle and are unable to participate in the stochastic heating at the sheath boundaries. This was analyzed in [3.134], where the electron distribution function for a collisionless plasma was found. The authors also examine the case when the sheaths are collisionless but there are some rare collisions in the plasma at $L \gg d$.

It should be noted that in a collisionless plasma, in which the electrons practically do not collide with the atoms, electron–electron Coulomb 'collisions' become essential, contributing to the spectrum maxwellization. The plasma is sustained by rare ionizing collisions of energetic electrons that gained energy via both the stochastic heating and the spectrum maxwellization. Besides, such energy balance conditions of the electron gas are possible when the energy from the stochastic heating is largely carried away to the electrodes by energetic electrons capable of penetrating through the sheaths. A low-pressure discharge with stochastic heating of electrons was consistently simulated in [3.75] and other works we consider below.

3.13 Numerical simulation of low-pressure RF discharges

A low-pressure RF discharge is such a complicated phenomenon that approximational analytical theories, including those mentioned above, can describe only some aspects of the process and explain the nature of individual effects or point out the basic mechanisms, but they can hardly provide a complete and comprehensive picture. For this reason, of special value are numerical simulations, especially self-consistent approaches which take some general external conditions and certain microscopic characteristics, like cross sections or transport properties, and yield all discharge parameters and their distributions. There is extensive literature on simulation of low-pressure RF discharges, and in recent years there have been a few reviews of the subject [3.120, 3.135, 3.136]. Even a specialized theoretical issue on simulation problems has been published [3.106]. Here we will briefly discuss the available techniques with the aim to systematize the material and give appropriate references.

3.13.1 The hydrodynamic approximation

A common and well-elaborated technique is to use a hydrodynamic or drift-diffusion approximation. In some cases it provides satisfactory results for moderately low pressures, when the electron path lengths are small compared to the characteristic dimensions of the system—the sheath thickness and the gap size, though the method has limitations due, in particular, to the neglect of nonlocal effects. The approximation is based on equations similar to (2.1) and (3.80), which we will not write here again. One normally gives either the instantaneous electrode voltage, $V = V_a \sin \omega t$, expressed as an integral of the instantaneous field over the gap, or the current density $j = j_a \sin \omega t$, the latter being a sum of charge and displacement currents. In the general case, the external circuit equations are given. Such calculations were illustrated and discussed in Section 2.1 for moderate-pressure discharges and in Section 3.10 for the case of electronegative gas discharge. Sometimes, equations like (2.1) and (3.80) are supplemented by a hydrodynamic equation for the electron energy [3.19], taking into account the electron temperature dependence of transport coefficients. Various modifications of the hydrodynamic approximation have been employed in [3.24, 3.84, 3.92–3.94, 3.96–3.103, 3.137–3.143].

A simple and effective algorithm of RF discharge simulation in the drift-diffusion approximation suggested in [3.96] reduces the computation time by two or three orders of magnitude as compared with conventional approaches. A method allowing for the nonlocal character of the electron distribution function (EDF) was employed in [3.24, 3.97, 3.102, 3.139–3.142]. The distribution function is found by various modifications of the Monte-Carlo method and then used to evaluate transport coefficients, which are substituted into the hydrodynamic equations. For this purpose, in [3.139, 3.140] the cumbersome Monte-Carlo method was replaced by the solution to the Boltzmann equation in a conventional two-term approximation.

3.13.2 The two-group model

Even with account of the EDF nonlocality in the calculation of kinetic coefficients, the hydrodynamic approximation fails to describe the presence of fast electron beams and two gas ionization sources—by the plasma electrons and by the energetic electrons accelerated in the sheaths. For the same reason, it fails to describe the spatial structure of the γ-mode in which ionization by fast electrons is so essential. A simple way of treating this ionization is to apply the two-group approximation, in which the plasma electrons are described in a hydrodynamic approximation while the beams by another method using a kinetic equation for a strongly anisotropic EDF. This has been done in [3.110, 3.109, 3.144, 3.145]. The method of monoenergetic beams employed in [3.110] is to some extent similar to the method used in [3.146] for an electrodeless discharge, which is sustained by high repetition rate high voltage pulse rather than by a sine-wave field. But in [3.146] electron beams were examined more thoroughly with account of the electron energy spectrum and relaxation. In spite of the large number of simplifying assumptions, the calculations made in [3.110] have shown a good agreement with the experiments of [3.107] discussed in Section 3.11.

3.13.3 The Monte-Carlo method

At low pressures, when the path lengths for electrons and ions are comparable with the sheath size and, the more so, with the gap size, the hydrodynamic approximation becomes inapplicable. Nor is the conventional two-term approach applicable to the Boltzmann equation for the EDF [3.136]. The most suitable procedure for studying the discharge behavior is then the Monte-Carlo method of solving the kinetic equation [3.135, 3.147–3.152]. Such calculations can provide a possible scenario of motion of charged particles in the field with allowance for elastic and inelastic collision, including ionizing ones. In conventional calculations, one follows the trajectories of each particle; but this, even with the optimum algorithm of the so-called null collision method [3.148–3.151], leads to great computation expenditures. So the Monte-Carlo method is largely used either for giving the field distribution [3.150] or for finding the EDF in the combined models [3.24, 3.97, 3.102, 3.141, 3.142] discussed above. Self-consistent calculations, including the solution to the field equation, require large expenditures even in the case of a dc glow discharge [3.144, 3.145], so they are performed only to calculate the cathode sheath. The Monte-Carlo method has been employed for the calculation of the distributions of electrons [3.152] and ions [3.55] in an RF sheath, for two-dimensional modelling of a symmetric RF discharge at $p = 0.1$ Torr and $f = 13.56$ MHz in a typical plasmachemical reactor [3.153]. Here the method was applied only to electrons, while ions were treated in the hydrodynamic approximation.

3.13.4 The super-particle method

Self-consistent simulation of low-pressure RF discharges has considerably advanced, owing to the application of modifications of the super-particle method widely used in modern physics [3.154–3.156]. These modifications are the particle-in-cell (PIC) approach and the cloud-in-cell (CIC) approach close to the first one. The basic idea underlying the super-particle method is to substitute the real laboratory plasma by a model computer plasma of pseudo-particles with a large mass and charge, each of which consists of numerous real particles. The quasi-particles move according to the classical mechanics laws in a self-consistent electric field and are involved in collisions with neutral molecules, as common electrons and ions. Since the density of quasi-particles in a 'computer' plasma ($n' \sim 10^3$–10^5 cm^{-3}) is much smaller than the charge density in a 'laboratory' plasma ($n \sim 10^8$–10^{11} cm^{-3}), it is much easier to implement a fully self-consistent algorithm, which can provide the quasi-particle trajectories and collisional processes, using the Monte-Carlo method. So, starting with the work [3.157], the PIC approach combined with the Monte-Carlo collision method (so-called PIC-MCC) has been actively used for self-consistent simulation of low-pressure RF discharges [3.75, 3.158–3.161]. The results obtained fit the experiment well, so they can serve as a basis for constructing simplified analytical theories.

The major advantage of the PIC-MCC algorithms over commonly used MCC lies in a considerable reduction of the amount of computations and memory storage down to a level accessible to modern computers. Therefore, there may be a temptation to decrease the quasi-particle density in order to save the computer time. But in this case the computer plasma may acquire properties qualitatively different from those of a laboratory plasma. In particular, an essential parameter of the plasma—the number of particles in a Debye sphere that characterizes the degree of its ideality—sharply decreases, giving rise to pseudo-collisions and higher thermal noise in the calculations [3.154–3.156]. To avoid this, it is necessary to increase the density of model quasi-particles and to design better algorithms.

3.13.5 The convective scheme

The so-called convective scheme (CS) offers a promise for low-pressure discharge simulation. It was first suggested in [3.164] for self-consistent kinetic modeling of a glow-discharge cathode region and later successfully used for a self-consistent, completely kinetic simulation of a low-pressure RF discharge [3.123, 3.165]. The CS is, to some extent, similar to the PIC method; but in contrast to the PIC-MCC algorithms, it finds the distribution function of electrons and ions in a spatially nonuniform, unsteady electric field without a probable scenario of collisional events involving random numbers, as in the MCC. This greatly reduces the amount of computations. Its other merit is the absence of time-step restriction imposed by the Courant criterion. A detailed description of the CS algorithm can be found in [3.164, 3.165]. Note that in the PIC and PIC-MCC computations, one 'follows' the quasi-particle motion and collisional processes in fixed Eulerian coordinates; whereas in the CS computations, one 'observes' the 'Lagrangian' cells in the phase space (coordinate, velocity) with a variable number of electrons or ions in a cell due to collisions. The latter are taken into account in a way similar to that for the collision integral in Boltzmann's kinetic equation.

We would like to illustrate this with the results of a completely self-consistent calculation [3.123, 3.165] using the convective scheme. We take for comparison two of the variants of helium discharge at $f = 13.56$ MHz, $p = 0.1$ Torr, $L = 4$ cm, and applied voltage amplitude $V_a = 500$ V. These variants differ only in the secondary emission coefficient γ. Table 3.1 gives the current density amplitude j_a, the constant plasma potential in the gap middle $\overline{V}_p$, the maximum sheath thickness d_{max}, the maximum plasma density n_{max}; the maximum q_{max} and cycle-average $\overline{q}$ ionization rates, the mean electron energy $\overline{\varepsilon}$, and field amplitude E_a are all for the gap middle.

The most impressive thing here is the good agreement of nearly all parameters for both values $\gamma = 0.1$ and $\gamma = 0$, although we are certainly not dealing with the α-mode, because the mean electron energy is quite low. Independence of the results of the presence or absence of secondary emission indicates a minor role of this process and of γ-electrons. In fact, we deal with a typical third mode involving electron-sheath collisions, when electrons acquire considerable energy due to the stochastic heating in collisions with the oscillating sheath boundaries (Section 3.11.1). We have pointed out the specific distribution functions of electrons in this case found experimentally [3.121] and supported by calculations in the super-particle approach [3.122]. Note that within the convective scheme, there

TABLE 3.1

γ	0.1	0.0
j_a (mA cm^{-2})	2.7	2.6
$\overline{V}_p$ (V)	221	222
d_{max} (cm)	1.1	1.1
n_{max} (10^{10} cm^{-3})	1.0	0.9
q_{max} (10^{14} cm^{-3} s^{-1})	9.0	8.3
$\overline{q}$ (10^{14} cm^{-3} s^{-1})	5.5	5
$\overline{\varepsilon}$ (eV)	0.54	0.58
E_a (V cm^{-1})	1.9	1.8

may be a strong computational diffusion in space and energy, resulting in serious errors in the calculations of the rate constants of ionization and excitation of atoms by electron impact. A procedure for avoiding this drawback has been suggested in [3.166].

3.14   Magnetron RF discharge

At present, there is an active interest in RF discharges operating in an external constant magnetic field normal to the RF field and current. The magnetic field is commonly localized near the loaded electrode. This design provides a higher ionization degree at lower RF electrode voltage, decreases the energy of ions bombarding a target placed on the loaded electrode for treatment and increases the ion flux from the plasma, thus increasing etching efficiency. A magnetron discharge can be effectively sustained at much lower gas pressure, nearly down to 10^{-4} Torr, and this provides a better orientation of ion fluxes that penetrate through a sheath practically with no collisions. The ratio of the ion flux to the flux of chemically active neutral radicals becomes larger. These factors improve the etching quality of submicron anisotropic structures, making the magnetron discharge a promising tool for high precision treatment of semiconducting materials.

Figure 3.34 shows schematically the arrangement of a magnetron discharge in a technological reactor [3.167] designed for ion etching of SiO$_2$, Al and other materials. RF voltage is applied, as usual, to the smaller lower electrode with a target to be treated on it. A samarium-cobalt constant rectangular magnet is placed in vacuum under the electrode. The dashed line indicates the direction of magnetic lines in the discharge chamber above the electrode. One can see some sites on the target surface with the magnetic field directed horizontally along the surface, i.e., normal to the RF field. Since the region with the horizontal field is quite limited

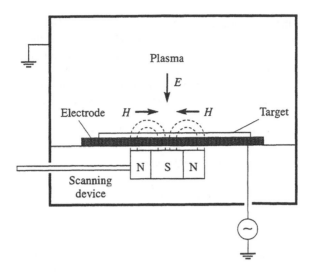

FIGURE 3.34
Schematic magnetron discharge employed in [3.167]. See text.

and does not cover the whole electrode, the magnet is moved along the surface by a special scanning device. Thus, for some time, the whole near-surface region of the discharge is subjected to the action of the field transverse to the current. In some experiments [3.168–3.171, 3.13, 3.74], magnetron discharges were excited in cylindrical systems with coaxial electrodes with the magnetic field along the cylinder axis.

The idea of using crossed constant magnetic and RF electric fields is that under the action of the magnetic field normal to the current flow, electrons oscillating together with the sheath boundary wind up around the magnetic lines at a Larmour frequency

$$\omega_H = \frac{eH}{mc} = 1.76 \times 10^7 \, H \, [\text{Oe}] \, \text{s}^{-1} \tag{3.85}$$

If the magnetic field is strong enough and the Larmour frequency high, the oscillation amplitude of an electron along the RF electric field sharply decreases. In a sense, the magnetic rotation of electrons plays the same role as collisions with atoms, preventing the electrons from achieving the free oscillation amplitude. The Larmour frequency ω_H acts as the collision frequency ν_m. But since the oscillation amplitude of electrons that make up the sheath boundary serves as the scale of sheath thickness, the thickness and the voltage fall across the loaded electrode are decreased, decreasing the self-bias, ion energy, and the RF voltage needed to maintain the plasma.

To see that it is this effect that the magnetic field has on oscillations and to understand the reasoning for the choice of the respective magnetic strength, consider the motion of electrons in crossed fields. Direct the x-axis along the electric RF field ($E \equiv E_x$) normal to the electrode surface and the z-axis along the magnetic field ($H \equiv H_z$). We should now introduce the Lorentz force in the motion equation (1.1) and deal with harmonic quantities in the complex form, as in Section 1.5.3, but not in the trigonometric form adopted in Section 1.2.1. The final formulas in this case have a more compact form. So we have

$$m\dot{\boldsymbol{v}} = -eE_\mathrm{a}e^{\mathrm{i}\omega t} - \frac{e}{c}[\boldsymbol{v} \times \boldsymbol{H}] - m\nu_\mathrm{m}\boldsymbol{v} \qquad (3.86)$$

Project equation (3.86) onto the x- and y-axes

$$m\dot{v}_x = -eE_\mathrm{a}e^{\mathrm{i}\omega t} - \frac{e}{c}v_y H - m\nu_\mathrm{m}v_x$$

$$m\dot{v}_y = \frac{e}{c}v_x H - m\nu_\mathrm{m}v_y \qquad (3.87)$$

In order to solve this set of equations with respect to v_x and v_y, it is more convenient to pass over to a new variable $v_x \pm \mathrm{i}v_y$. This is easy to do by adding the equations and subtracting them from one another, having multiplied the second one by a factor i. The resultant linear equations can be solved very easily. The solutions describing the oscillations of an electron induced by the oscillating electric field have the form

$$v_x \pm \mathrm{i}v_y = \frac{eE_\mathrm{a}e^{\mathrm{i}\omega t}}{m[\nu_\mathrm{m} + \mathrm{i}(\omega \mp \omega_H)]}$$

From this pair of equations we find the electron oscillation velocity v_x normal to the electrode surface, and then from the equation $\dot{\xi} = v_x$ we obtain the value of oscillatory displacement ξ

$$v_x = \frac{eE_\mathrm{a}e^{\mathrm{i}\omega t}}{2m}\left(\frac{1}{\nu_\mathrm{m} + \mathrm{i}(\omega - \omega_H)} + \frac{1}{\nu_\mathrm{m} + \mathrm{i}(\omega + \omega_H)}\right)$$

$$\xi = -\frac{\mathrm{i}eE_\mathrm{a}e^{\mathrm{i}\omega t}}{2m\omega}\left(\frac{1}{\nu_\mathrm{m} + \mathrm{i}(\omega - \omega_H)} + \frac{1}{\nu_\mathrm{m} + \mathrm{i}(\omega + \omega_H)}\right) \qquad (3.88)$$

These formulas describe the general case of an arbitrary relation between the frequencies ω, ν_m and ω_H. When reduced to the trigonometric form, formulas (3.88) transform to (1.2) for $\omega_H = 0$ ($H = 0$).

In the general case the expressions for the amplitudes and phases v_x and ξ are very cumbersome. Without writing them out, we turn to the low-pressure conditions of interest, when $\nu_\mathrm{m} \ll \omega$ and the electrons oscillate without collisions. At zero magnetic field and $\nu_\mathrm{m} = 0$, equations (3.88) yield the amplitudes of velocity and displacement (1.3) $u_\mathrm{a} = eE_\mathrm{a}/m\omega$, $a = eE_\mathrm{a}/m\omega^2$. The aim of applying a magnetic field is to reduce these amplitudes. Indeed, taking $\omega_H \gg \omega$ and making expansion in the brackets of equations (3.88) with respect to ω/ω_H at $\nu_\mathrm{m} = 0$, we can find the collisionless motion amplitudes in a strong magnetic field

$$u_{aH} = \frac{eE_a\omega}{m\omega_H^2} \qquad a_H = \frac{eE_a}{m\omega_H^2} \qquad (3.89)$$

For instance, at $f = 13.56$ MHz and $H = 200$ Oe, as in [3.167], $\omega_H = 3.5 \times 10^9$ s^{-1} is 40 times that of ω, and the magnetic field reduces 1600 times the amplitude of electron collisionless oscillations at the same electric field E_a. The ion trajectories in magnetic fields of this scale are not distorted, since the ion cyclotron (Larmour) frequency turns out to be much lower than ω.

At moderately low pressures but strong magnetic fields, $\omega \ll \nu_m \ll \omega_H$, electrons travel, according to equations (3.88), along the electric field $E = E_a \sin \omega t$ in the mobility regime at drift velocity

$$v_d = \frac{eE\nu_m}{m(\nu_m^2 + \omega_H^2)} = \mu_{e\perp}E \qquad \mu_{e\perp} = \frac{e}{m\nu_m} \frac{1}{1 + \omega_H^2/\nu_m^2} \qquad (3.90)$$

where the mobility transverse to the magnetic field is by a factor of $1 + \omega_H^2/\nu_m^2 \gg 1$ smaller than common mobility. The transverse diffusion coefficient for electrons decreases as much. Therefore, full electron fluxes entering the drift-diffusion approximational equations for collisional motion of an electron gas are reduced in the direction transverse to the magnetic field.

Magnetron discharges have been calculated in [3.172–3.174]. For instance, in [3.172] fluxes of electrons and ions were calculated in a hydrodynamic drift-diffusion approximation with account of electron magnetization. In order to find the electron mobility and diffusion coefficient, the Boltzmann equation was solved in a two-term approximation in [3.173]. The commonly used drift-diffusion approximation was taken to be valid for ions. On the contrary, in [3.174] ion kinetics was simulated by the particle-in-cell approach, while electrons were approximated by the usual hydrodynamic approach with transport coefficients depending on the local electron energy, which was found by solving the energy balance equation.

Experiments and numerical computations show that a magnetic RF discharge exhibits all characteristic properties of a common low-pressure RF discharge, and that its specific features are associated with the presence of a constant magnetic field. In particular, by reducing the electron fluxes, the transverse magnetic field makes it difficult for them to escape the plasma and enter the electrodes. So, the time for which the electrons stay in the discharge rises, which lowers the requirements on their ionizing ability.

Due to the escape of some electrons from the gap, positively charged sheaths are also formed near the electrodes. In a strong transverse magnetic field, the electron mobility may become lower than the ion mobility, the oscillation amplitude of ions in the sheath exceeding that of electrons. In this case the electron current to the electrode is determined by diffusion, which is, in spite of electron magnetization, quite large because of the great excess of electron temperature over the ion temperature. The electron current varies little over a cycle [3.173]. Under certain conditions, stochastic heating of magnetized electrons at the sheath boundaries is also possible [3.175].

An asymmetric magnetron discharge possesses all basic properties of an asymmetric RF discharge: self-bias in the presence of a blocking capacitance in the circuit [3.13, 3.74, 3.169–3.171] and the battery effect when the external circuit is capable of passing direct current [3.172, 3.176]. Since the sheath is much thinner in the presence of a magnetic field than at zero field, the self-bias in the first case is much lower; this was clearly observed in experiment. The energy spectrum of ions bombarding the electrode in a magnetron discharge is very similar to that in a common discharge, but the ion energies vary with the applied magnetic field, because it determines the sheath voltage fall. For calculations of ion spectra in a magnetron discharge, the reader is referred to [3.170, 3.177].

4

Experimental Methods and Measurements

In the previous sections, we have focused entirely on the nature of physical processes occurring in RF capacitive discharges, on the discharge structure and characteristics, referring, where necessary, to experimental findings. But we have not discussed experimental designs and techniques for measuring various discharge parameters. Extensive knowledge and understanding here are, however, necessary not only for the experimentalist but for the theoretical physicist, too. The latter must know how particular data were obtained in order to be able to assess their validity. In this section we will be concerned with experimentation in discharge physics, in particular in moderate-pressure discharges. The importance of understanding experimental techniques can be illustrated with this example. It followed from observations made in the 1950s–1960s that the constant plasma potential, being several hundred volts at low pressures, decreased with increasing pressure down to a few volts at $p \gtrsim 1$ Torr. The wrong conclusion was drawn from this fact as for the nature of moderate-pressure RF discharges, because the role of electrode sheaths responsible for the constant potential was underestimated. Later analysis of the details of the RF discharge showed that the previous measurements had contained significant methodological errors, which did not affect much the low-pressure potential but radically distorted the potential data on moderate-pressure discharges. More accurate measurements showed the plasma potential to be quite large, 100–200 V, and the sheaths to be as well-defined as at low pressures. Moreover, sheath parameters, in particular their thicknesses, were found to be ambiguous.

4.1 Voltage measurements and current-voltage characteristics

The current-voltage characteristic (CVC) of an RF discharge is a major experimental parameter, whose theoretical interpretation may provide an insight into the discharge behavior (Sections 1.8, 2.2, 3.2). Taking CVC data is reduced to

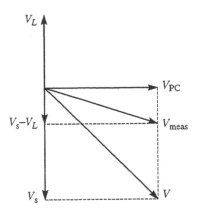

FIGURE 4.1
Vector diagram of complex voltages.

measuring electrode voltages and discharge currents while varying the amplitude of one of these parameters, V_a or i_a, or of the power applied to the discharge. Of equal importance is knowledge of the plasma voltage fall, i.e., the electrode voltage minus the voltage falls across the sheaths and electrode coatings, if any. It was the knowledge of the plasma CVC that helped to understand the physical reason for the normal current density in the α-discharge of moderate pressure (Sections 1.8, 2.2, 2.3).

When measuring electrode voltage V and phase shift between the 'sine-waves' of V and current i, one should take a special care to avoid regular errors due to the inductance of line accessories which supply current to the electrodes. If this error source is ignored, especially when the RF voltmeter is connected to the wrong place, the result may be greatly distorted. This is illustrated by a vector diagram of RF voltages in Figure 4.1 for the case when active conductance of the sheaths can be neglected. The diagram shows the various components of the electrode voltage V. It consists of the sheath voltage V_s (naked electrodes) and the positive plasma column voltage V_{PC}, $\pi/2$ phase-shifted if the plasma displacement current is neglected. If the voltage to be measured V_{meas} also includes the voltage drop due to the line inductance V_L, then V_L is subtracted from the V_s vector, thus decreasing V_{meas} relative to V. If the inductance between the point of contact of the measuring instrument and the operating surface of the active electrode is such that $|V_L|$ is comparable to the sheath voltage $|V_s|$, the error in the electrode voltage measurement may be very appreciable. Generally, $|V_s| \sim 100\text{–}300$ V. At current i of a few amperes typical of moderate-pressure discharges at frequency 13.56 MHz and line inductance $L' \approx 1$ μH, their impedance is $Z_L = \omega L' \approx 80$ Ω while $V_L = \omega L' i$ may also be a few hundred volts as V_s. The measurement error will then be large. Therefore, L' should be reduced as much as possible; and when this

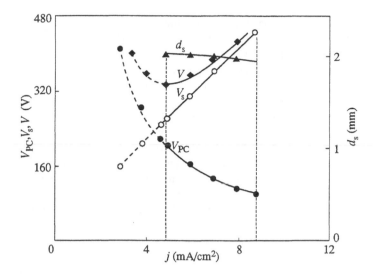

FIGURE 4.2

The CVC in an α-mode. The voltages across electrodes V, sheaths V_s, and plasma column V_{PC} as well as sheath thickness d_s were measured as a function of current density j for a discharge in air, $p = 30$ Torr, $f = 13.56$ MHz. The curve sections marked by a dashed line were measured for coated electrodes (see text).

cannot be done any more, one should connect a voltmeter to the electrodes through additional conductors arranged in such a way as to avoid discharge current flow through them.

The 'harmful' effect of inductance can be turned to advantage by measuring with it the plasma voltage [4.1]. The principle of this procedure follows from the above vector diagram in Figure 4.1 or from the respective formula for the amplitudes

$$V_{meas} = \left[V_{PC}^2 + (V_s - V_L)^2 \right]^{1/2} \tag{4.1}$$

This formula reflects the fact that the voltage falls on the sheath capacitive resistance and the external side inductance have opposite signs. By connecting a voltmeter at different points along the current-carrying line, one finds the minimum reading. From Figure 4.1 and equation (4.1) it corresponds to the exact compensation by the inductance for the sheath capacitive resistance. Then $V_{meas} = V_{PC}$.

Figure 4.2 illustrates measurements made in this way of the plasma CVC $V_p(j)$, the total CVC of the α-discharge $V(j)$ as well as the sheath CVC $V_s(j)$ [4.2]. The latter has been found by reducing the gap to an almost complete disappearance of the positive column when the opposite sheaths nearly come in contact. The voltage fall in the plasma is then small, and all the measured electrode voltage may be considered to fall in the sheaths. In contrast to the method of external inductance variation, when the sheath voltage does not show being compensated by the inductance voltage and the measured electrode voltage V coincides with V_{PC}, here V_{PC} vanishes and V coincides with V_s.

However, a serious problem arises requiring the application of a special technique. The point is that the α-discharge we are discussing exhibits the normal current density effect. At given pressure p, gap length L and gas, the current density cannot be lower than a definite normal value $j_{n\,\alpha}$. In Figure 4.2, however, there are points referring to $j < j_{n\,\alpha}$. They are connected with a dashed line to indicate that these states are normally unfeasible. Measurements for these 'unfeasible' current densities were made at the same p and L values but with coated electrodes. The principle of this technique will become clear if one refers to Section 2.9.1. When the electrodes are insulated from the plasma with dielectrics of thickness δ and permittivity ε, the normal current density of the α-discharge is the smaller, the lower is the total sheath plus dielectric capacitance; i.e., the larger is their effective thickness $d_\alpha + 2\delta/\varepsilon$, where d_α is the sheath thickness. By varying the dielectric thickness δ (and the material, i.e., ε) at constant L and p, one can measure V_{PC} at current densities j smaller than normal density $j_{n\,\alpha}$ with naked electrodes but the same positive column length. As the distance L between the electrodes is made smaller (until the positive column entirely disappears) to measure $V_s(j)$, the normal current density decreases according to equation (2.4). Therefore, in this case there is no obstacle to realizing j of interest. The discharge should just be transformed to an anomalous modification with the necessary current density j by increasing the discharge current.

4.2 Probe measurement of constant space and plasma potentials

4.2.1 Connection of a blocking element to the probe circuit

In order to measure the plasma potential relative to the ground in a dc discharge, when the potential does not vary in time, a probe is introduced at the site of interest and insulated from the ground and charged bodies. The probe takes on the potential V_{pr}, which is lower than the potential V_p at this point by the floating potential value $|V_f|$, which is determined by the plasma electron temperature (Section 3.7). By measuring the probe potential, one can find the plasma potential as $V_p = V_{pr} + |V_f|$. If the plasma potential contains a high frequency component, as is usually the case for the RF discharge, the problem of reliable measurement of plasma constant (average) potential $\overline{V}_p$ becomes much more complicated. The commonly used insulation of the probe from the ground does not extend to alternating current capable of flowing through the probe into the ground, since any insulation possesses a capacitive conductance. A positive space charge sheath is always formed around the probe for the same reasons as it arises near an electrode. This sheath possesses a finite time-varying impedance. Now an RF probe current flows through the sheath and, in addition to the common floating potential, an additional alternating voltage falls across the sheath. But due to the sheath impedance nonlinearity because of its variation during a cycle, the cycle-average value of the additional voltage turns out to be non-zero. It has the same sign as the floating potential, because the alternating voltage component of the probe sheath enhances the effect produced by the velocity difference of electron and ion transport through the sheath, giving rise to a floating potential near the body.

This is easy to see if one applies equation (3.49) to the probe, keeping in mind that this formula has been deduced for a collisionless sheath and may generate an error at moderately low pressures. Qualitatively, however, it can hardly be expected to play a trick when extended to a collisional sheath. Suppose a plasma has a potential $V_p = \overline{V}_p + \Delta V_p \sin \omega t$ relative to the ground. The probe potential can be represented as $V_{pr} = \overline{V}_{pr} + \Delta V_{pr} \sin \omega t$. In equation (3.49) $\overline{V}_p$ is the average plasma potential relative to a body with no dc current. In our case this is $\overline{V}_p - \overline{V}_{pr}$. In equation (3.49) ΔV_p is the amplitude of the RF plasma potential component relative to an insulated body. Here the role of ΔV_p is played by $|\Delta V_p - \Delta V_{pr}|$. Naturally, we are interested to find relatively small deviations of the probe potential from the plasma potential, so we take the value $x = e(\Delta V_p - \Delta V_{pr})/kT_e$ to be small, $|x| \ll 1$, and make use of the appropriate expansion of the Bessel function $I_0(x) \approx 1 + x^2/4$. The resulting expression for deviation of the average probe potential from the average plasma potential is

$$\overline{V}_p - \overline{V}_{pr} = |V_f| + \frac{T_e}{4}\left(\frac{\Delta V_p - \Delta V_{pr}}{T_e}\right)^2 \qquad (4.2)$$

where T_e is expressed in volts. We see that the correction for the alternating voltage effect always has the same sign as the floating potential, i.e., it further decreases the average probe potential with respect to the constant plasma potential.

If, in spite of the nonlinearity of the probe sheath impedance Z_{pr}, one makes evaluations in terms of common concepts of the linear theory of electricity, one can state that the additional voltage fall in the sheath is proportional to Z_{pr} and to the RF current through the probe, $\Delta V_p - \Delta V_{pr} \approx i_{pr} Z_{pr}$. Therefore, to reduce the error of constant plasma potential measurement, one should minimize the probe current, for instance, by connecting to the probe circuit a blocking element highly resistant to the RF current.

The idea of introducing, in the probe circuit, a choke with high inductive resistance follows not only from the theory above but also from a general physical reasoning; it was suggested as far back as the 1930s [4.3]. This technique has been used in some experiments [4.4, 4.5]. A schematic diagram of measurements and a refined equivalent electric circuit constructed in [4.6] are shown in Figure 4.3. Here L_b is a blocking element (in the experiments above, a choke) with impedance Z_b, C is a capacitor with a constant voltage (constant probe potential $\overline{V}_{pr}$ to be measured), and C_{str} is the stray probe and blocking element capacitance to the ground which is always present. Constant voltage between the plasma at the

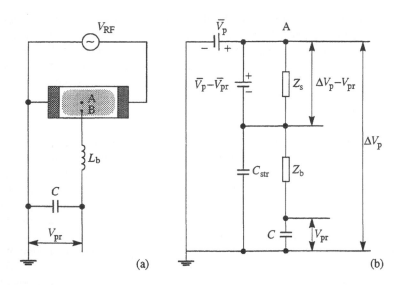

FIGURE 4.3
(a) Schematic diagram of probe measurements of constant plasma potential and (b) its equivalent electric circuit.

point of the probe location A and the grounded electrode, which is established during the discharge process, is represented schematically as a dc battery with an emf equal to the constant plasma potential $\overline{V}_\mathrm{p}$. Similarly, the source of constant voltage between the plasma and the probe $\overline{V}_\mathrm{p} - \overline{V}_\mathrm{pr}$ is also represented as a battery with the respective polarity. The sign ΔV_p indicates alternating voltage from the generator applied to the operating electrode, and $\Delta V_\mathrm{p} - \Delta V_\mathrm{pr}$ stands for alternating voltage between point A and point B, the latter representing the probe body.

It would seem that by using a blocking element with a very large impedance, one could completely suppress the probe current, reducing the difference between $\overline{V}_\mathrm{p}$ and V_pr to the inevitable floating potential. This, however, is not the case. The presence of stray capacitance sets limitations on the reduction of RF current through the probe, making an increase in Z_b above $(\omega C_\mathrm{str})^{-1}$ meaningless, because in this case the stray capacitance will have a shunting effect. Therefore, the only way of increasing the sensitivity of probe measurement of $\overline{V}_\mathrm{p}$ is to minimize the stray capacitance. This can be achieved by a good choice of the blocking element, by the design and appropriate location of the probe, by choosing the right dimensions of both, etc. The inevitable measurement error can be identified from equation (4.2). Denote the probe circuit impedance as $Z_\Sigma = (Z_\mathrm{b}^{-1} + \omega C_\mathrm{str})^{-1}$ (capacitance C to be used for measuring $\overline{V}_\mathrm{pr}$ can be ignored in Z_Σ, since it can always be made large enough to possess no resistance). Clearly, $\Delta V_\mathrm{pr} = i_\mathrm{pr} Z_\Sigma \sim i_\mathrm{pr} Z_\mathrm{b}$. Then $(\Delta V_\mathrm{p} - \Delta V_\mathrm{pr})/\Delta V_\mathrm{pr} \sim Z_\mathrm{s}/Z_\Sigma$. But for a symmetric RF discharge with no blocking capacitance in the circuit, we approximately have, on

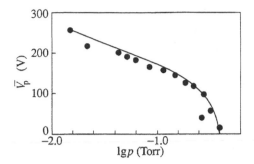

FIGURE 4.4
Probe measurement of constant plasma potential in hydrogen, $f = 68$ MHz, $L = 10$ cm [4.4]. The drop of $\overline{V}_p$ at pressures above 0.5 Torr resulted from the measurement procedure and is invalid.

the assumption of ideal plasma conductance and ignoring the floating potential, $\Delta\overline{V}_p \approx \overline{V}_p \approx V_a/2$, where V_a is the amplitude of RF voltage applied to the operating electrode (Section 3.1.2). By minimizing the deviation of the probe potential from the plasma potential, we tend to the equality $\Delta V_{pr} \approx \Delta V_p \approx \overline{V}_p$. Hence, the measurement error for $\overline{V}_p$ defined by the second term in equation (4.2), in case it is small, has the order

$$\frac{\overline{V}_p - \overline{V}_{pr} - |V_f|}{\overline{V}_p} \sim \frac{\overline{V}_p}{4T_e}\left(\frac{Z_s}{Z_\Sigma}\right)^2 \qquad (4.3)$$

Thus, to reduce the error, one should try to block, in the probe circuit, RF current with impedance Z_b—which is not only much larger than that for the probe sheath but also meets the more rigid requirement $Z_\Sigma/Z_s \gg (\overline{V}_p/4T_e)^{1/2} \approx 5$ at $\overline{V}_p \approx 200$ V and $T_e \approx 2$ eV. A detailed discussion of probe measurement errors, as applied to a collisionless sheath, is given in [4.7].

The primary and necessary condition for probe measurement of constant RF plasma potential—a good choice of the blocking element—was not fulfilled in experiments [4.4, 4.5] for moderate pressure. So, the results obtained were in error, and we present them as an illustration of the importance of this problem (Figure 4.4). The measurements made at low pressures were reasonable and agreed with the results obtained by other techniques, say, by measuring positive ion energies at the electrode. But at $p \gtrsim 1$ Torr, the measured constant potential drops to a very low value of about one volt, which contradicts both theory and more recent measurements. This is evidence for failure of RF probe current blocking at elevated pressures. It will be shown in the next section that the probe potential may become even negative in this case.

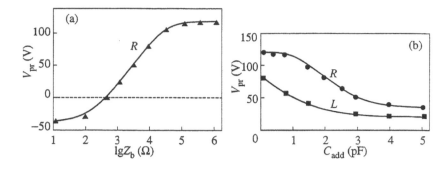

FIGURE 4.5
Probe data versus (a) impedance Z_b of a blocking element (low-capacitance highly ohmic resistor) and versus (b) additional capacitance C_{add} with a choke as blocking element (curve L) and with the above resistor (curve R). The α-mode in air, $f = 13.56$ MHz, $p = 10$ Torr, $L = 3$ cm, probe in the gap middle.

4.2.2 Experimental check on the choice of a blocking element and measurements

The validity of the above arguments (Section 4.2.1) can be tested experimentally by controlled measurement of the blocking element impedance Z_b as well as by parallel connection of additional measurable capacitance C_{add}, which acts in a way similar to natural, stray capacitance. The discharge was excited at 13.56 MHz in a quartz tube of 6 cm diameter filled with air at 10 Torr. Flat disk electrodes were arranged at a distance $L = 3$ cm, and the probe was inserted at the gap center [4.6]. The measurements are presented in Figure 4.5. The blocking element in Figure 4.5(a) was made as a low capacitance highly ohmic resistor (a choke is of little use for this purpose at moderate pressures, because it possesses an appreciable stray capacitance to the ground). At small Z_b, the measured constant potential is negative, indicating a huge gain in the floating potential due to RF current through the probe. In the 'absence' of considerable resistance Z_Σ to RF current in a disconnected probe circuit, the probe acts as an additional electrode connected through the blocking capacitance C (which disconnects the circuit and is to be measured). From equation (4.2), $\Delta V_{pr} \approx i_{pr} Z_\Sigma \ll \Delta V_p = \overline{V}_p$. Optical effects around the probe are identical to those near RF discharge electrodes. As Z_b is increased, the probe constant potential rises, becomes positive and gradually reaches saturation, which, judging by the value $\overline{V}_{pr} \approx \overline{V}_p \approx 100$ V, can be taken to be an approximation to the sought constant plasma potential.

Still, this encouraging result needs to be verified. Saturation of $\overline{V}_{pr}$ may be considered not only as an approximation to the true plasma potential (minus the floating potential, $\overline{V}_p - |V_f|$) but as a result of the shunting effect of stray capacitance. Shunting restricts the influence of the blocking impedance Z_b, because the equivalent impedance Z_Σ eventually stops growing in spite of the Z_b growth.

This was checked by introducing, parallel to Z_b, a controllable capacitance C_{add} simulating the effect of C_{str}. At large values of C_{add}, which shunts the same highly ohmic low-capacitance resistor, the measured probe potential was low; but at sufficiently small C_{add} the potential $\overline{V}_{pr}$ also reached saturation, becoming equal to the value with no C_{add} [curve R in Figure 4.5(b)]. When a choke was used as a blocking element, the reduction of the shunting capacitance to zero did not result in the probe potential saturation [curve L in Figure 4.5(b)]; this potential remained substantially smaller than $\overline{V}_p - |V_f|$. This proves that a choke possessing a high stray capacitance does not serve its function. It is quite likely that this was the source of measurement errors made in [4.4, 4.5] at $p \gtrsim 1$ Torr. The sheath impedance became larger, so the measurement accuracy (determined by the Z_s/Z_Σ ratio) decreased.

Figures 4.6 and 4.7 illustrate measurements of constant plasma potential as a function of applied RF voltage [4.1]. All curves in Figure 4.6—to be more exact, straight lines, because the $\overline{V}_p$ versus V_a plot turns out to be linear—refer to the α-discharge. The lower points in the curves correspond to a normal α-mode, while the others correspond to an anomalous α-mode, since the RF voltage starts growing only after the current has spread all over the electrode. In addition to the normal and anomalous α-modes (lower branches), Figure 4.7 shows data for a γ-mode. One can see how the RF voltage and the plasma constant potential change after the $\alpha-\gamma$ transition (dashed jumps). It is interesting that the RF voltage drops at the transition while the constant potential remains practically unchanged and even slightly rises. The lower points of the upper lines correspond to a normal γ-mode. The growing constant potential and RF voltage corresponding to the upper sections of the curves describe an anomalous γ-mode.

The probe method was also used to measure the constant potential distribution along the discharge axis in the sheath and positive column. These data were presented in Figure 1.14 in Section 1.8. Note the low peaks of $\overline{V}_p$ in the regions of brighter α-glow near the sheath-plasma boundaries. They are likely to be due to the peaks of local plasma density n. Diffusional polarization of the plasma induces a dc field directed away from the n peak towards the gap center.

4.2.3 Measurement of maximum constant plasma potential

Measurement of constant potential distribution within a sheath is a fairly complicated task because of the small sheath thickness, especially at high frequencies. For example, at $f = 81$ MHz, a γ-sheath at moderate pressures is a few fractions of a millimeter. But for many applications, knowledge of the maximum value of constant potential is quite sufficient. To measure this value, there is often no need to place the probe near the sheath or in the discharge middle. When the RF probe current is high, the probe and the blocking element become overheated, breakdowns occur along the surface of the latter and the influence of stray capacitance is great. It may suffice to seek a constant potential maximum on the discharge periphery or even outside it. The following observation pointed to this

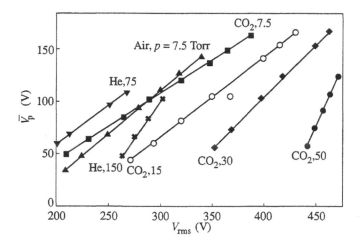

FIGURE 4.6
Probe measurements of constant plasma potential $\overline{V}_p$ versus rms electrode voltage V for
an α-mode, $f = 13.56$ MHz; gas and pressure in Torr are given near the curves.

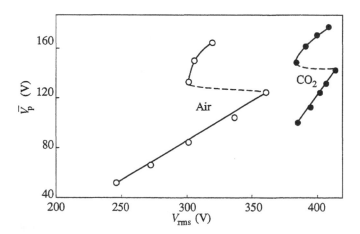

FIGURE 4.7
Probe measurements of constant plasma potential $\overline{V}_p$ versus rms electrode voltage V for
α- and γ-modes, $f = 13.56$ MHz, $p = 7.5$ Torr. Dashed lines, abrupt changes at $\alpha-\gamma$
transitions.

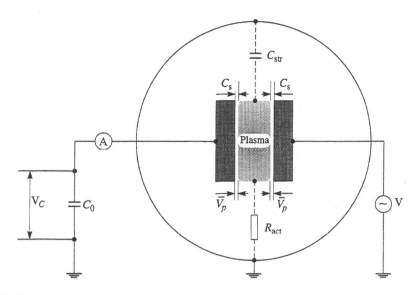

FIGURE 4.8

Schematic diagram of the discharge chamber in which constant voltage across the capacitor C_0 was observed. V, an RF generator; the electrodes are shaded; C_{str}, stray plasma capacitance to the chamber; R_{act}, stands for active conductivity of a discharge-free gas; A, RF current meter.

possibility [4.1]. A discharge was excited in the middle of a large metallic chamber of 60 liter volume between water-cooled metallic disks of 10 cm in diameter. A capacitor C_0 was connected between one of the electrodes and the ground to calibrate the RF current meter A (Figure 4.8). The discharge was symmetrical to avoid a common negative self-bias across the electrode. In spite of this the capacitor was charged to a constant negative potential $V_C < 0$ on the plate connected to the electrode. The value of V_C varied with the electrode RF voltage amplitude and was about 100 V. The effect was observed in He and N_2 at $p \sim 50$ Torr, but in CO_2 at $p > 5$ Torr the capacitor voltage was insignificant.

One might suggest that capacitor C_0, connected in series to stray capacitance C_{str} between the plasma and the grounded chamber walls, would become charged together with C_{str} (Figure 4.8). The polarity of the C_0 voltage turns out to be correct. One could be sure, however, that $C_0 \gg C_{str}$. Indeed, estimation gives the sheath capacitances $C_s \ll C_0$. But the stray capacitance $C_{str} \ll C_s$, otherwise the current would close on the grounded chamber walls rather than the grounded electrode. Therefore, $C_{str} \ll C_s \ll C_0$. But in this case most of the constant potential $\overline{V}_p$ falls on the small stray capacitance C_{str}, and only its small portion of about $\overline{V}_p(C_{str}/C_0)$ falls on C_0. In contrast, the measured voltage $V_C \approx -\overline{V}_p$ is much larger.

This effect was satisfactorily explained on the assumption of finite active conductivity in the region between the discharge plasma and the grounded chamber walls. In this case the grounded capacitor plate is charged positively, through the gas conductor on the periphery, up to the constant potential in the plasma bulk. The constant plasma potential connected to the ground through the peripheral conductor now acquires the potential of the ground, i.e., it becomes zero. Factors inducing a constant positive (relative to the electrodes) potential $\overline{V}_p$ are still active, therefore the constant electrode potential acquires the value $-\overline{V}_p$ relative to the ground and to the discharge plasma.

This hypothesis was tested in a discharge excited in a quartz tube, whose potential was non-zero, unlike the potential of the grounded metallic chamber. In this case there was no constant voltage on C_0 connected to the electrode through one plate and to the ground through the other (Figure 4.8). This is quite natural, because the grounded plate has now become electrically insulated from the plasma. This situation could not be changed by a purposeful increase of the plasma stray capacitance to the ground by placing the discharge tube into a grounded metallic screen. However, when a small hole was made in the tube wall to introduce a thin probe connected to the ground through a low-capacitance resistor of 1 MΩ, a constant voltage appeared on the capacitor C_0 and had about the same value and polarity as in the case of a metallic chamber.

These experiments lead us to the conclusion that there is a weakly conducting medium, denoted in Figure 4.8 as active 'resistance' R_{act}, between the grounded metallic chamber and the RF discharge in its middle. The capacitor C_0 is charged up to a certain voltage V_c, close in magnitude to $\overline{V}_p$, through this resistance. Absence of this effect in an electronegative CO_2 discharge seems to be due to the loss of electrons through attachment outside the discharge space. But if the space at a distance of about 10 cm from the plasma volume possesses a certain conductivity, the probe introduced into this space will acquire its potential (of course, up to the floating potential value, which here will be smaller because of the lower electron temperature outside the discharge volume).

Measurement of maximum constant plasma potential $\overline{V}_{p\,max}$ now reduces to placing the probe onto an equipotential surface corresponding to $\overline{V}_{p\,max}$. In practice, the measurements are made as follows: a probe is moved along the discharge periphery, and the maximum readings of a highly ohmic voltmeter— for instance, of an electrostatic voltmeter protected against RF interference—are taken. The probe in this case is not a floating probe but is connected to the ground through a very large resistance. The value of $\overline{V}_{p\,max}$ thus obtained coincides, within the accuracy of a few percent, with $\overline{V}_p$ measurements made by introducing a floating probe in the plasma.

We would like to mention that there is a way of increasing the probe measurement accuracy by using resonance RF chokes as blocking elements [4.8]. This technique, however, is limited to low frequencies $f \lesssim 1$ MHz due to a low Q-factor and the impossibility to principally eliminate stray capacitances. For probe measurements in RF plasma, see also [4.8a-4.8c].

Note that the experimental data that confirm the existence of self-bias in a moderate-pressure asymmetric discharge between coaxial cylindrical electrodes and of the battery effect are an independent evidence for a high constant potential in moderate-pressure discharges [4.6]. These effects were discussed above in Section 3.5.2.

4.3 Active dc probing of an RF discharge

Among a great variety of methods available for the study of the spatial structure of RF capacitive discharges, dc probing is known as a simple and very informative technique. When combined with calorimetric studies of dissipated power, it permits unambiguous identification of discharge operation modes as well as determination of such integral sheath characteristics as thickness d_s and ohmic resistance R_s, of the electric field strength in the plasma E_p, and other parameters [4.9, 4.10].

The principle of this technique is that a source of controllable constant voltage U is connected to RF electrodes if they are naked, otherwise to additional electrodes introduced in the gap; and the current-voltage characteristic of the circuit, including the RF discharge portion to be examined, is taken. The linear fraction of the probing circuit CVC is found experimentally in order to determine the ohmic resistance of the object placed between the probing electrodes. If one wants to exclude the influence of the space charge sheaths that always arises at the probing electrode surfaces, one should take the probing circuit CVCs at various distances between the additional electrodes (see below).

Active dc probing of an RF discharge is an attractive technique, because the discharge RF current i and the probing current I can be easily separated by using inductance coils and capacitors.

4.3.1 Longitudinal probing

Consider a way of measuring d_s and R_s suitable for both discharge modes. It is more profitable to use for this aim plane naked water-cooled electrodes with a variable distance L between them. Since we need to find integral sheath characteristics, L should be chosen as small as possible to eliminate the influence of the plasma column on the measurements but to preserve the discharge intact.

At first, one needs to measure the electrode RF voltage V and the discharge current i with a simultaneous registration of the electrode area S covered by the discharge. An RF generator is connected to the electrodes through a blocking capacitor C. A stabilized source of constant controllable voltage U and a direct current meter I are connected to the same electrodes via inductance coils L_b blocking the RF current. Owing to the application of L_b to block the RF discharge current and of C to block the probing direct current, these currents are completely separated. RF current i is determined by V and by the total impedance of the gap, while direct current I depends on U and on the active conductivity of the discharge. We have already mentioned that the influence of the positive column on the gap conductivity can be minimized by an appropriate choice of the interelectrode spacing. One can then consider that a combination of the measured values US/I will determine the sheath ohmic resistance R_s per unit area, and VS/i will give the total impedance Z_s of these sheaths also per unit area.

Keeping in mind that the sheath impedance Z_s is determined by active R_s and capacitive resistances of the sheaths $1/\omega C_s$, and using the measured values of R_s and Z_s, one can easily calculate the effective sheath capacitances C_s per unit area. The latter is expressed as $C_s = \varepsilon_0/d_s$, where $\varepsilon_0 = 8.85 \times 10^{-14}$ F cm^{-1} is vacuum dielectric permittivity and d_s is an effective value close to the cycle-maximum sheath thickness. If the sheath thickness in a symmetric α-discharge varied around an average value in a strictly sine-wave law, the value of d_s would be exactly equal to the maximum sheath thickness (Sections 3.3, 3.4). In this way, d_s can be measured in various operation modes, and its dependence on the gas pressure and frequency can be found, if one makes the measurements at various p and ω.

The effective value of d_s is related to the experimentally measured parameters by the expression

$$d_s = \frac{\varepsilon_0 \omega}{(Z_s^{-2} - R_s^{-2})^{1/2}} \qquad (4.4)$$

The value of d_s calculated from equation (4.4) is consistent with sheath thicknesses determined by other methods from constant potential distribution $\overline{V}(x)$ and from glow intensity distribution [4.9]. The registered wave form of the voltage $V(t)$ and of the current $i(t)$ is close to the sine-wave shape. This, of course, is indicative of a linear character of the circuit as a whole in spite of a nonlinear characteristic of each individual sheath (the latter is indicated by a high constant voltage arising between the plasma and the electrodes). However, one should identify with caution the measured values of d_s with the average or maximum sheath thicknesses, because the wave forms of voltage and current are not sensitive enough to the laws of sheath thickness variation in a symmetric discharge (Section 3.4).

When analyzing methodological errors in the measurements, we should primarily take into account specific features of the α- and γ-modes of the RF discharge. In this particular case, they reveal themselves in the fact that γ-sheaths, like the cathode region of a dc glow discharge, may exist independent of a positive plasma column. This readily provides conditions with $V \approx V_s$ by making the interelectrode spacing smaller. The major source of positive ions for α-sheaths is the plasma column, more exactly, the column portions adjacent to the sheaths. In other words, α-sheaths are regions with non-self-sustained active conductivity, for which Townsend's criterion (2.14) is invalid, in contrast to γ-sheaths. Ionization occurs in the plasma phase; this circumstance leads to an error in the d_s values calculated with equation (4.4) giving overestimations, because V in an α-mode always exceeds V_s due to the contribution of the positive column RF voltage. In experiment, the excess of V over V_s must be controlled by measuring the phase-shift φ between V and i. Taking into account that the active sheath conductivity is much smaller than the plasma conductivity, it is easy to show that, say, at $\varphi > \pi/3$ the relative excess is $(|V| - |V_s|)/|V| < 15\%$.

Another source of errors in measurements of RF current density j and d_s in the α-mode is associated with measurement of i. When the interelectrode gap is only partly filled with plasma in the direction transversal to the current, and when the gap itself is small, the shunting effect of the RF current flowing outside the discharge becomes essential. This effect can be minimized by smoothly increasing i to a level when the whole gap is filled up by the discharge plasma. The area of the current coincides with the electrode area, $S = \pi r_{el}^2$, where r_{el} is the electrode radius.

Measurements in a dc probing circuit have some specificity. The expression for active sheath resistance $R_s = US/I$ is valid only if (i) the probing current passing through the gap has no influence on the sheath structure; and (ii) no emf arises between the electrodes during the RF discharge operation (Section 3.5.2), that is, the discharge serves as a passive load for the probing current source. Experiments show that both conditions are generally violated. In particular, at $U > 10$ V the probing circuit CVC becomes nonlinear, indicating an effect of probing voltage on the sheath structure, especially in the α-mode. A decrease in U below 5 V sets more rigorous requirements on the suppression of RF induction in the measuring circuit. The second condition could not be met, because even at zero voltage at the probing voltage terminals the probing current in the γ-mode was as high as a few milliamperes. This is indicative of a constant emf of a few volts arising between the RF electrodes. To avoid an error, I should be measured twice in this case: at $U \neq 0$ and $U = 0$. The true value of I is taken to be the sum of the currents registered, if they flow in the opposite directions, or their difference, if they flow in the same direction.

To illustrate the capabilities of the method described, Figures 4.9 and 4.10 present measurements of the sheath thickness, ohmic resistance, and impedance of the sheaths as well as the discharge current density in α- and γ-modes in helium and air at various pressures and at frequency 13.6 MHz.

We will now discuss in some detail the experimental evidence for the normal current density effect in both discharge modes. A direct way based on discharge cross section measurement at various currents is often hard to employ, because the sheath contours are far from being regular, especially in the γ-mode. Moreover, the discharge column may continuously move across the electrode surface. Therefore, it is reasonable to make use of active probing, namely to employ the measurements just described to build a plot of the measured total sheath resistance $U/I = R_0$ as a function of the reciprocal value of RF current i^{-1}. Indeed, by definition, $R_0 = d_s/\sigma_a S$, where σ_a is active conductivity of the ionized gas in the sheath. On the other hand, $i = jS$, hence $R_0 = d_s j/\sigma_a i$. If the experimental points R_0

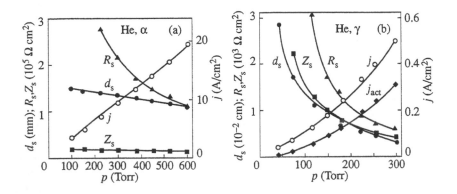

FIGURE 4.9
Parameters of normal α- and γ-discharges with naked electrodes in helium at $f = 13.56$ MHz obtained from active longitudinal dc probing: rms normal density of discharge current j, active current density component j_{act} across the electrode (γ-mode), sheath thickness d_s; ohmic resistance R_s and impedance Z_s per unit sheath area.

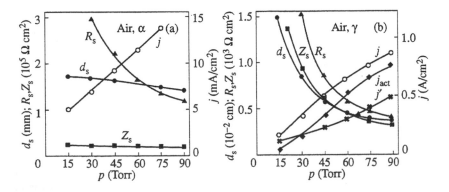

FIGURE 4.10
The same as in Figure 4.9 for air; j', normal current density in a discharge with uncooled electrodes, the difference with j being due to the gas heating.

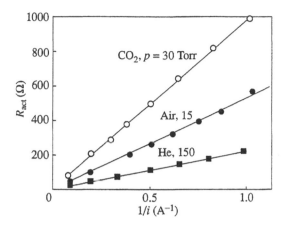

FIGURE 4.11
Experimental evidence for the normal current density effect in a γ-discharge at $f = 13.56$ MHz: total discharge resistance R_{act} to dc probing current depends linearly on reciprocal RF current i (rms values).

and i obtained by varying RF current lie in the straight line $R_0 = \text{const}\, i^{-1}$ in coordinates R_0, i^{-1}, this will mean that d_s and σ_a of the sheath and the discharge current density j are constant, and that the area S occupied by the RF current grows in proportion with i. In other words, there is the effect of normal current density.

Results typical of the γ-mode are presented in Figure 4.11. One can see that the experimental points lie in the lines $R_0 = \text{const}\, i^{-1}$ with a good accuracy. The constant in this equality varies with the gas and electrode material, V in each run remaining practically unchanged. Thus, the experimental data confirm the idea that the values of j, V_s, and d_s in a γ-mode with partly covered electrodes ($i < j\pi r_{el}^2$, $j = j_{n\,\gamma}$) are independent of i; that is, there is the normal current density effect. The technique considered can be applied if i is much larger than the RF current outside the discharge. Otherwise, the latter must be taken into account.

The method of longitudinal probing of an RF discharge can be made more informative if it is combined with a controllable variation of the distance L between the electrodes [4.6]. Let us measure, as described above, the ohmic resistance per unit area $R = R_0 S$, varying L. Let us maintain a constant discharge current density. As a result, we obtain the experimental dependence $R(L)$. Assuming the gas temperature to be also constant, we can suppose that the parameters and ohmic resistance R_s of the sheaths do not change. Then, introducing the ohmic resistance per unit area of the plasma $R_p = (L - d_s)/\sigma$, where σ is its conductivity, we write

$$R(L) = \frac{R_0}{S} = R_s + R_p = R_s + \frac{(L - d_s)}{\sigma} \qquad (4.5)$$

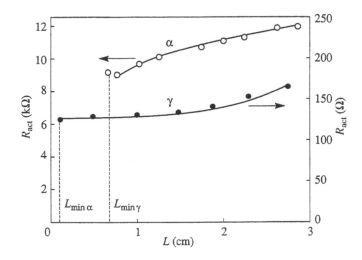

FIGURE 4.12
Measured resistance R_{act} in α- and γ-discharges in air as a function of the gap size L; $f = 13.56$ MHz, $p = 10$ Torr, the electrode and discharge column area 95 cm^2. L_{min} are the minimum L values at which the respective discharge mode could be excited under the conditions described.

By differentiating equation (4.5), we can find the plasma conductivity from the measured dependence $R(L)$ with the expression

$$\sigma = \left(\frac{\mathrm{d}R}{\mathrm{d}L}\right)^{-1} \tag{4.6}$$

Representative measurements made for totally covered electrodes, that is, at $S = \pi r_{el}^2 = $ const, are shown in Figure 4.12, where the curve slopes characterize the conductivity. An experimental error may arise from the gas temperature variation with increasing interelectrode spacing because of poor heat removal. The experimental technique can be improved by cooling the discharge with gas flow, creating conditions for heat removal to the side walls or by making allowance for the temperature effect in the heat conduction problem.

The curves in Figure 4.12 end at certain minimum distances between the electrodes, $L_{min\,\alpha}$ and $L_{min\,\gamma}$. These distances are close to the α- and γ-sheath thicknesses, and the respective resistances can be considered as ohmic sheath resistances. It is essential that $R_{s\,\alpha}$ is nearly by two orders of magnitude larger than $R_{s\,\gamma}$. On the other hand, $R_{s\,\gamma}$ is rather close to the cathode sheath resistance in a normal dc glow discharge, other conditions (gas, pressure) being equal. This experimental fact was taken as a direct evidence for a close similarity between the γ-mode and the dc glow discharge [4.9]. Moreover, the curve for $R(L)$ for the γ-discharge is a nearly horizontal line with a very slow slope down to $L \simeq 1.5$ cm.

The low plasma resistance is accounted for by the fact that the γ-sheath is followed by the regions of 'negative glow' and 'Faraday dark space' with fairly high conductivities, much higher than that of the positive column. The latter shows a steeper slope of $R(L)$ for the γ-mode at $L \simeq 2$–2.5 cm.

4.3.2 Transverse probing

When a discharge is probed by direct current applied to RF electrodes, the resistance to this current is largely exerted by the electrode sheaths, which possess a lower active conductivity than the plasma. This conclusion can be drawn from a comparison of measured σ_a values for the sheaths and the plasma conductivity σ values that can be obtained from the positive column CVC with respect to the RF current as well as from transverse probing data.

Using transverse dc probing, one should insulate the major (RF) electrodes from the probing direct current. Additional electrodes are introduced into the plasma, and direct current I is passed through them at constant voltage U across these additional electrodes. The plasma exerts resistance $\Omega = U/I$ to the probing current. Unfortunately, plasma conductivity σ cannot be derived directly from this quantity, since the configuration of the current lines between the new electrodes is complicated and poorly understood. It is known, however, that the current lines diverge from the additional electrodes; and at a certain distance from them, they run through the whole plasma cross section $S_\perp$, which is constant and measurable.

In order to eliminate the resistance at the unknown sites with diverging lines in the vicinity of the additional electrodes (including the space charge sheaths around them), one can resort to one of the two ways. First, two measurements can be made with the same electrode pair at two different distances l and l' between them, achieving identical current I in both cases by varying the electrode voltage, the minor electrodes being moved along the major electrode surface. Suppose U and U' are the respective electrode voltages. Subtract, one from the other, the expressions of the type $U = \Omega I$, where Ω is the total plasma resistance to the probing current with account of its spatial distribution. Since the probing currents are identical in both measurements, unknown components, corresponding to the sites with complicated but identical current distributions near the electrodes, are thus eliminated from Ω. We find that

$$U' - U = (\Omega' - \Omega)I = \left(\frac{\Delta l}{\sigma S_\perp}\right)I \qquad \Delta l = l' - l \qquad (4.7)$$

and can now determine σ.

In the other approach, instead of moving the two electrodes, one introduces in the plasma three identical fixed electrodes (1, 2, and 3) separated by different, sufficiently large distances l_{12} and l_{23} such that there are sites between electrodes 1–2 and 2–3 where the current is known to pass through all known plasma cross sections $S_\perp$. Designate the lengths of these sites as l'_{12} and l'_{23}. If we manage to get identical probing current I between the electrode pairs 1–2 and 2–3, then,

TABLE 4.1
Comparison of two RF discharge modes in air at $p = 30$ Torr and in helium at
$p = 200$ Torr, $f = 13.56$ MHz, $L = 1$ cm

Discharge		j	j_a	j_{dis}	$\overline{E(0)}$	d_s	Z_s^{-1}	σ_a/d_s
mode	Gas	(mA cm^{-2})			(kV/cm)	(cm)	(Ω^{-1} cm^{-2})	
α, normal	air	7	0.4	6.9	0.92	0.23	2×10^{-5}	1.7×10^{-6}
	He	6.01	0.3	6	0.79	0.11	5×10^{-5}	3.4×10^{-6}
α, before	air	15	2	14.9	1.9	0.20		
$\alpha - \gamma$ trans.	He	19	2.3	18.9	2.5	0.10		
γ, normal	air	300	140	260	34	0.008	12×10^{-4}	6.2×10^{-4}
	He	300	150	255	33	0.006	1.6×10^{-3}	1.1×10^{-3}

Designations: j, rms density of discharge current; j_a and j_{dis}, active and rms reactive components in the sheaths; $\overline{E(0)}$, constant field in the right at the electrode; d_s, sheath thickness; Z_s^{-1}, reciprocal total impedance per 1 cm^2 area; σ_a/d_s, reciprocal sheath resistance per 1 cm^2 area; σ_a, used in the sense of active sheath conductivity.

evidently, $l'_{12} - l'_{23} = l_{12} - l_{23} = \Delta l$. If Ω_{div} is the resistance of sites with diverging current lines at each electrode, we can write that

$$U_{12} = 2\Omega_{div} I + \frac{l'_{12} I}{\sigma S_\perp} \qquad U_{23} = 2\Omega_{div} I + \frac{l'_{23} I}{\sigma S_\perp} \tag{4.8}$$

By subtracting the equalities from one another, we obtain a formula

$$U_{12} - U_{23} = \frac{\Delta l I}{\sigma S_\perp} \tag{4.9}$$

similar to (4.7) from which σ can be found using all measured values.

Measurements have shown that the plasma conductivity of an α-mode determined by transverse probing exceeds, by two or three orders of magnitude, that found by longitudinal probing and corresponding to active sheath conductance near the RF electrodes. This is an independent evidence for the existence of weakly conducting sheaths in moderate-pressure α-discharge.

To conclude this section, master Table 4.1 is presented to illustrate the measured characteristics of the two moderate-pressure discharge modes.

4.4 A method for studying the transverse discharge structure

So far, we have been concerned with the longitudinal discharge structure including electrode sheaths separated by discharge plasma. A discharge column, however,

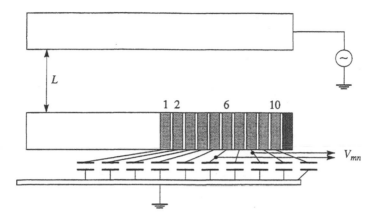

FIGURE 4.13
Experimental design with a sectioned electrode made up of insulated metallic rings for measurement of radial distributions of the constant potential component.

has limited transverse dimensions. If a discharge does not cover all electrode surfaces and operates in a normal mode—which always happens when the current is not very strong—then there is a periphery where the discharge plasma or the ionized gas of the electrode sheaths come in contact with a weakly ionized gas followed by an unionized gas. In the unionized gas between the electrode peripheries, there is also a field induced by the electrode potential difference; but it is clear that the longitudinal potential distribution on the periphery and in the adjacent zero-current zone would differ from the distribution $\varphi(x)$ along the discharge axis. Therefore, the discharge has not only a longitudinal but also a transverse structure. There is little experimental information on the transverse (radial) structure, and the discharge is considered, quite reasonably, as being one-dimensional, when the column diameter greatly exceeds the interelectrode distance. Still, the parameters of a radial discharge structure are also of interest.

Here we describe an experimental technique for the study of RF discharge radial behavior. This technique has the advantage of leaving the discharge and the plasma unperturbed. It is shown schematically in Figure 4.13.

One of the electrodes is a solid metallic disk, and the other represents a set of contacting but insulated metallic rings. For the RF component of the potential to be identical for all the rings and for the constant component to establish spontaneously with the discharge radial structure, large capacitances C are connected between the rings and the ground so that the capacitor resistances $(\omega C)^{-1}$ are very small. The RF potential component can then be assumed to be close to zero in all the rings. RF voltage is applied to the solid electrode. One can measure the constant potential difference between any rings, or the potential difference U_{mn} between a ring of radius r_m and the last ring r_n, which is the largest and is actually

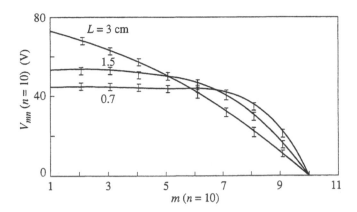

FIGURE 4.14

Measured radial distributions of the constant potential component across a sectioned electrode made up of 10 rings. Discharge in air, $f = 13.56$ MHz, $p = 7.5$ Torr. The curves are for various L values.

located outside the discharge. This procedure provides the radial distribution of the constant potential component. It has been shown experimentally that a discharge with sectioned electrodes and large individual blocking capacitances does not differ in general characteristics from a solid electrode discharge.

Typical measurements for α-discharges filling up all sections are presented in Figure 4.14. Remarkable is the fact that the discharge in a short gap filled mostly by the sheaths with nearly no room for a positive column is quite uniform radially at the center, with the uniformity being perturbed only on the periphery. In case of a sufficiently long gap with a plasma in its middle, the current spot on the electrode is nonuniform in the direction normal to the current. Of course, the radial potential distribution measurements should be supplemented by measurements of the RF current density distribution by registering the RF current through each section. In any case, the co-existence of α- and γ-modes in the same gap, no doubt, evidences for a radial nonuniformity of the discharge. It is possible, for instance, that the α-sheath voltage near the column axis exceeds breakdown voltage, while on the periphery it does not. This situation is also stimulated by greater gas heating around the axis, promoting current flow in this region.

Transverse nonuniformity also produces multichannel discharge structures. Experiments show that, as the current increases in a heavy inert gas α-discharge, one high current density filament arises at first, then another, and so on. It has been found that the discharge CVC in such cases has a nonmonotonic character; the appearance of each plasma filament is accompanied by an abrupt decrease in the RF electrode voltage. Then the voltage grows with current until the next jump occurs due to the appearance of another filament (Figure 4.15). Sometimes, plasma filaments in the gap become well ordered. Similar effects have been observed for

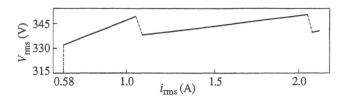

FIGURE 4.15
Abrupt changes in the discharge CVC accompanying the appearance of a plasma filament in a mixture of air and Xe, 1:6; $p = 35$ Torr.

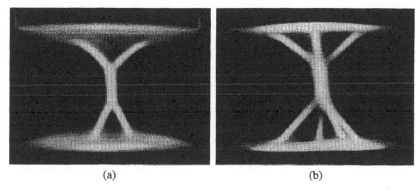

(a)   (b)

FIGURE 4.16
Branching of the plasma filament in a γ-discharge near the electrodes, $f = 13.56$ MHz; air: Xe $= 1{:}6$, $p = 35$ Torr; (a) for $i = 2.8$ A, (b) for $i = 5.7$ A.

a γ-discharge. Phenomena associated with transverse discharge nonuniformity can all be studied using the sectioned electrode technique and its modifications. Structural nonuniformity sometimes transforms a γ-discharge into an arc mode, even with partly current-filed electrodes, and produces branching of the filament near the sheaths (Figure 4.16). Other effects may also occur.

We have silently accepted so far that the surface of a solid metallic electrode is equipotential. This is true, provided that the electrode diameter is much smaller than the wavelength corresponding to RF frequency; for instance, for $f = 13.56$ MHz the wavelength $\lambda = 22$ m is indeed fairly large. But at higher frequencies, $f \approx 100$ MHz with $\lambda \approx 3$ m; and in the standing wave mode on an electrode with a characteristic size of a few dozen centimeters, as well as in laser technologies, there is the problem of nonuniform distribution of RF voltage $V(y)$ along the electrode. There are various methods available for leveling such nonuniformities, in particular, by connecting inductance coils at various points of the interelectrode space, by maintaining the running wave mode of a steady RF discharge, etc. The above analysis of the spatial discharge structure offers still another approach to leveling plasma parameters under the conditions of nonuniform $V(y)$. It is necessary that the dielectric thickness $\delta(y)$ depend on the y-coordinate along the electrode and be chosen such that the discharge current density with the given distribution $V(y)$ be independent of y. In practice, this can be implemented as follows. The dielectric coating surface that faces the discharge chamber interior is made plane, while that in contact with the electrode is shaped according to $V(y)$; the electrode surface must be shaped exactly as the coating. Then, in spite of the nonuniform RF voltage distribution along the electrodes, the fall of the discharge voltage including the sheath and the plasma voltages will be the same along the whole electrode.

4.5   Optical methods for the study of spatial discharge structure

Discharge photography and intensity distribution measurements of intrinsic optical radiation are the most common and accessible ways of studying the spatial discharge structure. The basic concept underlying these approaches is that an optical radiation source at a given point in space is the stronger, the higher is the electron density at the point and the higher is the field that imparts energy to the electrons, increasing their temperature and the radiative ability of the gas. By looking at dark and light spots in a photograph, one can unmistakably identify those where the influence of both factors (n_e and E) is stronger. A mere comparison of the two photographs in Figure 1.11 showing, seemingly one and the same discharge (since the RF voltage in the two cases differ but slightly), can help reveal their structural differences. These visible distinctions acquire a quantitative character from comparison of the respective glow intensity distributions along the gap (Figure 1.12). Such observations alone may give impetus to speculations about a possible existence of two RF discharge modes—an idea we have referred to repeatedly throughout this book. This idea is also supported by the photograph in Figure 2.30 showing the coexistence of both modes in the same gap, one mode being quite unusual, because the sheath thickness is larger than in a standard γ-mode in Figure 1.11 (this is a dielectric-stabilized 'subnormal' γ-discharge; see Section 2.9).

Why electrode sheaths always look like dark regions in photographs is quite understandable. In the cathodic phase, sheaths have almost no electrons, though the field is very strong; in the anodic phase there are many electrons, but the field is weak and this phase is quite short. Differences in the mechanisms underlying the dynamics of sheaths and adjacent plasma regions have also been demonstrated by spectroscopic studies. The spectral structure of bright regions adjacent to dark γ-sheaths coincides with that of the negative glow region near the cathode sheath in a dc glow discharge. On the contrary, the spectrum of the bright glow adjacent to dark α-sheaths is similar to that of a positive column. In air, the former is white blue, the latter reddish.

Some information on discharge behavior, in particular on its spatial structure, can be obtained from time-resolved sheath and plasma radiation. These approaches have long been used in RF discharge experiments [4.5, 4.11–4.13]. For instance, it has been shown that the radiation intensity of a plasma column is modulated with a double frequency. This is natural because the oscillating field amplitude reaches a maximum twice over a cycle, and at these moments the field imparts most of its energy to the electrons (direction of the field is, clearly, of no importance). In contrast to this situation, the sheath glow, although weak, pulsates at frequency ω. Double frequency modulation has an amplitude by two orders smaller [4.11]. This is also clear since the sheath thickness and the field reach their maximum values only once over a cycle. Besides, it was pointed out that a glow might appear at one and the same spot twice over a cycle (in the α-mode) or once (in the γ-mode) [4.12].

At low pressures, glow pulsation with frequency ω may result from beams of fast electrons emitted by electrodes and accelerated by a strong field. This may happen even when the Townsend criterion for the sheath is not fulfilled and there is no breakdown in it. At moderate pressures, application of optical methods with time and space resolution can provide unambiguous identification of α- and γ-modes, especially when glow oscillograms are synchronized with oscillograms for the plasma current and field [4.13]. Optical methods have proved very useful in studying the effects of dielectric coatings and their thickness on the stabilization of γ-'subnormal' states (see Section 2.9). They were also used to observe variation in the γ-sheath thickness as a function of the dielectric thickness. Moreover, it was established by optical studies that in a 'subnormal' mode, the spectrum of bright regions adjacent to dark sheaths is similar to the negative glow spectrum in a dc glow discharge [4.14]. Naturally, conventional plasma diagnostic techniques can be used along with optical and other methods mentioned above.

4.6 Laser-induced fluorescence and laser-optogalvanic spectroscopy

In recent years, novel methods of laser spectroscopy, namely laser-induced fluorescence (LIF) and laser optogalvanic (LOG) spectroscopy have been introduced. In both cases, tunable dye lasers excited by pulsed excimer or nitrogen lasers are used. An extensive list of publications on LIF and LOG methods can be found in a review [4.15]. Techniques based on LIF and LOG spectroscopy provide more detailed and reliable information on the plasma under study than the classical methods of optical emission spectroscopy. They are based on registration of radiation (fluorescence) of resonance-excited atoms and ions induced by a tunable laser. This process can be considered as radiation scattering at frequencies close to the resonance frequency for one of the atomic transitions.

LIF is much more effective than most optical emission techniques. Its important advantage is that the experimentalist can stimulate fluorescence of a certain plasma component in a given discharge region at a given moment by regulating the laser radiation, that is, by choosing the necessary wavelength and intensity of the exciting photon flux. LIF provides information on local plasma parameters, because observations are made in the direction normal to the probing beam with a high spatial and time resolution (10^{-4} cm and 10^{-7} s). Fluorescence becomes saturated at a high probing radiation intensity. If fluorescence constants are known, one can find the concentrations of fluorescent components. But even if one does not know the constants, LIF permits determination of relative concentration distributions of the plasma components and their identification.

The LIF technique was developed in [4.16–4.18] and combined with measurement of Stark fluorescence line broadening to register local electric field. In Section 3.10, we described the application of LIF and Stark effect of radiation from BCl radicals produced by BCl_3 molecular dissociation in order to measure instantaneous electric field distribution in electrode sheaths of RF capacitive discharges. These measurements were effective only in the sheaths where the field was relatively high, the minimum field registered being 75 V cm^{-1}.

A technique similar to LIF and based on Rydberg atom spectroscopy, in which atoms are excited by a probing laser radiation, permits registration of lower fields, down to 10 ± 1 V cm^{-1} [4.19, 4.20]. A Rydberg atom is an atom with an electron excited to a high energy level characterized by the principal quantum number n. With growing n, the atom acquires properties similar to a hydrogen atom. Radiation at the transition from a Rydberg state to the ground state makes Stark effect measurements of the field strength more sensitive.

The LOG method [4.21] is based on the effect of a probing laser pulse on ions. Laser-excited ions change their mobility, and this leads to current perturbations to be registered. Generally speaking, a LOG signal can also be affected by a change in the charge number balance due to a change of the diffusion or recombination coefficient of excited ions. But these effects appear to be minor as compared to the mobility variation, since the lifetime of excited states is much less than the characteristic times of diffusion and recombination. On the other hand, the lifetime of an excited state is comparable with the time between elastic collisions of ions and charge exchange events, so that the variation in the ion mobility and drift velocity in the field may be quite appreciable. With the known ion mobility change (increase or decrease), the current perturbation sign depends on the sign or direction of the electric field, opening a possibility for experimental registration of field reversal. The latter occurs in the negative glow region and the Faraday dark space in a dc glow discharge, as well as in an RF γ-discharge, if we have in mind the average constant field (for this effect in a moderate-pressure γ-discharge, see Section 2.5). LOG-spectroscopy was used in [4.18, 4.22, 4.23] to register negative ion density distributions in RF discharges operating in electronegative gases. Probing laser radiation initiated local photodetachment of negative ions, leading to the current perturbations registered in the experiments.

Application of LIF- and LOG-spectroscopy has contributed to our understanding of the longitudinal structure of cathode regions in a dc glow discharge [4.24–4.26]. Distributions of ions and metastable atoms were obtained, and the effect of electric field reversal in the negative glow region and the Faraday dark space was registered. The LOG-signal greatly exceeded the noise level. Mixtures of Ar and N_2 were studied at a pressure, say, 0.3 Torr and a gap of 3 cm when there is no room for a positive column. Probing radiation was used to excite molecular ions N_2^+ from the ground to a short-living state $N_2^+ (B^2 \Sigma_u^+)$ possessing a higher mobility. Registration of a LOG-signal, which indicated the field reversal, was accompanied by registration of LIF in the transition of excited molecules from this state to the ground state. The laser radiation parameters were such that the number of excited ions $N_2^+ (B^2 \Sigma_u^+)$ was saturated, which permitted determination of the density distribution of N_2^+ ions.

4.7 Excitation and control of an RF discharge

Papers on RF discharges do not normally describe the excitation procedures, RF generators, ways of matching the generator or the discharge load; nor is there information on how the experimentalist controls the discharge and obtains desirable parameters. In schematic diagrams, the generator or the connection point of a supply are simply marked with a conventional symbol. This seems to imply that the experimentalist already has all the necessary skills, while a beginner acquires them through his own experience and contacts with experienced researchers. As for theoretical physicists, they can do nothing better than go on being ignorant or extract some information from their experimenting friends. This section is intended for those readers who do want to get a general idea of the RF discharge experiment and are ready to take some radio-engineering concepts for granted.

4.7.1 Self-excited tube generators

Excitation and steady-state operation of an RF capacitive discharge (especially in the moderate-pressure range) with the active current component density up to 10^2 A cm^{-2} and discharge volumes up to 10^3 cm^3 require powerful RF sources from several units to a hundred kilowatts. For this purpose, self-excited vacuum-tube generators are commonly used. The frequency and amplitude of RF oscillations in this case are mainly determined by the generator performance characteristics.

The majority of available modifications of the tube generator can be reduced to a general diagram shown in Figure 4.17, which is commonly known as a three-point design, according to the major tube electrodes—a cathode, an anode, and a control grid—to which the circuit elements are connected (hereafter the electrodes

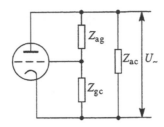

FIGURE 4.17
Generalized three-point generator circuit.

are designated with the subscripts c, a and g). The design in Figure 4.17 shows only the alternating current and voltage components, so there is no dc source of the anode. The Z impedances (Z_{ac}, Z_{gc}, Z_{ag}) are generalized complex resistances $Z = R + iX$, with their active R components being usually very small compared to the reactive X components. So, we can assume with certainty that $Z \approx X$. The oscillation frequency is determined by the natural frequency of the circuit formed by three reactive resistances X, that is by the condition

$$X_{ac} + X_{gc} + X_{ag} = 0 \qquad (4.10)$$

in which the resistances with inductive and capacitive, X_L and X_C, have opposite signs.

To induce oscillations, this system must possess a positive feedback. This means that the grid potential relative to the cathode must have a maximum negative value when the anode potential relative to the cathode has a maximum positive value. According to the directions of the circuit current flow in the elements Z_{ac} and Z_{gc} in Figure 4.17, two resistances, X_{ac} and X_{gc}, must have identical signs. For the condition of (4.10) to be satisfied, the third resistance X_{ag} must have an opposite sign. These requirements can be met in three ways, hence there are three basic generator designs. As a rule, for frequencies $f \gtrsim 1$ MHz, one of the three X elements is an inter-electrode capacitance of the vacuum tube. The three designs differ in which of the electrodes is connected to the other two external elements, and this electrode is grounded.

When two external elements, namely the inductive resistances X_{gc} and X_{ac}, are connected to the cathode, and the internal element is the capacitance X_{ag} between the grid and the anode, this is a design with a common, grounded cathode [Figure 4.18(a)]. Such designs are usually used in self-excited generators with an operating frequency up to 13.6 MHz. The design with a common grounded anode [Figure 4.18(b)]—where X_{gc} is the internal capacitance between the grid and the cathode, X_{ac} is the capacitive resistance and X_{ag} is the inductive resistance—is employed in the frequency range $f \approx 13-150$ MHz. The common grid design [Figure 4.18(c)] is used for frequencies over 150 MHz, but this is a less popular

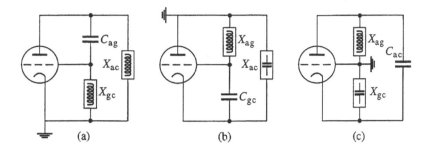

FIGURE 4.18
Versions of a self-excited three-point generator used in discharge practice: (a) with a common cathode, (b) with a common anode, and (c) with a common grid; C_{ag}, C_{gc} and C_{ac}, tube capacitances between respective electrodes. Inductive and capacitive reactive resistances are shown.

modification in discharge practice than the other two.

The RF generator yields the largest useful power and possesses maximum efficiency, if the equivalent resistance of the load R_{el} coincides with the internal tube resistance

$$R^* = \frac{\xi E_a^0}{\alpha_1 i_{a\,max}}$$

Here E_a^0 is constant voltage of the anode source; $\xi = U/E_a^0$ is the anode voltage utilization coefficient; where U is the anode oscillation voltage, $i_{a\,max}$ is maximum current through the anode; and α_1 is so-called Berg's coefficient, which varies with the anode current cutoff angle. The details can be found in handbooks on self-excited generator calculations.[1] If $R_{el} = R^*$, the generator has a normal load. The mode with $R_{el} > R^*$ is overvoltaged, and the generator is underloaded. If $R_{el} < R^*$, the mode is undervoltaged and the generator is overloaded. In either case, the useful power and the efficiency are less than in the normal operation mode. However, underloading does not affect much the generator performance, while overloading is undesirable because it leads to an increase in the anode current and losses. Generally, loading ranges from $0.75R^*$ to $1.3R^*$.

4.7.2 Real RF-generator design for discharge studies

In a simple modification, a discharge chamber, which serves for RF current as a capacitor formed by electrodes, can, in principle, be connected directly to the anode circuit, say to the element Z_{ac} in Figure 4.17. In fact, this variant is

---

[1] The physics of maximum power is similar to that (though much simpler) of a dc circuit. If $\mathcal{E}$ is the source emf, R^* is its internal resistance and R_l is the resistance in the external circuit, then the direct current is equal to $\mathcal{I} = \mathcal{E}/(R^* + R_l)$. The load releases the power $P = \mathcal{I}^2 R_l = \mathcal{E}^2 R_l/(R^* + R_l)^2$. The function $P(R_l)$ has a peak at $R_l = R^*$.

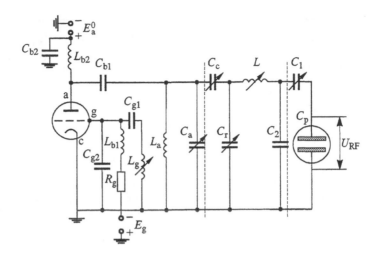

FIGURE 4.19
Real self-excited RF generator with a matching element and discharge load,
$f = 13.6$ MHz. Vertical dashed lines separate the generator, matching element and load.
See text.

unacceptable due to incompatibility of the requirements on frequency stability
and approximate constancy of the load equivalent resistance. It is clear that the
capacitor capacity should change due to discharge excitation, and a new element—
plasma—appears in the circuit, which possesses a resistance. When the discharge
operation mode changes, both capacitance and ohmic resistance also change,
affecting the resonance frequency and the load equivalent resistance. For this
reason, the self-excited generator and the load are coupled indirectly through a
matching element, say, a π-circuit.

A schematic diagram of a 3 kW generator used by one of the authors of this
book to study 13.56 MHz moderate-pressure discharges (see above) is shown in
Figure 4.19. The generator is a common cathode modification based on a three-
point principle (Figure 4.18). The tube internal capacitance between the anode and
the grid served as one of its elements (not shown in the diagram). The element X_{gc}
was a contour made up of inductance L_g and capacitances C_{g1} and C_{g2} possessing
a total inductance response. The element X_{ac} was an untuned contour (L_a, C_a),
also with inductive resistance. These elements comprise the self-oscillating part
of the scheme shown on the left of the left-hand vertical dashed line in Figure 4.19.
The blocking capacitance C_{b1} on the right of the anode blocks the way to direct
current excited by the anode source. The blocking capacitance and inductance,
C_{b2} and L_{b2}, in the power supply section E_a^0 prevent RF current from flowing to
the source. Bias E_g can be applied to the grid through the blocking inductance
L_{b1}, and the resistor R_g.

The π-matching element, formed by variable regulating capacitance C_r and inductance L as well as fixed capacitance C_2, is coupled to the generator through variable coupling capacitance C_c. The load is positioned on the right of the right-hand vertical dashed line. It is formed by capacitance C_p between the electrodes in the absence of discharge, or by the impedance of all resistances inherent to an interelectrode discharge, as well as by variable capacitor C_1 connected in series to the discharge chamber. The latter is necessary to control the load, compensating for the working capacitance variation during the excitation and various changes in the discharge operation. Capacitance C_1 represents an air gap between one of the working capacitor plates and an auxiliary plate, which can be moved to vary the air gap size and, hence, C_1.

One can easily see that by varying C_1 it would be impossible to simultaneously maintain a constant frequency and equivalent resistance of the oscillation contour, if the contour with a certain inductance L, series capacitances C_1 and C_p and with active plasma resistance R_p were directly connected to the anode circuit of the generation. Indeed, the intrinsic frequency of the contour $\omega = \sqrt{LC_e}$, where $C_e^{-1} = C_1^{-1} + C_p^{-1}$, would then determine the generation frequency and the load equivalent resistance would be $R_{el} = L/C_eR_p = \rho^2/R_p$, where $\rho = 1/\omega C_e$. Suppose C_p and R_p have increased owing to the discharge excitation. For R_e to remain constant, ρ should be increased; but for this, C_1 should be decreased by making the air gap wider. But at a constant contour inductance L, any variation in C_e would lead to changing frequency. This is the reason why multicontour designs of the type shown in Figure 4.19 are used to maintain a constant generation frequency.

Regulation of RF voltage U_{RF} across the working capacitor electrodes—or, to be exact, variation of the oscillatory power applied to the discharge—is performed by changing the capacitance of the regulating capacitor C_r in the matching element. Radical alterations in the whole system operation, for instance, replacement of the gas-discharge chamber or electrodes, require a proper choice of other elements of the matching system, too: inductance L, capacitance C_2 as well as capacitance C_1.

It is sometimes reasonable to use resonance circuit properties and include the discharge gap into a parallel contour, to which RF power is applied. It may also be necessary to connect a series ballast impedance to stabilize the discharge. The stabilizing action of the ballast impedance will be effective if its value exceeds the discharge gap impedance. To save the power, the ballast should be made reactive, and this is one of the advantages of an RF discharge over a direct current discharge. But even if the ballast element is reactive (for instance, a capacitor), all the same higher voltage must be applied to the ballast and the discharge chamber connected in series. The power dissipated in the discharge appears to be appreciably smaller than that applied to the RF-generator output. To avoid this, the discharge gap together with the ballast impedance could be connected to the parallel contour and tuned in resonance with the RF generator frequency. If Q is the quality factor of the contour, the current in each branch will exceed, by a factor of Q, the current applied to the contour. Such a circuit can be supplied by a generator of Q-times lower oscillation power than with a series impedance connected to the generator output circuit, as described above.

Thus, the generator circuit of a three-point design is separated from the discharge load. A load is connected to the generator through a coupling capacitance C_c and a π-contour. Owing to the relatively small value of C_c and, hence, to its large reactive resistance, changes in the load affect but little the operating generator frequency given by equation (4.10), i.e., largely by the parameters of the three-point scheme on the left of the left-hand dashed line in Figure 4.19. The matching element serves to make equivalent load resistance consistent with the generator resistance. By doing so, one achieves an approximate condition of $R_{el} \approx R^*$ by choosing proper parameters of the π-circuit and regulating them during the work with the discharge.

4.7.3 Gap impedance change with discharge excitation

The problem of parameter matching becomes acute when load characteristics change on discharge excitation. This is especially so for pulsed or periodic-pulsed operation modes, since the conditions necessary for a gap breakdown to occur and for maximum power contribution to the plasma are essentially different because of a considerable increase in capacitance and active conductivity of the gap after a plasma has been produced. The fact of gap impedance change on discharge excitation seems quite obvious. It was observed experimentally by Townsend in 1929 and studied recently [4.27, 4.28], but its interpretation suggested in [4.28] appears questionable.

It was pointed out in the latter work that, according to realistic RF discharge simulation data [4.27], the appearance of a plasma in the interelectrode gap contributes, to the supply circuit, supplemental capacitance ΔC, whose value depends on frequency f and applied power, i.e., on the discharge current and operation mode. The general conclusion that the value ΔC seems to increase the empty gap capacitance C_0 is correct; but this, in the authors' opinion [4.28], indicates a parallel connection between ΔC and C_0.

However, the real physical reason for the gap capacitance increase, as was shown in Sections 1.5.4 and 1.5.5, are the introduction of a plasma conductor in the gap middle and the sharp reduction of the distance between the 'electrodes,' which are now represented by the actual electrodes and the conducting plasma between them. Prior to breakdown, the gap capacitance was determined by the distance between the actual electrodes L, but now it is determined by the total sheath thickness $2\bar{d} \approx d_{max}$. In other words, instead of one small capacitance C_0, we now deal with two large variable series capacitances of the sheaths. Their equivalent capacitance is, in a good approximation, constant in time (in a symmetric discharge) and corresponds to the maximum thickness of one sheath

d_{max}. We have $C_0 = S/4\pi L$ and

$$\Delta C = \frac{S}{4\pi d_{\text{max}}} - \frac{S}{4\pi L} \qquad \frac{\Delta C_0}{C_0} \approx \frac{L}{d_{\text{max}}}$$

with the plasma capacitance neglected.

Thus, a circuit with a series connection of a new capacitance (Figure 1.7) is more advantageous and physically valid than a parallel connection of the previous and new capacitances, as is believed in [4.28].

4.7.4 Gap voltage control by varying output contour regulating capacitance

Providing an optimal operation mode of an RF generator and regulating RF voltage applied to the discharge chamber can be implemented more readily by a proper choice and regulation of the output contour elements than by controlling RF power through its supply circuits. However, in this case some ambiguities arise associated with the shape of the resonance curve of the working contour. Consider experimental measurements of RF electrode voltage U_{RF} as a function of regulating capacitance C_r [4.1]. The measurements were made for both steady-state discharge and discharge-free conditions. The latter data provide the intrinsic resonance characteristic for the output contour. Its quality factor Q can be varied by varying the π-contour parameters L and C_2.

Under discharge-free conditions when the resistance r of the output contour losses is constant, the quality factor is

$$Q = \frac{\sqrt{L/C_\Sigma}}{r} \qquad C_\Sigma = \frac{C_r(C_0 + C_2)}{C_r + C_0 + C_2}$$

where C_Σ is the total equivalent capacitance of the output contour, provided the air gap in the capacitor C_1 is made very small and the capacitance C_1 is much larger than the empty gap capacitance C_0. By decreasing L_p with a simultaneous increase of C_Σ through increasing C_2 ($C_2 < C_r$), one can reduce Q of the output contour in the absence of discharge, retaining its resonance frequency $\omega_r = (LC_\Sigma)^{-1/2}$ constant. This is very essential since we can then have a smoothly descending slope for U_{RF} (near the resonance) against regulating capacitance C_r.

The discharge gap shunting by a large extra capacitance $C_2 > C_p$, exceeding the growing capacitance C_p between the working electrodes after the plasma has appeared, leads to a lesser detuning of the output contour (discharge excitation is accompanied by the gap capacitance growth by a value $\Delta C = C_p - C_0$, see Section 4.7.3). The upper value of C_2 is, however, limited by the danger of excessive reduction of the U_{RF} variation range by changing C_r too much.

Figure 4.20 presents measured dependences of U_{RF} for two C_2 values, 50 and 200 pF, at atmospheric pressure in the chamber when no discharge can arise. The chamber was a quartz tube of 8 cm in diameter and 90 cm long. Two RF electrodes in the form of duralumin plates of $15 \times 60 \text{ cm}^2$ each were mounted outside the

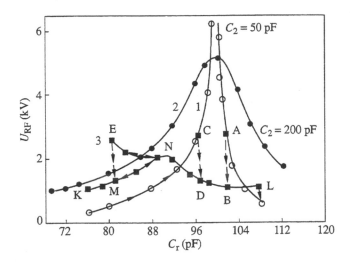

FIGURE 4.20
Experimental electrode voltage U_{RF} plotted against regulated capacitance C_r. Curve 1, no discharge, additional capacitance $C_2 = 50$ pF; curve 2, no discharge, $C_2 = 20$ pF; complex curve 3 with capitals, with discharge. See text.

tube along its longitudinal axis. The $U_{RF}(C_r)$ curves have an asymmetric bell-like shape, characteristic of resonance curves. Resonance curve 2 corresponding to the larger capacitance C_2 is broader and lower than curve 1. This indicates a poorer Q-factor of the output circuit including the discharge-free chamber. Generally, the growth of C_2 and C_Σ should be accompanied by a decrease in the induction L for the plasma resonance frequency to remain constant. In practice, however, L can be left unchanged. A small frequency variation within 1% is considered permissible.

Complex curve 3 KMENDBL with a two-valued fragment in Figure 4.20 was taken at air pressure 10 Torr and $C_2 = 50$ pF in the same chamber as for steeper resonance curve 1 (no discharge). Points A and B indicate a breakdown and the appearance of a discharge. At points K and L the discharge extinguished. The appearance of a discharge and the presence of plasma load in the chamber brings about some specific features in the behavior of the $U_{RF}(C_r)$ curves, indicating complicated processes in the discharge.

Let us take $C_2 = 50$ pF, evacuate the chamber down to 10 Torr and decrease C_r, starting from a maximum possible value: about 108 pF. At first, the RF electrode voltage rises along resonance curve 1, corresponding to the discharge-free condition. At point A, the voltage reaches a breakdown threshold value, and a discharge is initiated at some place in the chamber. RF voltage U_{RF} drops to a value corresponding to point B. Further decrease in C_r makes the discharge space expand along the tube, and its luminosity increases. These changes signal the expansion of a normal α-mode. Starting approximately from point D, the α-discharge fills up the chamber and becomes anomalous. As C_r decreases, the voltage grows; and in the vicinity of point N the $\alpha-\gamma$ transition occurs, bringing to life a γ-mode. The transition is not accompanied by a voltage drop, because the electrodes are separated from the discharge by a thick stabilizing layer of dielectric material of the tube.

At point E, a jump-like change in the discharge characteristic occurs: the voltage drops to a value of point M, the luminosity becomes weaker, and the discharge space along the tube shrinks. This is a reverse $\gamma-\alpha$ transition. If one continues to decrease C_r, the α-discharge will extinguish at point K. But if C_r is increased after the change at EM, the discharge, being an α-mode, approaches the state N on the lower branch of the $U_{RF}(C_r)$ curve along the section MN.

If C_r is increased starting from a minimum possible value—which is 76 pF, defined by the capacitor design—U_{RF} will rise monotonically along the resonance curve for the discharge-free condition, up to point C on the left-hand branch. A breakdown occurs at point C, and the voltage drops to a value of point D (an anomalous α-discharge is initiated).

As C_r is increased further, U_{RF} decreases down to point L, where the discharge extinguishes. At the section BL, there is a normal α-mode.

One of the reasons for an abrupt drop of U_{RF} after the discharge initiation is the increasing capacitance in the gap from C_0 (no discharge) to C_p (with plasma). At the moment of breakdown, for example at point C, the gap capacitance rises so much that this signals the transition from the inductive resonance curve branch to the capacitive one. Therefore, further rise of C_r can only decrease U_{RF}, which is a fact supported by experimental observations. A change in the opposite direction, that is, a decrease of C_r after the breakdown, leads, on the contrary, to resonance recovery, providing higher U_{RF} across the electrodes.

Without knowledge of the two discharge modes, one can hardly explain satisfactorily the two-valued fragments EN and MN in the $U_{RF}(C_r)$ curve. There have been attempts to account for this feature invoking the concepts of circuiting effects when U_{RF} growth is accompanied by increased dissipated power in the plasma [4.29]. But this interpretation leaves aside the questions of what exactly determines the discharge operation mode and why the effective capacitance is larger in one case than in the other. Besides, the question arises why we cannot get to the section NM with decreasing C_r, if we start from point B. Specific behavior of the $U_{RF}(C_r)$ curve can be understood only if one invokes the concepts of two modes of RF discharge operation and respective transition mechanisms. But then it should be kept in mind that, as C_r and the operating discharge change, the gap capacitance C_p always varies. Each value of C_p produces its own resonance curve. The 'travel' along the curve LNEK or KMNL is accompanied by 'jumping' from one resonance curve to the other. The problem of decoding this travel is not a simple task, since it requires plotting the whole family of resonance curves for various C_p values.

5

Application of RF Capacitive Discharges for Gas Laser Excitation and Plasma Technology

The principal applications of RF capacitive discharges today are excitation of an active medium in gas lasers, primarily in CO_2 lasers, and materials treatment by bombarding targets with ions accelerated in space charge sheaths. The latter has been repeatedly described in review articles and books. The former, however, has received very little attention, though about a quarter of all modern commercial CO_2 lasers of various power are excited by RF discharges. This work is normally done on empirical and engineering levels without a clear understanding of the physics, which impairs progress in laser construction. For this reason, this chapter will be primarily concerned with RF discharge application to laser construction, and only in the last section will we briefly consider materials treatment to help the layman to get a general idea of technological processes implemented by means of RF discharges.

5.1 RF discharge and gas lasers: A brief history

Historically, it was the RF capacitive discharge that was first used in the 1960s to create an active medium in helium-neon [5.1] and CO_2 [5.2] lasers. This was probably due to the following discharge features well known at that time. First, RF discharge plasma does not have enough time to decay over a field oscillation cycle; therefore, one can produce a plasma with a nearly steady density, $n_e(t) \approx$ const. Second, the electric current from the plasma can be closed on the electrodes by displacement currents, RF current densities acceptable for laser generation, $j \gtrsim 10$ mA cm^{-2}, being quite feasible. Consequently, the electrodes can be mounted outside a dielectric discharge chamber, thus simplifying the laser design and excluding a contact between metallic electrodes and an active laser medium. This made the design convenient from all points of view.

Nevertheless, methods for RF excitation of gas lasers could not be developed at that time. The physics of RF discharges remained unclear, and the principal advantages over a dc discharge were not evident, so there were no grounds to consider the RF discharge as a preferable one. The dc glow discharge, on the contrary, was understood much better and possessed, as it still does, the unrefutable merit of requiring no complex power supplies and load coupling as the RF discharge. For these reasons, RF discharges were largely used as a supplementary means of maintaining spatially uniform plasmas with the appropriate laser parameters when a self-sustained dc discharge failed to do so. Powerful laser operation requires the use of large currents, high pressures and current densities; but these requirements decrease the uniformity of a dc glow discharge, which is also liable to contraction (filamentation), reducing the laser generation to zero. A detailed description of these effects and of the measures for controlling contraction in CO_2 lasers as well as the discharge arrangement can be found in [5.3]. We would like to note that contraction usually occurs when the output energy density in the discharge plasma $w = jE$ exceeds a certain critical value. This value is not very large in dc plasmas, only several watts per cubic centimeter. In most cases, this limit is lower than the permissible limit of energy release, at which the gas temperature rises to 250–300°C. This, in principle, excludes generation in a CO_2 laser irrespective of the way of exciting the active medium. In fact, it is the problem of glow discharge stabilization and achievement of the highest possible energy input limit that are the key aspects of the power gas laser construction. In contrast, this limit is achievable in an RF discharge, because if it is designed correctly, its instability occurs at a much greater energy input.

Regarding the history of RF discharge application, note that a combination of an RF discharge transverse to the gas flow and a longitudinal (along the flow) dc discharge in a high flow rate CO_2 laser provided a more than 3-fold power increase in a continuous operation [5.4]. A longitudinal dc discharge excited in a large plane channel at a gas flow velocity of 140 m s^{-1} provided a maximum laser power of 8 kW, while a combination with a transverse RF discharge yielded 27 kW. These are high values even for modern lasers. Serious disadvantages, however, were a large size of the laser setup and a complicated electrode system.

The use of a combined discharge in a coaxial system with diffusional cooling (conduction removal of energy) provided a CO_2 laser power of 210 W m^{-1} (per unit resonator length) [5.5] against the commonly obtained 50 W m^{-1} in tube dc-excited lasers. At zero RF field (normal to the cylinder axis), the laser generation ceased completely, because the longitudinal (along the cylinder axis) dc discharge could not provide a uniform filling of the long annular gap between the cooled coaxial cylinders of 2.85 and 4.65 cm in diameter, which limited the discharge space. At the same time, the use of a transverse RF discharge alone did not give a satisfactory result, either [5.5]. The advantage of the coaxial geometry for a laser with diffusional cooling had been observed earlier [5.6]. An attempt to create a steady tube laser with only transverse RF excitation and electrodes arranged externally along the tube was unsuccessful [5.7]. The maximum unsaturated gain obtained in that work did not exceed $k_0 \approx 0.1 \text{ m}^{-1}$ at partial CO_2 pressure

$p_{CO_2} = 0.1$ Torr. At higher CO_2 pressures, the probing signal was absorbed instead of being enhanced, which was indicative of inversion disappearance. But it was shown in [5.8] that under the same external conditions ($f = 13.56$ MHz and the same discharge tube diameter) and at $p_{CO_2} > 5$ Torr, the gain was as high as $k_0 \approx 1$ m^{-1} and in a combined discharge even $k_0 \approx 2$ m^{-1}. This variation was due to the appropriate discharge mode used.

The results of later experiments also seemed, at first glance, ambiguous and controversial. For instance, it was stated in [5.9] that an RF discharge transverse to the resonator optical axis was unsuitable for pumping a CO_2 laser. Other authors [5.10–5.12], on the contrary, presented data indicating a good promise of a transverse RF discharge for pumping CO_2 lasers with diffusional cooling. As a matter of fact, there is no controversy here, because the result depends on the choice of the appropriate operation mode and RF frequency. The difficulties arise from the fact that even under the same external conditions—the gas composition, gap geometry, field frequency, and even electrode voltage—a moderate-pressure RF discharge can operate in two strongly differing modes (which were discussed in the previous chapters). The α- and γ-modes essentially differ in the x-coordinate distribution of basic characteristics that influence the laser effects—the plasma density and the RF field amplitude. It is these differences that cause the various performance characteristics of RF-excited gas lasers [5.12–5.14].

Owing to the purposeful studies and empirical trials of the past decade, the promise of RF discharges for gas laser excitation has become quite evident. Some of these systems or respective investigations have been described in the literature [5.14–5.26], and many of such lasers are currently produced commercially. Among these is a variety of RF-excited CO_2 lasers of low power, from a few tens to a few hundred watts, and there are powerful lasers of one kilowatt and more.

5.2 Arguments in favor of RF laser excitation

One should not stay under the delusion that RF excitation methods are perfect. Their demerits are the complexity and high cost of power supplies, the difficulty of matching the RF generator and the discharge load, as well as the necessity to protect the personnel from harmful RF field by screening the whole setup. These, however, are purely technical problems. As for the physics, RF discharges, no doubt, possess advantages over dc glow discharges. So, when choosing an optimal laser design for a particular application, one is faced with a comprehensive task involving physical, technical and economic aspects. In spite of the above limitations, RF-excited gas lasers find an increasingly wider application, since an appropriate choice of parameters may considerably improve the emitter quality; in particular, to decrease the emitter size and weight, to simplify their design, control and maintenance, to increase their durability and reliability, and to achieve an easier adjustment when changing from one laser operation regime to another.

5.2.1 Feasibility of slab systems with diffusional cooling

Commonly, only 10% of the input energy is transformed into laser radiation energy (record values are as high as 20%). The remaining 80–90% of the dissipated electrical power change into heat contributing to the gas heating. However, a temperature above 550–600 K is damaging to the laser generation, so a primary task in creating CO_2 lasers is to provide effective heat removal. There are two ways of removing heat, according to which all CO_2 lasers are subdivided into two groups: (i) the heat is removed via heat conductivity to the walls of the discharge chamber (so-called diffusional cooling), and (ii) the heat is removed from the chamber by a fast flow of laser mixture through it. The latter way is known as convective cooling, and the respective lasers are known as high flow rate lasers. The first of the two cooling procedures proves to be sufficient and effective for small and moderate power lasers, usually below 1 kW; the second one is commonly used in high power lasers and involves serious technical problems. It is necessary to make a closed contour including a compressor or another gas pump arrangement, coolers, etc. (These issues are briefly discussed in [5.3], but for details the reader should refer to books on gas lasers.) The first way is attractive because of its simplicity and reliability. In many cases, natural cooling turns out to be sufficient. Tube lasers with diffusional cooling and a longitudinal dc glow discharge (Figure 5.1) have found a very wide application. We will discuss the other types of lasers in Section 5.6 and now focus on diffusion-cooled lasers, which clearly show the advantage of RF excitation.

Diffusion-cooled tube systems have some principal limitations associated with the temperature limit $T_{max} \approx 600$ K for laser generation. Indeed, if power jE, or $\langle jE \rangle$ in the RF case, is released in a 1 cm^3 plasma volume per second, and if the tube radius is R and the wall temperature $T_0 \approx 300$ K, then the equation for the

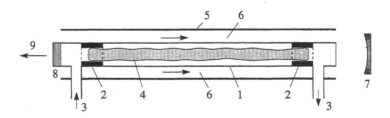

FIGURE 5.1
Diffusion-cooled dc CO_2 laser of medium power: 1, discharge tube; 2, annular electrodes; 3, low flow rate laser mixture; 4, discharge plasma; 5, external tube; 6, cold circulating water; 7, opaque mirror; 8, outlet semitransparent mirror; 9, outgoing radiation.

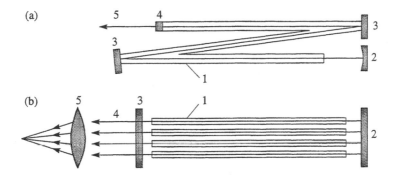

FIGURE 5.2
Multisectional tube lasers for 1 kW power and more. (a) Series connection of the tubes: 1, discharge tubes; 2, opaque mirror; 3, turning mirrors; 4, semitransparent outlet mirror; 5, outgoing beam. (b) Parallel array of tubes in an optical resonator: 1, discharge tubes; 2, opaque mirror; 3, semitransparent outlet mirror; 4, outgoing beams; 5, focusing lens.

gas heat balance approximately is

$$jE\pi R^2 = 2\pi R\lambda \left(\frac{dT}{dr}\right)_{r=R} \approx (2.4)^2 \pi \lambda (T_{max} - T_0) \qquad (5.1)$$

where λ is heat conductivity, and the factor 2.4 is typical of the characteristic diffusion length $\Lambda = R/2.4$ for an infinite cylinder (Section 1.4.3). A value close to this appears in equation (5.1) from an approximate calculation of the temperature radial profile and from the heat flux to the wall entering into equation (5.1) [5.3]. Since the heat conductivities of laser mixtures vary within narrow ranges [at $T \approx 300-600$ K, $\lambda \sim (2-5) \times 10^{-4}$ W cm^{-1} K^{-1}] and $\Delta T = T_{max} - T_0 \approx 250-300°$C, energy $iE = j\pi R^2 E$ larger than a certain limit cannot be put into unit length of the tube; and hence, one cannot obtain from it a large laser power. Normally, these limits are about 1 kW m^{-1} of electrical power and 50–70 W m^{-1} of laser power. These values are independent of the type of discharge mode or its parameters and tube radius. For example, if we increase radius R and longitudinal dc current i, we have at the same time to decrease the pressure and field E ($E/p \approx$ const) to avoid exceeding the limit $(iE)_{max} \sim 1$ kW m^{-1}.

There are two ways of gaining laser power in a tube system. One way is to increase the tube length by connecting several tubes in series to the resonator [Figure 5.2(a)]. The length of a single tube l is limited by a reasonable voltage value of about 10 kV proportional to l. The other way is to use many parallel tubes to a common resonator [Figure 5.2(b)]. Both ways are employed in practice: the first one has become conventional, the second one was patented in [5.27] and used in lasers [5.28, 5.29]. Each design has its own drawbacks; for instance, in the second design the beams from individual tubes are incoherent, decreasing the radiation quality.

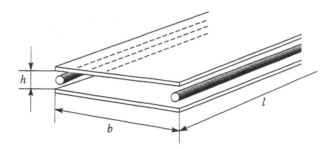

FIGURE 5.3
Experimental design for a longitudinal discharge in a slit gap.

A simple idea may help to surpass the above limit for power per unit length of the discharge volume and of the resonator. This idea was, probably, first stated clearly in [5.8] and is as follows. For effective heat removal to be feasible ($\lambda dT/dr \sim \lambda \Delta T/R$), a small radius tube (small in two directions) is not at all necessary. It would suffice to make a slab, which would be plane, coaxial and small only in one dimension h. The other dimension—the slab width b—can be made as large as necessary, keeping the length l, corresponding to the tube length, constant. In this way, one can essentially increase the laser volume hbl, at the same l and the same density of the heat flux to the walls (if the slab height $h \approx 2R$) and greatly exceed the respective tube volume $\pi R^2 l$. By limiting the power density, one can achieve a much higher total power of the system at the given length l. The gas heat balance equation for the slab geometry (5.6) is written below.

5.2.2 Infeasibility of a dc discharge in the slab geometry

All attempts to fill up a large area slab gap by a dc plasma have failed. As an illustration, we present here experimental data obtained by one of the authors of this book. The discharge chamber was formed by two long dielectric stripes of length l spaced at a distance $h \lesssim 1$ cm (Figure 5.3). Water-cooled copper tubes extending along the length l of the stripe edges served as the electrodes. The gap size coincided with the stripe width $b = 10$ cm. The chamber was filled with the laser mixture $CO_2:N_2:He=1:1:8$ at p from 1 to 30 Torr. At $p > 1$ Torr, the gap was not filled by a uniform plasma in spite of complete filling of the cathode by current (when it was strong enough). Due to contraction, the current passed through one or several plasma filaments, whose transverse size did not exceed the slab height h. In this geometry, even RF excitation at the frequencies 13.56 and 40 MHz failed to fill up the entire space with a uniform plasma (cf. Section 2.5.6). The 'longitudinal' discharge in the gap, with the current along the large dimension b, was ineffective because of the positive column contraction.

In a 'transverse' dc discharge between plane electrodes separated by a small distance $h \lesssim 1$ cm, when the current was directed along the small dimension h, it covered practically the same area of the cathode and of the anode—the areas growing with current strength. As a result, a normal uniform glow discharge was produced. But the voltage fall in the positive column, which was even invisible (in the same mixture 1:1:8 at $p = 15$ Torr and $h = 1$ cm), was very small compared to normal cathode fall $V_n \approx 380$ V. So application of a transverse dc glow discharge in a narrow gap between the electrodes with an effective heat removal is energetically meaningless, because most of the electrical power is released in the cathode layer which contains a small amount of electrons and cannot contribute to the lasing.

5.2.3 Effectiveness of transverse RF discharge in the slab geometry

RF voltage application to large plane electrodes separated by a small distance can perfectly solve the problem of creating a highly efficient laser medium, if one uses knowingly the properties of the discharge. First of all, this should be the α-mode. We will list here the main advantages of the RF α-discharge.

1. There are practically no useless power losses in the reactive electrode sheaths, in contrast with the dc discharge and the RF γ-mode, where the losses are considerable and active ionic current is fairly strong at elevated pressures. So an α-gap is readily filled with quite a uniform plasma.

2. The electron density distribution along the x-coordinate, along the smaller slab dimension, is much more uniform than in tubes with longitudinal dc current, where the radial profile of n_e is determined by the Bessel function $J_0(2.4\,r/R)$ with a rapid decrease towards the periphery. This is profitable for effective use of the laser medium. Profitable for lasers is also the output energy distribution along the x-axis and the current flow as compared to the radial distribution in the case of dc current in the tube. We emphasize this fact because the local excitation intensity of the laser medium is approximately proportional to the released energy density here. The radial distribution $w = jE$ in the tube has the Bessel profile $n_e(r)$, while the function $w(x)$ in the α-mode has a double hump shape with a maximum on the periphery (Figure 2.12 in Section 2.3.3). The peripheral gas layers are also involved in lasing, and the large temperature gradients near the plasma edges stimulate heat removal to the electrodes or their dielectric coatings. This is well supported by measurements of unsaturated gain $k_0(r)$ in a tube of 3 mm radius in the case of longitudinal direct current and transverse RF current [5.30]. In the first situation, $k_0(r)$ decreases towards the periphery like $n_e(r)$ and $w(r)$, while in the second it rises, with the cross section average gain being about 1.5 times larger in the RF excitation.

3. There is a possibility to choose an appropriate plasma electron density by regulating the α-current density, choosing a suitable frequency. The problem of plasma parameter regulation by choosing suitable frequencies is of great importance for laser construction and will be considered individually in Section 5.3.

4. The function of the dielectric coating of RF electrodes reduces not only to the trivial possibility of removing the electrodes from the discharge chamber

but also to controlling the plasma parameters, using various subnormal γ-modes intermediate between the α- and γ-modes (Section 2.9). Dielectric coating can be used for space modulation of the discharge parameters and, vice versa, for equalizing them in the direction normal to the current. The latter is especially important at very high frequencies, when the wavelength is comparable with the system size and the nonequipotential character of the electrode surface has an effect.

5. Since the α-mode operation does not depend on the electrode material and on whether the electrodes are covered with dielectric or not, there is a possibility to choose the material and dielectric only in terms of technological criteria and requirements on the optical resonator. The latter is especially important for the waveguide laser operation.

6. A small gap size in a transverse RF discharge permits the application of small RF voltages of a few hundreds of volts instead of ten kilovolts, like in dc tube lasers.

7. A small gap also facilitates laser initiation, because the values of ignition voltage and of steady operation voltage in RF discharge are quite close. In contrast, in dc tube lasers the excitation and stabilization of the laser operation present a problem, because the breakdown voltage in a long gap exceeds the operating voltage of the glow discharge in it (for meter-range tubes the differences may be as large as $V_{\text{break}} \approx 20$ kV and $V_{\text{steady}} \approx 8$ kV). Of course, low RF voltage in the slab geometry is not inherent only to the RF discharge. In a dc discharge between closely spaced electrodes, the voltage is not very high, of the order of cathode fall of several hundreds of volts; but in a slab gap in this case of a transverse dc discharge, most of the power is released in the cathode layer and lost.

8. RF excitation permits an easy regulation of laser radiation power from zero to a maximum value by simply changing the RF current strength. However, in dc excitation this is difficult to do due to the necessity to use large ballast resistances in the external circuit to stabilize the discharge.

9. RF excitation permits a variety of slab geometries, and the slab may not necessarily be made plane but have an annular or a more complicated shape (see below). In this case the positive column, which serves as an active laser medium, follows all the electrode and sheath curvature, since the sheaths and the plasma in the α-mode are tightly bound.

10. The technological parameters and reliability of the system elements are very high, because there are no fragile parts, for example glass tubes. The elements can be made of solid metals and covered by dielectric materials.

11. Another advantage of an RF slab laser over a tube laser is the possibility of a rapid replacement of the working mixture in the discharge space by letting it flow sideways into large volumes connected with the gap. If a tube is used, the mixture should be pumped along it.

12. To stabilize an RF discharge, one can use reactive ballast resistances, or just dielectric coatings, in which the power is not lost. However, about 30% of the power is lost in ballast resistances when a dc discharge is used.

It should be stressed that even RF excitation may fail if one ignores the difference in the discharge operation modes. For example, if a γ-discharge is excited in a slab gap, there will be no lasing at all or its efficiency will be very low, since most of the gap will be filled by negative glow and by the beginning of the Faraday dark space where the field is weak and insufficient to excite the laser mixture.

For an effective excitation, the values of E/p must be more or less definite, and they do not always coincide with the values realizable in the positive column, providing ionization balance of the plasma. The latter values are normally higher than the optimal values for lasers. In the case of an RF α-discharge, this problem is less acute than in a dc discharge.

Many of the advantageous properties of RF discharges manifest themselves not only in slab systems but in conventional tube systems, too. A γ-discharge may also be used advantageously. With an appropriate use of a transverse RF discharge in tube lasers, one can improve their characteristics, as compared to those of a longitudinal dc discharge. On the average, both excitation procedures can be considered equally feasible in tube laser excitation. However, a transverse RF discharge still has no rivals in the operation of very promising slab systems, although studies on combined discharges and magnetic stabilization (see below) are still in progress.

5.3 Frequency dependence of discharge and active laser medium parameters

We emphasized in Section 5.2.3 that the possibility of regulating plasma parameters by choosing the frequency is an essential advantage of RF laser excitation. Therefore, it is important to understand the physical mechanism of frequency dependence of conductivity σ, electron density n_e and plasma RF field E_p. There have been several attempts to clarify the frequency effect on plasma parameters [5.19, 5.30, 5.15, 5.31], but the existence of electrode sheaths has been entirely ignored. In [5.15, 5.31] the frequency effect is attributed to a difference in kinetic plasma processes in RF and dc discharges. Naturally, this approach failed to provide a satisfactory explanation of the experimental dependences of σ, n_e, and E_p on frequency.

In fact, the basic mechanism underlying the frequency effect on plasma parameters is determined by the existence of electrode sheaths in the α-mode, which form with the plasma an integral and inseparable system. Analysis of this system only [5.32, 5.14, 5.33] permits an adequate interpretation of the frequency dependence of σ, n_e, and E_p as well as some other experimental facts that puzzled the authors of [5.31] (see the end of this section). As for kinetic processes, the ionization rate may vary with frequency in certain frequency and pressure ranges, when the field frequency ω is comparable with the frequency ν_u of establishing the electron spectra and mean energy (Section 1.4.1). If, however, $\omega \ll \nu_u$ or $\omega \gg \nu_u$, the ionization rate becomes frequency-independent [equations (1.35) and (1.36)]. Plasma fields calculated from these limiting expressions do not differ much (Figure 1.3). One may assume then that the effect, if any, of kinetic processes on the frequency dependence of RF plasma parameters is not as great as that of the sheaths, which in no case should be ignored.

Section 2.3 discussed in detail the descending character of the current-voltage characteristic (CVC) in the positive column (PC). In the molecular laser mixtures $CO_2 + N_2 + He$ and in air, this is primarily associated with heating and thermal expansion of the gas in the discharge space. Phenomenologically, the descending CVC of the PC can be described by a simplified approximational formula for voltage across a PC of length L_1, $V_{PC} = CpL_1/j^m$, where C and m are constant in the given conditions. Since the voltage fall across the α-sheaths, through which displacement current flows, is approximately equal to $V_s \approx 4\pi d_\alpha j/\omega$, where d_α is the sheath thickness, the electrode voltage fall V as a function of j [formula (2.3)] has a minimum. The current density $j_{n\,\alpha}$ corresponding to this minimum and representing the least possible (normal) current density is determined by equations (2.4), (2.36), and (2.37) together with the minimum voltage $V_{\alpha\,min}$ necessary for an α-discharge operation. In the last two formulas, m is taken to be unity, $m = 1$, but account is taken of a possible dielectric coating of thickness δ.

In the examination of normal current density effects in Section 2.3, and in the implementation of subnormal γ-states with dielectric-coated electrodes in Section 2.9, it was important to take into account only the existence of an α-CVC minimum. Hence, one could take there $m = 1$ to simplify the formulas. But at $m = 1$, the power released per 1 cm^2 PC, jV_{PC}, turns out to be independent of the current, which does not permit the analysis of gas heating effects. Indeed, energy release per unit plasma volume $w = jE_p = jV_{PC}/L_1$ grows with growing current density; therefore, one should take $m < 1$ in the formula for the PC CVC. The power approximation of the PC CVC is closer to the real dependence if the power index m lies within $1/2 < m < 1$.[1] At $m \neq 1$ and with account of the dielectric coating, equations (2.36) and (2.37) take a more general form

$$j_{n\,\alpha} = \left(\frac{\sqrt{m}CpL_1\omega}{4\pi(d_\alpha + 2\delta/\varepsilon)} \right)^{1/(m+1)}$$

$$V_{\alpha\,min} = \beta(CpL_1)^{1/(m+1)} \left(\frac{4\pi(d_\alpha + 2\delta/\varepsilon)}{\omega} \right)^{m/(m+1)}$$

$$\beta = \left(\frac{1}{m^{m/(m+1)}} + m^{1/(m+1)} \right)^{1/2} \tag{5.2}$$

---

[1] A simplified physical model of the PC CVC with gas heating is considered in [5.3]; but, unfortunately, its result is the inverse function $j(E)$ instead of the function $E(j)$ necessary for finding the CVC minimum. The former cannot be inverted analytically because of the temperature dependence of the gas thermal conductivity. But even if we neglect the temperature dependence, the expression for $E(j)$ is too complicated, so we use a power approximation.

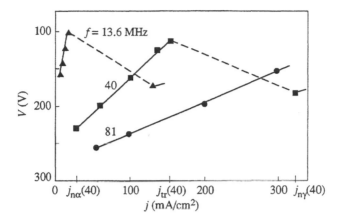

FIGURE 5.4
Current-voltage characteristics of an RF discharge at various frequencies: air, $L = 0.75$ cm, $p = 15$ Torr, brass electrodes. Solid lines, anomalous α-discharge with the extreme left points corresponding to a normal α-mode and the extreme right point to the $\alpha-\gamma$ transition. Dashed line is directed to a normal γ-mode [5.33].

The α-sheath thickness for a moderate-pressure discharge is, in order of value, equal to the electron drift oscillation amplitude $d_\alpha \approx v_d/\omega$, where the electron drift velocity v_d corresponds to the RF field amplitude in the plasma, $v_d = \mu_e E_{pa} = (\mu_e N)(E_{pa}/N)$. The mobility $\mu_e \sim N^{-1}$. We now deal with molecular density N instead of pressure, since the gas temperature T during the heating depends on the power input into the discharge and on the current, so that the fixed pressure $p = NkT = \text{const}$ cannot serve as a measure of density on which μ_e and the rates of plasma processes actually depend. At moderate pressures and in molecular gases, the E_{pa}/N ratio is largely determined by the necessity to maintain the ionization–recombination balance. Except for conditions when $\omega \sim \nu_u$, this ratio does not strongly depend on the field frequency, n_e and j (Section 1.4.5). Incidentally, in electronegative gases—for instance, in the laser mixtures CO_2+N_2+He and in air—the effective coefficient of the electron–ion recombination is, with account of attachment, an order of magnitude larger than the real value [5.3]. It follows from the foregoing that one can approximately assume $v_d(\omega, N) \approx \text{const}$, from which the α-sheath thickness $d_\alpha \sim \omega^{-1}$ and is independent of the gas density. These relationships have been confirmed by direct measurements (Figure 2.24).

Thus, according to equation (5.2), for naked electrodes the normal (minimum) current density in an α-discharge is $j_{n\alpha} \sim \omega^{2/(1+m)}$, and the minimum voltage across partly current-filled electrodes is $V_{\alpha\min} \sim \omega^{-2m/(1+m)}$. For example, for the convenient and quite reasonable value of $m = 1/2$, we have $j_{n\alpha} \sim \omega^{4/3}$,

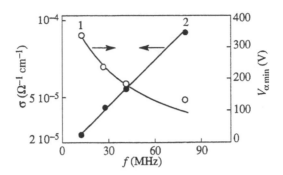

FIGURE 5.5
Frequency dependence of minimum rms RF voltage (curve 1) and of conductivity
(curve 2) in a normal α-discharge: discharge in CO_2, $p = 30$ Torr, $L = 0.9$ cm (curve 1);
discharge in air, $p = 10$ Torr, $L = 1$ cm (curve 2) [5.33].

$V_{\alpha\,\text{min}} \sim \omega^{-2/3}$ ($w = jE_p \sim \omega^{2/3}$). These results follow exclusively from the
recognition of the existence of electrode sheaths and of the descending character
of the PC CVC, irrespective of plasma kinetics. They are consistent qualitatively
with experiment [5.33] (Figures 5.4 and 5.5).

The plasma current, which at moderate pressure largely represents conduction
current, coincides with the sheath displacement current and with the discharge
current j. If $\omega \ll \nu_\text{m}$, which is valid at moderate pressures, the plasma conductivity
from equation (1.18) is $\sigma \sim n_e/N$ and does not directly depend on frequency.
It follows from $j \approx \sigma E_p \sim n_e(E_p/N)$, from a weak frequency dependence of
E_p/N and from the proportionality $j_{n\,\alpha} \sim \omega^{4/3}$, that the plasma density and
conductivity in an α-discharge grow with growing frequency: $\sigma \sim n_e \sim \omega^{4/3}$.
The result is derived from the 'sheath concept' and also qualitatively agrees with
experiment (Figure 5.5).

After the electrodes are totally covered by current, the α-mode becomes anomalous, and the electrode voltage grows with discharge current i and its density
$j = i/S$, where S is the area of the electrodes. This growth goes on until the
current and plasma densities reach critical values j_tr and n_tr, at which the $\alpha-\gamma$
transition occurs. The limiting values of the α-mode, as well as 'normal' ones,
increase with frequency. From equation (2.17), where $n_{cr} \equiv n_\text{tr}$ and with account
of $A \approx d_\alpha \sim \omega^{-1}$, we find that $n_\text{tr} \sim \omega$. Physically, this relationship is due
to the fact that with increasing ω, the α-sheath thickness d_α is reduced; so, for
the sheath voltage fall $V_s \approx 2\pi e n_+ d_\alpha^2$ to reach a breakdown value V_t, the ion
density in the sheath n_+ must be larger. Indeed, on the right-hand branch of the
Paschen curve, the breakdown voltage grows with d a little slower than $V_t \sim d$
[see equation (2.15)]. Therefore, for breakdown of the sheath d_α, the ion densities
in it must be $n_+ \sim V_t/d_\alpha^2 \sim d_\alpha^{-1}$. But at moderate pressures the α-sheath ion
density is of the order of the plasma density (Section 2.1), so the plasma density

at which an $\alpha-\gamma$ transition occurs is $n_{tr} \sim d_\alpha^{-1} \sim \omega$.

Thus, with frequency increase, the range of plasma density variation (from n_n corresponding to normal current density to n_{tr} corresponding to an $\alpha-\gamma$ transition $n_n \leq n_e \leq n_{tr}$) shifts towards higher frequencies nearly in proportion to the frequency and becomes wider. This is also in agreement with the experimental data of Figure 5.5. Though secondary emission processes do not play an essential role in the α-mode, the $\alpha-\gamma$ transition parameters, which serve as the upper limits of the j and n_e ranges favorable for laser pumping, do depend on these processes. Hence, the practically important conclusion is that the breakdown sheath voltages can be increased by coating the electrodes with materials possessing poor emission properties. In other words, the α-current can be raised to a higher value without the risk of transition to an unfavorable γ-mode, that is; we can increase laser power without increasing the frequency.

The fact that the plasma current and density in a normal α-discharge are determined by the behavior of the sheath–plasma system but not by the properties of the RF plasma alone, is evident from the following experiment [5.17]. As is clear from equation (5.2), $j_{n\alpha}$ and $V_{\alpha\,min}$ vary with the electrode coating thickness. The experimental discharge chamber was formed by two quartz plates of thickness $\delta > d_\alpha/\varepsilon$. A metallic coating was deposited on one side of a plate and served as an electrode. Normal current density was measured in the same gas at the same values of ω, p, and L (gap length); but in one case the metallic coating was facing the plasma and corresponded to $\delta = 0$ in equation (5.2), and in the other case it was turned away from it with $\delta \neq 0$. In the first case of naked electrodes the current density $j_{n\alpha}$ was higher, as follows from equation (5.2). If the plasma and current densities of an α-mode were determined exclusively by plasma processes, they would hardly depend on whether a dielectric coating is present or not.

Considering available experimental data, various researchers have pointed out that the plasma field E_p and the E_p/p ratio decrease with increasing frequency at a given pressure. It is clear from the above that this is due to the descending PC CVC and to the current density rise with growing frequency: $E_p/p \approx V_{PC}/pL_1 \approx C/j_{n\alpha}^{1/2} \sim \omega^{-2/3}$ ($m = 1/2$), but not to some special kinetics of the RF plasma. Though the E_p/p ratio drops with increasing ω, the E_p/N ratio responsible for the change in the mean energy (temperature) of electrons and in the rates of kinetic processes is less sensitive to the frequency. As ω increases, j and $w = jE_p$ grow, and the gas temperature rises decreasing its density. So, the field E_p necessary to maintain the ionization decreases, too.

Similarly, the 'sheath–plasma' concept can be used to explain the experimentally registered dependences of electrode RF voltages for waveguide [5.31] and slab [5.17] CO_2 lasers on the gas pressure and frequency at fixed input power (Figure 5.6). Let us substitute into equation (2.3) with $m = 1$ (this is now permissible) the current density expressed through the output energy density $w = jE_p$

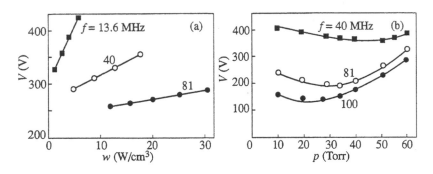

FIGURE 5.6
Dependences of rms RF voltage (a) on the input power density and (b) on the pressure at
fixed power at various frequencies; (a) air, $p = 30$ Torr, $L = 0.6$ cm; (b) laser mixture
$CO_2:N_2:He=1:1:3$, $L = 0.3$ cm, $w = 20$ W cm^{-3} [5.17].

and $d_\alpha = v_d/\omega$. We will obtain the electrode RF voltage in the form

$$V = \left[\left(\frac{E_p}{p} L_1 \right)^2 p^2 + \left(\frac{4\pi v_d w}{(E_p/p)\omega^2} \right)^2 \frac{1}{p^2} \right]^{1/2} \tag{5.3}$$

Suppose a discharge fills up the whole volume Ω and, therefore, is anomalous,
as it was in the experiments. Then for varying pressure and fixed power, $w =
P/\Omega =$ const, but j and E_p change. (In a normal discharge, E_p, the current area
on the electrode, plasma volume Ω and w would change, while j would remain
constant, $j_{n\,\alpha}$.) The function $V(p)$ from (5.3) has a minimum

$$V_{min} = \frac{(8\pi v_d w L_1)^{1/2}}{\omega} \qquad p_{(min)} = \frac{1}{(E_p/p)\omega} \left(\frac{4\pi v_d w}{L_1} \right)^{1/2} \tag{5.4}$$

The frequency dependences of V_{min} and $p_{(min)}$ qualitatively agree with experiment,
and even numerical values are found to be correct if one takes for evaluations the
reasonable values of $E_p/p \sim 10$ V cm^{-1} Torr^{-1}, $v_d \sim 10^7$ cm s^{-1}, as well as the
experimental values of ω, $w = 20$ W cm^{-3} and $L_1 \approx 0.3$ cm. A formula of the
type (5.3) can be used for experimental evaluation of E_p/p in the plasma from
measured voltage and power.

5.4   Selection of designs and parameters of transverse RF-excited CO_2 lasers with diffusional cooling

Here we discuss some specific features of these lasers, restrictions on their parameters and the quality of various schemes. The researcher is inevitably faced with

these problems in designing a laser. In our discussion, we will concentrate on various aspects of discharge arrangement, but this is only part of the matter. The construction of a laser requires the solutions to two principal problems: (i) creation of an effective laser medium, which is directly related to discharge arrangement, and (ii) transformation of the medium excitation energy to laser radiation, which reduces to making a good optical resonator. The problems of the optics are not directly related to the major subject of the book and will be mentioned briefly in Section 5.5.

5.4.1 Restrictions on field frequency in slab and capillary CO_2 lasers

A CO_2 laser with a transverse RF excitation, when the RF field is directed along the small slab or tube dimension, can operate only in a certain frequency range that depends on the small size of the gap h. The upper and lower frequencies are limited. We have already emphasized that the α-mode is the most favorable form of the RF discharge for this type of laser. The α-sheath thickness is nearly inversely proportional to the frequency $d_\alpha \approx v_d/\omega \approx \text{const}/\omega$. For an α-discharge to be ignited easily in the gap, it is sufficient that its size $L = h$ should be larger than the maximum sheath thickness. But to have sufficient space for a plasma, which serves as the active laser medium, h must be quite large as compared to the sheath thickness. Since better heat removal conditions require small h, this improves the rigorous restriction on the lower working frequency of the RF field

$$d_\alpha \approx \frac{v_d}{\omega} \ll h \qquad \omega \gg \frac{v_d}{h} \tag{5.5}$$

If, as usual, the electron drift velocity corresponding to the plasma RF field amplitude is $v_d \approx 10^7$ cm s^{-1} and, for example, $h = 0.3$ cm, frequencies $f \gg 5$ MHz are necessary. Besides, low frequencies are not profitable, because the normal current density $j_{n\alpha} \sim \omega^{4/3}$ and the energy output $w = jE_p \sim \omega^{2/3}$ are too small for them, so the laser cannot generate high power at low frequencies. Thus, one should try to use as high frequencies as possible.

However, one must not go too far in one's attempts to increase the frequency, because very high frequencies are unacceptable for the RF way (but not for the microwave one) of exciting a discharge. The upper limit is determined by the maximum permissible gas temperature $T_{max} \approx 600$ K. Since the normal, that is, minimum possible, current density in an α-discharge grows with ω, the power input into the plasma at very high frequencies may turn out to be too great, leading to an overheating of the laser mixture. The upper frequency limit is also related to the small gap size. In the case of a plane slab the heat balance of the gas is approximately described by the equation

$$jE_p h = 2\lambda \left(\frac{dT}{dx}\right)_{x=\pm h/2} \approx \pi^2 \lambda \frac{(T_{max} - T_0)}{h}$$

$$jE_p h^2 \approx \pi^2 \lambda (T_{max} - T_0) \approx \text{const} \tag{5.6}$$

which is similar to the one for a tube (5.1). In equation (5.6) the 'thermal conduction' length, which characterizes the temperature gradient near the wall and the heat removal rate, is taken to be equal to the diffusion length $\Lambda = h/\pi$ (Section 1.4.3), from which it differs only slightly. Maximum permissible energy input is $w \sim h^{-2}$. For example, at $h = 0.3$ cm and $\lambda \approx 5 \times 10^{-4}$ W cm^{-1} K^{-1}, an energy input over 10 W cm^{-3} can lead to a considerable heating, though inputs as high as 100 W cm^{-3} may be achieved in an RF discharge. In practice, frequencies over 100–150 MHz may become unacceptable for such slabs.[2]

There are two other circumstances which make the application of very high frequencies undesirable. If a quarter wavelength $\lambda/4 = c/4f$ (75 cm for 100 MHz) is comparable to the length of the electrodes (tens of centimeters or, possibly, a meter), there is the effect of wave retardation, which gives rise to a longitudinal nonuniformity of the discharge and of the active medium. In principle, this difficulty can be removed using special technical procedures: connection of an array of induction coils at a certain distance from each other along the electrodes [5.34], sectioning of the electrodes [5.22], application of dielectric coatings of a variable thickness (Section 2.9), etc. But these measures increase the complexity of the laser construction. The other technical difficulty is associated with the matching of the RF generator and the discharge load.

5.4.2 Restriction on discharge current to avoid $\alpha-\gamma$ transition

Generally speaking, the maximum energy input w_{max} defined by equations (5.6) and (5.1) can be achieved not only by increasing the frequency, but also by increasing the discharge current and by implementation of a strongly anomalous α-discharge at a lower frequency. Here, however, there is the risk of transition to a γ-mode at a site on the electrode, where the inevitable nonuniformities and geometry distortions may provoke an $\alpha-\gamma$ transition. This happens quite often in practice, and we will discuss this effect in Section 5.4.3. Moreover, in a strongly anomalous α-mode, the power losses due to dissipation of the active current in the sheaths, which produces no lasing, appreciably increase. In this mode the portion of the active current in the discharge current increases and the sheath voltage rises, so their product defining the power losses proves to be substantial.

For this reason, it is more advantageous to deal with a weakly anomalous α-mode providing the filling of the whole chamber space without increasing much the current density and voltage above the normal values. Higher power (at a given h or R) should be achieved just by raising the frequency up to a maximum value.

The fact that only the α-mode is suitable for pumping CO_2 lasers by a transverse RF slab discharge is well illustrated by experimental data [5.12, 5.14]. The discharge was ignited in a CO_2-containing laser mixture, which filled a gap of

---

[2]Note that with respect to permissible input energy density $w = jE_p$ and the related frequency, a slab and a tube with $2R \sim h$ are more or less equivalent. The advantage of a slab is that the total power per unit length l of a stripe or a tube can be increased by making the gap width larger.

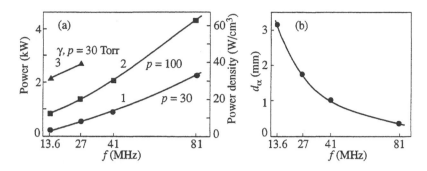

FIGURE 5.7
(a) The limiting powers and power densities (to be put into an α-discharge prior to its transition to a γ-mode) versus the field frequency [5.12]: laser mixture with CO_2, $p = 30$ Torr (curve 1) and 100 Torr (curve 2), power in the γ-mode at $p = 30$ Torr (curve 3) and for comparison (b) the α-sheath thickness versus frequency.

$h = 0.66$ cm between plane electrodes of area $S = 100$ cm^2. Maximum power P_{tr}, at which the α-mode changes into the γ-mode [curves 1 and 2 in Figure 5.7(a)] was measured as a function of frequency. Incidentally, these curves qualitatively illustrate the growth of the current density in the $\alpha-\gamma$ transition j_{tr}, to which the maximum power density w_{tr} is proportional. The latter, also presented in Figure 5.7(a), can be written as $w = P/Sh$, because in all cases the α-discharge filled up the whole gap. Figure 5.7(b) shows the decrease in the measured α-sheath thickness with growing frequency.

Calorimetric determination of the power was accompanied by registration of the enhanced probing signal from another CO_2 laser. At a maximum but subtransitional power of 2.1 kW at 81 MHz and $p = 30$ Torr, the gain k_0 was found to be 0.5 m^{-1}. This means that under the given conditions, the $\alpha-\gamma$ transition current is lower than the maximum current for the mixture heating up to T_{max}. In the case of overheating, there would be no gain. Curve 3 corresponds to the power input in a γ-discharge in the same chamber with the same cooling conditions for the same mixture and pressure of 30 Torr. But no gain was registered at the same power of 2.1 kW. Since in this γ-mode (but at 13.35 MHz) the discharge also filled the whole space and the temperature was, therefore, the same, the disappearance of inversion should not be attributed to thermal effects. One can conclude (Section 5.2) that the distributions of the plasma density and of the field in a slab γ-discharge do not stimulate the excitation of the laser mixture: much of the gap space is occupied by the inactive negative glow and the beginning of the Faraday dark space.

5.4.3 Adverse effect of $\alpha-\gamma$ coexistence in a slab

When the gap h is relatively large, say, a few centimeters, the $\alpha-\gamma$ transition and the initiation of a γ-mode can be clearly identified, and one can easily observe the discharge cross section reduction. But in the case of a comparatively small gap h, the situation becomes different. After the transition, an α- and a γ-mode may coexist in the gap, part of both electrode areas being occupied by one mode and the other part by the other mode (Section 2.7). At large h, there is no coexistence effect, because the operation voltage of the γ-mode—with its rather extensive regions of negative glow and Faraday space having a small voltage fall—turns out to be lower than the voltage necessary for an α-mode. In narrow gaps, most of which are occupied by extensive α-sheaths with a voltage fall smaller than the γ-sheath voltage, the operation voltage of a normal or slightly anomalous α-mode is often lower than for a γ-mode. Then, due to the electrodes being equipotential, the voltage at some electrode sites becomes that characteristic of a γ-mode. Since it now exceeds the α-operation voltage, the remaining electrode surface is covered by the α-current.

The coexistence of the two modes screens the $\alpha-\gamma$ transition, because all the space now seems to be filled by a glow; and one often fails to identify in it the characteristic γ-intensities and glow distribution. But the transition that has occurred leads to poorer CO_2 laser parameters in spite of the presence of α-sites. The screening of the transition often misleads researchers, who do not see the real reason for inferior laser operation.

This conclusion can be supported by the following experiment [5.12]. The discharge chambers for CO_2 lasers were quartz tubes of 170 cm long, in which an α-discharge was excited. In some cases, cooled electrode stripes of 1 cm wide were mounted internally at a distance $h = 1$ cm from each other; in others, the electrode stripes were mounted outside a 1 cm diameter quartz tube, the discharge space h being identical in both cases. The tubes were filled with the laser mixture CO_2:N_2:He=1:1:8 at $p = 20$ Torr. Figure 5.8(a,b) (curves 1, 2, and 3) shows the measurements of the radiation power as a function of discharge current i.

The laser generation begins at a certain RF current, its power then grows with the current, reaches a maximum and drops rather steeply (Figure 5.8, curves 1 and 2). In the case with naked internal electrodes (curve 1), the discharge cross section noticeably reduces when the generation ceases, the electrode area not being covered totally. In the case with external electrodes at the same frequency of 13.56 MHz (curve 2), the current area is not reduced, but a more careful examination shows that at some sites the discharge has changed into the γ-type. The sheaths there have become thinner, and there is a characteristic alternation of luminous and dark layers. Direct current probing (Section 4.3) reveals an increased active conductivity in the γ-sheaths with a small thickness. However, the existence of the typical α-pattern on the remaining electrode surface indicates a coexistence of the two modes. This is exactly the reason for the laser generation being preserved but with a lower power. At a higher frequency of 27 MHz (curve 3), when the $\alpha-\gamma$ transition current is higher, no visible change in the discharge occurs due to the transition; but the generation maximum is reached in the current range studied.

The fact that the generation decrease is not due to the gas overheating but to

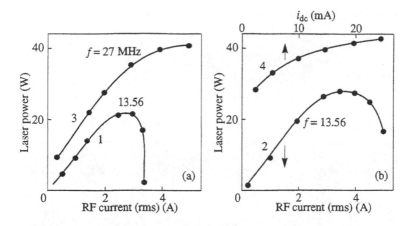

FIGURE 5.8
CO_2 laser generation power plotted against RF current at 13.56 MHz (curve 1) and
27 MHz (curve 3) for naked electrodes, $h = 1$ cm, $p = 20$ Torr; in a dielectric tube of
1 cm diameter with external electrodes: at $f = 13.56$ MHz (curve 2), radiation power in a
combined discharge at optimal RF current, $i = 3.5$ A, versus additional longitudinal
direct current i_{dc} (curve 4).

the change in the longitudinal and transverse discharge structure because of the
partial filling of the gap by a γ-mode is supported by the following experiment.
Additional direct current was passed along the tube with external electrodes at RF
current corresponding to the maximum generation in curve 2. The electrodes were
internal rings placed at the tube ends. In spite of the additional power input into
the plasma by direct current i_{dc} to increase the gas temperature, the laser power
only rises with i_{dc} (curve 4). It rises even at the additional dc input of 150 W,
which exceeds the 100 W RF power to be added to optimal RF power in order to
suppress the laser generation.

This is also clear from a comparison of photographs taken during a transverse
RF α-discharge in a tube and a combined discharge in which, in addition to
transverse RF current along the tube, there is a direct current in the same direction
[Figure 5.9(a,b)]. The middle of the combined discharge has a brighter glow
than in an RF discharge only. This indicates the localization region of the direct
current. It flows near the axis of the central column, increasing the plasma density
in it and thus creating additional excitation in the active laser medium. The latter
effect is manifested in the increased power of laser radiation evident from curve 4
[Figure 5.9(b)].

Thus, prevention of a γ-mode and of $\alpha-\gamma$ coexistence in a CO_2 laser by limiting
the RF current is a necessary condition for normal operation of slab and capillary
laser systems—but with an important reservation, which will be discussed in
Section 5.4.4, that the electrode width should be comparable with the gap width.

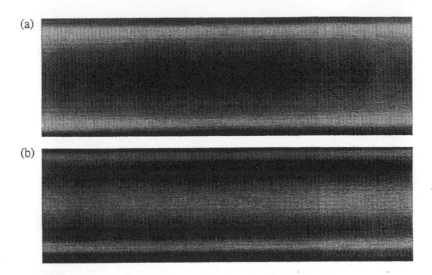

FIGURE 5.9
(a) RF α-discharge in air in a cylindrical tube of 4 cm in diameter with external
electrodes: $f = 13.56$ MHz, $p = 10$ Torr; (b) photograph of a combined discharge under
the same conditions with longitudinal direct current.

The effective use of advantages of transverse RF discharges has permitted
creation of compact slab CO_2 lasers with the discharge chamber sizes of $60 \times 10 \times$
0.2 cm^3 [5.22] and $77 \times 9.5 \times 0.2$ cm^3 [5.35] and radiation power exceeding 1 kW.
This was achieved in diffusional cooling conditions without fast gas pumping.

5.4.4   The use of combined γ- and α-modes in CO_2 lasers

Generally, in a simple geometry of the slab type with electrodes of the same size
as the chamber plane, the appearance of a γ-mode or the coexistence of an α-
and a γ-mode have an unfavorable effect on CO_2 laser generation (Section 5.5.3).
However, a reasonable application of features of various RF discharge types
permits finding such an electrode design in which unfavorable effects are used
advantageously [5.36]. A design of this kind avoids very high frequencies creating
specific difficulties (Section 5.4.1), but it provides an effective laser operation at
13.56 MHz—a frequency unsuitable for conventional slab and capillary systems
because of the inevitable γ-mode arising from increased current and power [5.37].

This design is based on the experimental fact of reduction of the Faraday dark
space (FDS) length in a discharge unconfined by lateral walls when the electrode
dimension transverse to the current is small. In this case, due to the sharp
reduction of the transverse dimensions of the negative glow (NG) region and the
FDS, and because of the fast diffusional expansion of the current stream, the

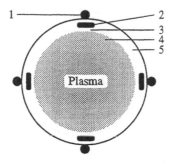

FIGURE 5.10
RF discharge in a dielectric tube showing a combination of the γ- and α-modes:
1, external wire electrodes; 2, negative glow regions; 3, the Faraday dark space regions;
4, common positive column; 5, α-sheaths.

electron density drops rapidly with distance from the electrode from a maximum value in the NG region to a much smaller value in the positive column (PC) (Section 2.5.4). Thereby the PC comes closer to the electrode. Taking advantage of this circumstance, one can fill up well, with the PC plasma, a dielectric discharge tube of a meter length and a 1 cm inner diameter, with four external electrodes along the tube made from a 1 mm diameter copper wire (Figure 5.10). RF voltage is applied to two of them, in any combination, and the other two are grounded. In the region adjacent to the electrodes, inside the tube, a typical γ-discharge arises at increased input power, which contains small, bright NG regions and more extended but also small regions of the FDS. The rest of the tube, except for the annular cylinder periphery, is filled up with the PC plasma. If the current is not very high, its increase is accompanied by spreading the γ-discharge along the tube. When the electrodes and the tube become totally filled with the current, an anomalous γ-mode occurs.

The dark peripheral ring at the inner tube surface represents a space charge sheath of the α-mode. The point is that some of the displacement current from the plasma is closed on the grounded object surrounding the tube, rather than on the electrodes, in such a way that the force lines of the RF field and of the RF current displacement near the tube surface between the electrodes have a component normal to the surface. It is the plasma boundary oscillations in this field that create an α-mode analog against the background of the peripheral ions. In this case one can speak of a 'coexistence' of the α- and γ-mode in the system. A multichannel laser based on this effect will be described in Section 5.4.5.

5.4.5 Multichannel laser systems

The idea of multichannel diffusion-cooled CO_2 lasers, in which many individual channels are connected 'in parallel' to a common resonator [Section 5.2.1 and

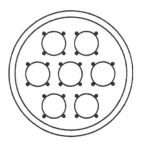

FIGURE 5.11
Multichannel laser combining 7 lasers shown in Figure 5.10.

Figure 5.2(b)], has recently attracted much attention again. This is due to certain progress in synchronization of individual channel operation, resulting not merely in summation of their generation powers but in formation of a single coherent radiation flux at the output, which has several important advantages (for the optical aspects of this effect, see Section 5.5). Due to its specific features, the transverse RF discharge may become a perspective means of exciting synchronized compact systems, taking into account the fact that the slab laser power has to be increased by increasing the stripe width, with a reasonable restriction of the stripe length to about one meter. But this approach also has limits. For this reason, it is more profitable to combine the best features of multichannel and slab systems.

A multichannel design of the conventional type, in the sense that the channels are optically independent of each other, was suggested in [5.17] (Figure 5.11). A single channel was shown in Figure 5.10. Seven parallel quartz tubes of 95 cm long with the inner 0.9 cm diameter were placed in a glass tube of 5.7 cm in diameter, through which transformer oil was circulated to cool the channels. The CO_2 laser mixtures were used at $p = 20$ Torr and $f = 13.56$ MHz. By varying the applied RF power, one could excite α- and γ-modes in the tubes (the latter mode in a way described in Section 5.4.4). Practically acceptable results (input into the plasma 6 W cm^{-3}, unsaturated gain $k_0 \approx 1$ m^{-1}, radiation power 200 W) were obtained only in the γ-mode combined with the α-mode. This design has advantages over similar dc systems [5.28]. These are the possibility to sustain a discharge in all channels simultaneously at a comparatively low electrode voltage, the absence of ballast resistances, and reliability associated with the external arrangement of the electrodes.

The new principle of designing multichannel CO_2 lasers arose from the unique properties of the RF α-discharge, which effectively reveal themselves if the frequency is properly chosen. This principle is based on the experimentally established property of the α-mode [5.13] to fill up with plasma the discharge gaps of arbitrary cross section. Some promising and partly implemented designs are shown in Figure 5.12. One of the features of such designs that have much promise is the possibility to create an intraresonator optical coupling between the channels

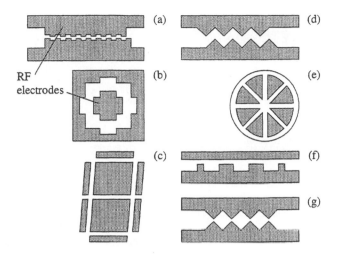

RF electrodes

FIGURE 5.12
Variants of multichannel slab laser systems.

(Section 5.5.2). Moreover, this type of laser opens up new technological options for monitoring the laser beam pattern in the nearby region by choosing an appropriate gap profile. A patterned laser beam can be used for labelling, stamping, etc. Such laser systems combining the principles of the slab geometry, multichannel design, and transverse RF fields have been realized in [5.23, 5.38–5.41].

Some of the multichannel systems contain wider 'working' channels with the slab size h, which are supposed to be filled by laser-active plasma, and narrow gaps h_1 between them designed for the optical coupling of the channels [see, for example, the jaw-type design in Figure 5.12(a,g)]. However, this system will not be able to operate as designed at any parameter values. It will work only if the α-sheath thickness $d_\alpha \approx v_d/\omega$ is such that $h_1 < v_d/\omega < h$. Then, an α-mode will be ignited in the working channels; but in the narrow gaps between the channels, a breakdown and steady regime will be difficult to induce. But if the chosen frequency is inappropriate and $v_d/\omega < h_1 < h$, the α-mode will be excited in the 'optical' gaps between the channels but not in the channels. Due to the equipotentiality of the electrodes, the voltage, which is established on them corresponding to a discharge in the smaller gap h_1, will turn out to be too low to sustain the discharge in the channel with $h > h_1$. Note that if one deals with a normal or a weakly anomalous α-mode at such ω and h that the operating voltage $V_{\alpha\,min}$ is less than the minimum voltage in the Paschen curve for the gas breakdown, then one should not be concerned about a transition to the γ-mode at acute angles where the field is concentrated. For this reason, the whole space becomes filled with an α-discharge.

In contrast with all variants with a 'parallel' electrical connection of channels,

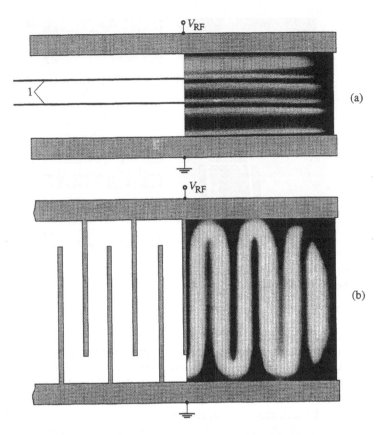

FIGURE 5.13
(a) Series and (b) parallel electrical connection of slab channels filled with an α-discharge. The photograph (a) shows well the typical discharge structure (cf. Figure 1.11 and 1.12).

for example, those in Figure 5.12, a 'series' electrical scheme of extended slab channels has been implemented experimentally [Figure 5.13(a)]. It shows certain advantages (see below). This scheme was suggested in [5.38], and it was probably the first design combining three important elements perspective for laser construction geometry, RF excitation and a multichannel system. Figure 5.13(a) illustrates its basic idea, which is as follows. In order to increase laser power, the discharge volume must be made larger. But one cannot increase infinitely the slab width and length. Besides, the gap capacitance rises with the electrode area, making the matching with the generator more complicated. On the other hand, one cannot greatly increase the gap size h at elevated pressures, because there is the upper limit ph (30–40 Torr cm for $f = 13.56$ MHz), above which no α-mode can exist (Section 2.2, Figure 2.9).

The value of h can be reduced several-fold by segmenting the interelectrode gap with plates, metallic or dielectric, running parallel to the electrodes. In the former case, the plates insulated from the electrodes acquire the space (floating) potential, and the gap equivalent capacitance reduces as much as the number of identical channels, as compared to a single channel capacitance, thus facilitating the matching. Moreover, the mechanical strength and rigidity of the evacuated discharge chamber are not decreased in spite of a strong compression by atmospheric pressure, since the chamber volume is enlarged without changing its area. Figure 5.13(b) [5.17] gives another illustration of a parallel electrical connection of the channels. One can see that the α-plasma fills up the complicated configuration of the space formed by two comb-type devices inserted into one another.

To conclude, we cite some major publications on waveguide (capillary) [5.42–5.63, 5.25, 5.34] and slab [5.64–5.75, 5.5, 5.12, 5.16, 5.26] CO₂ lasers with diffusional cooling and transverse RF excitation. The maximum radiation output per unit length of a capillary transverse CO₂ laser achieved so far is $1.1 \, \text{W cm}^{-1}$ [5.63a].

5.5 Optical resonators of waveguide and slab RF CO₂ lasers

5.5.1 Stable multipass and unstable resonators

Laser radiation power for a given discharge chamber design, discharge mode and input electrical power can be considerably increased by improving the optical resonator, in particular by increasing the radiation path length through the discharge using multipass schemes. A simple variant of an optical resonator of this type is a multipass stable resonator. Two extreme mirrors form a resonator, the intermediate mirrors serving to return the light beam into the active medium. The optical system in a stable resonator is such that the spatial field distribution in the beam does not change in its multiple passes, and, within the geometrical optics approximation, the electromagnetic field remains within the mirror limits in the transverse direction. Radiation can leave a stable resonator only due to the partial transmission of the reflecting elements themselves.

Multipass stable resonators are used in many technological dc-excited CO₂ lasers, including high power devices. They are also used in compact closed waveguide RF-excited CO₂ lasers. A laser [5.76] is made up of waveguide channels of 0.225 cm in diameter and 37 cm long. The total length of the active medium and the optical path can be increased by increasing the number of channels. Turning mirrors were used to form a V-shaped two-channel system, a Z-shaped three-channel system like the one in Figure 5.2(a), and a four-channel one. A one-channel laser yielded a maximum power of 31 W corresponding to $0.84 \, \text{W cm}^{-1}$. The power per unit length is close to maximum power achieved in a closed waveguide CO₂ laser by optimizing the laser mixture composition and the RF field frequency [5.11, 5.19]. In multichannel designs, the power per unit

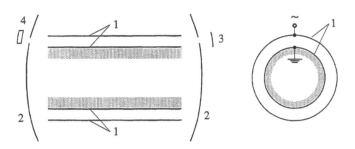

FIGURE 5.14
Multipass steady-state resonator in an annular coaxial laser [5.77]: 1, electrodes; 2, rotating mirrors with holes; 3, end and 4, semitransparent resonator mirrors.

length of the optical path is smaller due to diffraction losses on turning mirrors and to partial pumping of the power to other modes. For instance, in a four-channel laser, one-mode radiation power was 80 W, that is, 0.54 W per one centimeter of the optical path through an active medium at 800 W input electrical power, providing a 10% efficiency. The laser power was limited by deterioration of the mirror optical properties and even their damage. Generally, low beam durability of optical dielectric materials used for mirrors and windows is an essential drawback of stable resonators.

An annular (co-axial) cylindrical CO_2 laser [5.77] excited by a transverse RF discharge used a more complicated resonator design containing two turning large-radius spherical mirrors with a large curvature radius (Figure 5.14). A small hole was made at the edge of each mirror at a level close to the external cylindrical electrode radius. A nontransmitting and a partly transmitting mirror was placed behind the holes. The radiation is repeatedly reflected by the turning mirrors before it reaches—having been enhanced—the edge and passes through the holes to the resonator mirrors. A laser with a 2.3 cm external electrode radius and 26 cm electrode length along the cylindrical axis provided a 65 W continuous one-mode radiation with 9% efficiency. This power corresponds to the resonator optical length of 6 m, while its actual length is only 35 cm.

Another coaxial type of laser [5.26] with a 2.4 cm radius internal electrode and a 1.05 cm annular gap had a resonator of 2.4 m long formed by concave toric mirrors and a partly transparent outlet mirror with the curvature radii of 4.8 and 10 m, respectively (Figure 5.15). An axicon mirror converges the radiation, collected from all discharge space, into a beam coming through a small semitransparent mirror placed near the cylinder system axis. A maximum laser power of 350 W was obtained in this system with the mixture $CO_2:N_2:He=1:3:16$ at $p = 20-30$ Torr, RF power 5 kW, and $f = 27$ MHz.

Slab lasers use unstable resonators, too. An unstable resonator naturally provides multipass radiation without turning mirrors or other devices, which simplifies the design. The laser in [5.78] has a hybrid, stable–unstable resonator of a

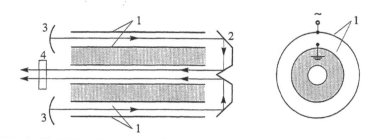

FIGURE 5.15
Coaxial resonator with an axicon reflector [5.26]: 1, electrodes; 2, axicon reflector; 3, toric concave mirror; 4, partly transparent outlet mirror.

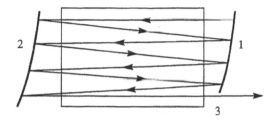

FIGURE 5.16
Slab resonator unsteady along the largest transverse dimension [5.78] (plane view): 1 and 2, concave and convex mirrors of the resonator; 3, outlet.

special design, which is stable for the small dimensions of the rectangular slab cross section to sustain a low-order waveguide mode and unstable for the large dimension of the slab of $38 \times 4.5 \times 0.225$ cm^3. The resonator is formed by spherical mirrors of curvature radii $R_1 = 5$ m and $R_2 = -4.22$ m, satisfying the confocality condition $R_1 + R_2 = 2L$, where $L = 39$ cm is the distance between the mirrors (Figure 5.16). Being successively reflected by the mirrors and enhanced, the radiation travels in the direction normal to the resonator axis (along the larger slab dimension) and eventually leaves the resonator through the exit hole. A maximum power of 240 W of one-mode radiation with 12% efficiency was obtained from the discharge mixture CO_2:N_2:He=1:1:3 with addition of 5% Xe at 60 Torr, electrical input power 2 kW, and $f = 125$ MHz. The whole laser head is placed in an evacuated capsule of $50 \times 15 \times 10$ cm^3.

5.5.2 Optically coupled multichannel laser systems

In recent years, multichannel module constructions have been studied intensively. If several (N) identical, optically uncoupled lasers produce each power P, the total radiation focused onto a target has the power NP. If, however, the laser modules

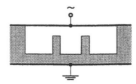

FIGURE 5.17
Three-element coupled waveguide resonator with a common transverse RF excitation.

are optically coupled so as to produce coherent radiation, the light wave amplitudes are summed up to give the resulting power N^2P. Various ways of coupling and stabilizing coherent modes have been considered in a recent review [5.79] of numerous investigations (see also [5.80]).

In a system of optically coupled identical lasers, as in the case of a single laser with a common resonator, there are intrinsic distributions of the field throughout the channel volumes, which possess a high quality factor. These may be called as super-modes or collective modes. The complicated configuration of the enhancing medium forms a spatial filter for the resulting field, providing a relatively small number of modes. The optical coupling can be most easily made using a common external mirror of the resonator, which would take some of the reflected radiation to the neighboring channels via diffraction spread [5.81, 5.82].

So-called 'nearest-neighbor' coupling was implemented in a set of semiconductor lasers [5.83], in two waveguide longitudinal dc-excited lasers [5.84] and later in RF-excited lasers [5.76, 5.85–5.87]. For instance, a compact CO_2 laser with optically coupled ceramic waveguide channels and a common transverse RF-excitation was created in [5.85]. The optical coupling of the channels is implemented via a narrow gap between the waveguide channels, through which radiation can partly travel to the neighboring channels (Figure 5.17). The optical coupling is based on choosing an appropriate distance between the channels and the gap size. A complete radiation synchronization was obtained for a system consisting of two and three channels. However, complete synchronization could not be achieved for a large number of channels. The reason is the complex dynamic behavior of a system of optically coupled waveguide channels, which reveals itself even in a two-channel scheme, as indicated by experiment and calculations [5.87].

The optical coupling suggested in [5.86] provided synchronized generation in one of the collective modes in six waveguide channels (also with transverse RF excitation). This was possible owing to a specific design of the waveguides capable of loss selection of modes. A set of six parallel waveguides of rectangular cross section is cut in halves, and one half is shifted transversely relative to the other by halfwidth of one channel (Figure 5.18). Minimum losses are characteristic of a mode which produces minimum scattering at the joint of the two halves, that is, the mode for which the electric field is equal to zero at the mid-channels

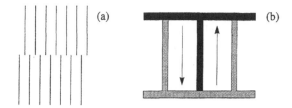

FIGURE 5.18
Six-element coupled waveguide resonator with staggered waveguide channels:
(a) ceramic waveguide channels (plane view); (b) front view illustrating selection of
collective modes.

of one half and on the walls of the other. The field distributions in both halves
of the waveguide sets turn out to be identical. This is the collective mode with
an antisymmetric field distribution relative to the waveguide middle in each half
(Figure 5.18). The radiation intensity of such a laser in the far zone has a double
hump structure.

For the optics of waveguide lasers, also see [5.88–5.90].

5.6 High flowrate CO_2 lasers excited by RF and combined RF-dc discharges

5.6.1 Specific power input limits in diffusional and convective cooling

It follows from the foregoing that an appropriate choice of RF discharge mode,
the frequency and the minimal size of the channel of capillary or slab configuration can provide an electrical power input into the plasma up to 100 W cm^{-3}
without exceeding the temperature limit $T_{max} \approx 550–600$ K admissible for CO_2
lasers, yielding up to 7–8 W cm^{-3} radiation [5.22, 5.35, 5.63a] and 1 kW total
power [5.22, 5.35]. It would seem that a substantial increase of the volume could
provide a much higher power of diffusion-cooled lasers. This is probably so, but
realistic promises of this approach in laser production do not go very far. The
matter is that high parameters are achievable only in tubes and slab channels of
exceptionally small diameters d or slab heights $h \approx d \approx 2–2.5$ mm. However,
one cannot infinitely increase the volume only by increasing the slab channel area,
even in a more compact geometry than the plane one. To try to create also highly
effective resonators is a very complicated task. Besides, the lasing quality, which
together with power is the focus of researchers' attention today, is, in this case,
not very high because of large diffraction divergence, at least in one direction:
$\Theta \approx \lambda_{CO_2}/h \approx 5 \times 10^{-3}$, where $\lambda_{CO_2} = 1.06 \times 10^{-3}$ cm.

Limitations on the discharge volume and lasing quality become, however, less rigorous, if one increases the small channel dimension, normal to the optical axis, from 2 mm to at least several centimeters. The gas overheating resulting then from the too slow diffusional cooling mechanism can be prevented by rapid pumping of the laser mixture through the discharge channel. Let l be the channel length along the flow and u the flow velocity. The power input limit into the discharge, at which the gas temperature reaches the admissible limit, is defined by an approximate equality

$$w = jE_p = \frac{\rho c_p(T_{max} - T_0)}{\tau_F} \qquad \tau_F = \frac{l}{u} \qquad (5.7)$$

where c_p is specific heat at constant pressure and ρ is the gas density proportional to the pressure.

If equation (5.6) is given the form of equation (5.7), the characteristic time of diffusional heat removal τ_T similar to the convective τ_F will take the form $\tau_T = \Lambda^2/\chi$, where $\chi = \lambda/\rho c_p$ is the thermometric conductivity, $\Lambda \approx h/\pi, R/2.4$. For example, at $p = 20$ Torr, when $\chi \approx 30$ cm^2 s^{-1}, the heat is removed through the channel walls of height $h = 6$ cm for $\tau_T \approx 10^{-1}$ s, then at the channel length $l = 30$ cm and velocity $u = 100$ m s^{-1} it will be carried out by the flow for $\tau_F = 3 \times 10^{-3}$ s. In principle, such a high flow rate would allow a 30-fold greater power per 1 cm^3 than heat removal to the walls.

Unfortunately, this is not completely feasible for large enough channel heights because of the positive column being liable to contraction [5.3] (see also Section 2.5.6). It is just the factor that sets limitations on specific energy input in high flowrate lasers operating on dc, RF or combined RF-dc discharges [5.3]. Perhaps, one can make the following general assertion: in capillary and slab lasers with transverse RF excitation, the limits on specific energy input and radiation power are set by the admissible gas heating, while in high flowrate lasers they are set by contraction, since the latter normally occurs before the limiting temperature for inversion in the CO_2 laser mixture is reached.

5.6.2   Stabilization of combined transverse RF-dc discharge

The first successful application of RF fields to high flowrate closed-cycle lasers [5.3] culminated in the design of a superhigh power (27 kW) CO_2 laser [5.4], which we briefly mentioned in Section 5.1. The discharge chamber was a large plane channel formed by thick dielectric plates of length $l = 53$ cm (along the gas flow and direct current) of height $h = 6.3$ cm and width $a = 244$ cm along the radiation direction. The resonator was designed in such a way that the radiation made many passes in a plane parallel to the channel plates (Figure 5.19). The dc electrodes were 360 cathode segments at the channel entrance and 4 parallel anode tubes at the exit. The RF voltage electrode plates were arranged above and below the dielectric walls of the channel. The power input into a mixture of CO_2:N_2:He=1:6.4:12.6 at $p = 30$ Torr and $f = 13.56$ MHz in a combined discharge prior to contraction was as high as 160 kW (2 W cm^{-3}), of which

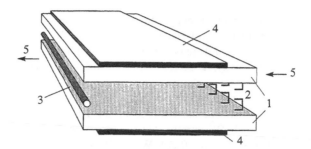

FIGURE 5.19
Large 27 kW high flowrate CO_2 laser with a combined discharge (RF plus dc
discharges) [5.4]: 1, dielectric plates; 2, rows of cathode elements; 3, one of the anodes;
4, RF electrodes; 5, direction of the gas flow.

60 kW were contributed by RF field and 100 kW by direct current. Without
RF field assistance, the maximum input was 60 kW (0.74 W cm^{-3}), after which
the discharge lost stability. The experiments preceding the laser design [5.91,
5.92] were done on setups of the same geometry but a smaller size ($a = 15$ cm,
$l = 46$ cm, h being nearly the same as the laser).

An attempt to apply RF voltage to the dc electrodes, that is, to initiate a
discharge with parallel dc and RF fields, failed because the contraction threshold
input energy could not be raised. Neither did local application of transverse
RF voltage to sites, thought by the authors to be the points of instability origin,
raise the stability threshold. It was only the application of transverse RF field
to the whole channel area that provided a higher contraction threshold of input
energy, up to 2.1 W cm^{-3}, slightly exceeding the limit in the large laser channel.
It was noted that nearly half of the RF power was dissipated in the 'resonance
absorption' sheaths of 0.6 cm thick, adjacent to the inner walls of the dielectric
plates. The authors of [5.91] erroneously identified the sheaths characteristic
of moderate-pressure RF discharges, which they seem to have observed, with
resonance absorption sheaths [5.93, 5.94] which represent quite a different thing
irrelevant to the conditions discussed. This point must be made quite clear, because
this interpretation was later cited by other researchers dealing with moderate-
pressure discharges.

Return to equation (1.50) for the impedance Z of a discharge gap with an RF
plasma. Suppose the gap length L to be much larger than the plasma electron
oscillation amplitude A, with which the average electrode sheath thickness coincides within a simplified theory. The absolute value of complex impedance (1.50)
at $L \gg A$ exactly coincides with the formula for $|Z|$ given in [5.94]. It is clear
from both that the resistance to the RF current has a maximum at plasma resonance
$\omega = \omega_p$, which is what is observed in the microwave range [5.93, 5.94]. But at
$f = 13.56$ MHz, the electron density $n_e = 2.3 \times 10^6$ cm^{-3} corresponding to the

resonance is by three orders of magnitude lower than the actual plasma density, while the collision frequency of electrons $\nu_m \approx 0.8 \times 10^{11}$ s^{-1} is three orders higher than $\omega = \omega_p$. Even if there is a resonance absorption in a narrow zone with such a low electron density, its excess over the usual, 'nonresonance' absorption is quite negligible, as $\omega = \omega_p \ll \nu_m$.

An elevated energy output near dielectric plates separating RF electrodes from a plasma is characteristic of the α-mode of the RF discharge (Section 2.3). Judging from the small RF current density of 2–3 mA cm^{-2}, which can be found from 60 kW RF power (0.74 W cm^{-3}) and an approximate value of $E/p \sim 10$ V cm^{-1} Torr^{-1}, the discharge described in [5.4, 5.91] operated in an α-mode. Although the large value of $ph = 190$ Torr cm in that laser goes beyond the range of α-mode operation between naked electrodes (Section 2.2, Figure 2.9), the presence of a thick dielectric layer between the electrodes and the plasma considerably extends the α-mode operation range [5.95] (Section 2.9).

There have been several hypotheses concerning the reasons for stabilization of a dc discharge by an RF field. Under the experimental conditions described in [5.4, 5.91], the frequency ω is noticeably less than the electron spectrum relaxation frequency ν_u, and the ionization rate sharply changes over a cycle (Section 1.4.1). In such cases ionization largely occurs during short moments when the resulting field (RF field plus constant field) has maximum values. According to [5.96], the stability of a combined discharge relative to ionization–overheating instability is high, because the instability develops only at the short moments of oscillating field maxima when the ionization occurs; during the rest of the cycle, the plasma recombines and is not subjected to instability. It was shown in [5.97], however, that such a stabilizing effect was quite weak in the case of a sine-wave field. The optimistic result of [5.96], which seems to have misled the author of [5.20] (see Section 5.6.3), resulted from the neglect of an important detail in the calculations (see [5.97]). The fact that stabilization of a combined discharge was observed in [5.91] only in crossed fields and that no stabilization was registered in parallel fields is a serious argument against the principal role of the mechanism [5.96], which is indifferent to whether the dc and RF fields are parallel or perpendicular to each other.

Lack of stabilization in parallel fields has drawn the attention to the suggestion made in [5.4] of a stabilizing effect of the resulting electrical vector 'sweep.' This suggestion has some physical grounds. Ionization–overheating instability develops by disturbing the plasma uniformity in the direction normal to the field vector, while the contracted current filament grows along the field. But if the field vector is not 'frozen to its place,' the filament slowly developing along the field has no time to follow its turns. This idea was developed in [5.98, 5.99] with respect to a discharge operating in a rotating RF field. In [5.100] the idea was implemented experimentally. RF voltage from two phase-shifted sources was applied to two pairs of mutually perpendicular plane insulated electrodes. The maximum power input into this discharge was up to 4 W cm^{-3}, though these experimental results are hard to be interpreted unambiguously. This kind of discharge excitation was not developed further, probably due to its complexity.

Assessing the applicability of combined discharges involving direct current, we would like to note that the 'cost' of RF energy is much higher than that of direct current. A combination of RF and dc fields would no doubt be profitable, if a high power dc discharge could be stabilized by a low power RF current. Such attempts, however, have failed in practice. In the setups used in [5.4, 5.91] the RF power was comparable with the dc power, and the total specific power input into the plasma was not large—only 2 W cm^{-3}. An input of 10–50 W cm^{-3} is possible only with a self-sustained transverse RF discharge in a chamber with a small interelectrode spacing. Indeed, since the publication of the work [5.4], the design of high power CO_2 lasers has been developing primarily through separate application of dc and RF discharges (Section 5.6.4).

As for combined discharges in channels of large height $h \approx 5-6$ cm, as in [5.4] where high specific energy inputs are not to be expected, more advantageous— from the above point of view—is a combined discharge suggested by one of the present authors [5.101] and implemented in a laser [5.102], in which the gas ionization and stabilization are performed by repeated low-power short electrodeless pulses, while most of the power input comes with direct current. The setup geometry and the field configuration are similar to those of [5.4] (Figure 5.19), but the cathode and anode were made as continuous tubes. The difference is that short pulses are applied to the electrodes insulated from the plasma. It was shown in [5.97] that their stabilizing effect considerably exceeds that of a sine-wave RF field. At present, improved modifications of this commercial CO_2 laser provide up to 8 kW continuous radiation and an average pulsed power of 4 kW at a pulse repetition rate of the order of 1 kHz.

5.6.3 A small gap as the major factor of RF discharge stability

In order to understand when and why an RF discharge possesses stability, special experimental investigations were performed [5.8, 5.103]. The data obtained showed that parallel RF and dc fields in a combined discharge should not be an obstacle to increasing its stability. By applying RF voltage to dc electrodes, as well as in the case of crossed fields, a steady discharge with a power input of 10 W cm^{-3} and more can be successfully created. This result greatly exceeds that obtained in [5.4, 5.91] in a high flowrate laser and that of [5.100] with a rotating field. The major factor in all these studies was the distance between the electrodes to which the RF voltage was applied. In the works [5.8, 5.103], the distance was chosen to be 6 cm, and the discharge was operating perfectly well, not only in crossed fields but also when both voltages were applied to the same electrodes. A stabilization effect was quite evident also in the experiments described in [5.4, 5.91] with $h = 6.3$ cm in a transverse RF field. But when RF voltage was applied to the dc electrodes, the interelectrode distance for this voltage was too large, 46 and 53 cm, so no stabilizing effect was observed. No doubt, the high energy inputs, up to 35 W cm^{-3}, achieved in the laser [5.104] using a transverse RF discharge in the flow, became possible owing to the small gap between dielectric coatings, only 2.5–4 cm (see below). The reason does not lie in the stabilization mechanism [5.96], to which an author of [5.20] refers and for which no 'interelectrode distance' exists at all, since an infinite homogeneous plasma is being considered.

It follows from the experiments [5.8, 5.17] that the stabilizing effect of an RF field on a combined discharge is primarily determined by the stability of the RF discharge itself under the given conditions. As a rule, when the pressure is several tens of Torr and the gap size $L = h$ is several centimeters, an RF discharge operates in a γ-mode. In this case a stable plasma can fill up the gap only if $L < L_{cr}$; and the higher the pressure, the smaller is L_{cr}. For example, in air at $p = 45$ Torr, $L_{cr} \approx 2$ cm (Section 2.5.6, Figure 2.22). If $L > L_{cr}$, the current can be raised only up to a certain value above which the discharge contraction occurs. But if $L < L_{cr}$, no contraction occurs in a normal γ-mode. It can be expected only after the plasma fills the whole gap space, as a result of the increased current and specific energy input in an anomalous mode. A probable physical interpretation of this effect was given in Section 2.5.6.[3]

Thus, when choosing optimal parameters of a laser system with the natural desire to put as much electrical power as possible into 1 cm^3 of plasma, one should rely on the undisputable fact established and confirmed by numerous and various experiments [5.13, 5.38, 5.17] that reduction of the interelectrode distance L contributes significantly to RF discharge stabilization and to raising the upper limit of energy input. This distance must decrease with increasing pressure but, of course, must be larger than the limit determined by the electrode sheath thickness [Section 5.4.1 and formula (5.5)]. In any case, the best results achieved so far were for the distances of a few centimeters and pressures of several tens of Torr. As for theoretical interpretation of this fact, there is still much to be clarified, and a convincing quantitative theory is still lacking.

5.6.4   Laser setups

A laser setup with a combined discharge [5.4] was described above. A reduction of the interelectrode distance to 2.5–4 cm allowed creation of an efficient high flowrate CO_2 laser excited by a transverse RF discharge [5.104] (Figure 5.20) with 4 kW continuous power and an average pulsed power of 6 kW at a pulse repetition rate from 0.1 to 15 kHz. The laser parameters are: the channel width along the

---

[3]The stabilization mechanism was attributed in [5.105] to the existence of highly conductive negative glow layers in the γ-mode, where the field is practically zero. This fact was used as a basis for the analysis of discharge stability. Further experiments [5.12, 5.17] by one of the authors of [5.105] showed, however, that the energy input into an α-discharge operating at $f = 81$ MHz and $h = 0.5$ cm could be brought up to 100 W cm^{-3}, which by two orders of magnitude exceeds the input limit calculated in [5.105]; though there are no highly conductive layers in the α-mode. So, the mechanism suggested in [5.105] should probably be discarded. Incidentally, the interpretation offered in Section 2.5.6 is equally applicable to the α-discharge.

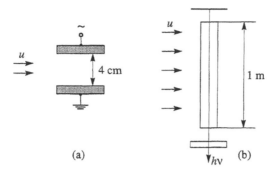

FIGURE 5.20
High flowrate laser with a transverse RF discharge [5.104]: (a) side view and (b) plane view.

optical axis 1 m, $u = 70-100$ m s^{-1}, $p = 45-75$ Torr, $f = 13.56$ MHz, the input power up to 35 W cm^{-3}, the radiation power up to 7 W cm^{-3}, and 13–20% of electrical power converted to radiation.

An RF laser with axial gas pumping has also been created. There are similar systems operating at dc fields. A longitudinal discharge similar to the one in Figure 5.1 is excited in a tube, through which a laser mixture is pumped at a high rate. Fast pumping provides a rapid heat removal and decreases, owing to flow turbulization, unfavorable effects of spatial nonuniformities on lasing quality. A transverse RF discharge (with external electrodes) in a tube with an axial flow improves the plasma homogeneity [5.106]. The laser setup described in [5.104, 5.104a] (Figure 5.21) provides 2 W cm^{-3} of radiation and even up to 10 W cm^{-3} by increasing the flow rate of the laser mixture. This points to great potentialities of tube systems with transverse RF excitation and fast axial gas flow. To promote this line of research in laser design, it appears reasonable to introduce into this system some elements of slab configurations, namely to use a coaxial geometry instead of a tube, which provides higher specific energy inputs at large total volumes. Adequate optical resonators have been designed for such systems [5.26].

In most of the available high flowrate lasers with frequencies of about $f \leq 27$ MHz and large energy inputs, the RF discharge operates in the γ-mode. The use of an α-mode might have a certain advantage, namely low power losses in the electrode sheaths, in which, in contrast with γ-modes, the active current component is very small. The difficulty, however, is that if one uses more effective elevated pressures $p \approx 50-100$ Torr, the gap between the electrodes is to be made smaller than 1 cm; otherwise, the discharge would become unstable. The gap, on the other hand, can be segmented into several plane channels by plates parallel to the electrodes, as in the experiment shown in Figure 5.13(a). Then the height of an individual channel h must be sufficiently small to fit the critical value pL_{cr} (Section 2.2) but must be several times the sheath thickness for a discharge to be

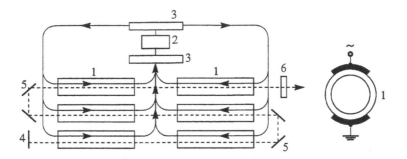

FIGURE 5.21
Six-channel RF-excited CO_2 laser with fast axial gas flow and a steady-state resonator
[5.104*a*]: 1, discharge tubes; 2, pumping device; 3, coolers; 4, end mirror; 5, turning
mirrors; 6, outlet semitransparent mirror. Solid lines with arrows, gas flow; dashed line,
beam path in the resonator. On the right, end view of a quartz discharge tube and of RF
electrodes.

ignited. Experiments conducted at $p \approx 100$ Torr, $u \approx 80$ m s^{-1}, $f = 81$ MHz,
and 1 cm length along the flow have been shown the feasibility of such a system:
the energy input into an α-discharge was as high as 100 W cm^{-3}. The difficulty of
arranging a resonator for such a system can be overcome by mounting it externally
downstream the flow. This is feasible, since the inversion in the lasing medium
persists for a period of time large enough to still go on at a distance of 10 cm from
the discharge zone, with $u \sim 100$ m s^{-1}.

To conclude, we mention some other reports on the spatial structure of high
flowrate RF discharges, as applied to designing CO_2 power lasers [5.107–5.121].

5.7   Lasers with alternative active media

It seems quite natural to try to apply RF discharge, so effectively used for CO_2
laser excitation, to other types of laser. Good results have been obtained with a CO
waveguide laser at the frequency 142 MHz [5.122, 5.123]. In the laser reported
in [5.122], two faces of a waveguide channel of a square cross section and the size
$38.6 \times 0.225 \times 0.225$ cm^3 were formed by massive aluminum plates that also served
as the electrodes. Two other channel faces were formed by dielectric Al-ceramic
slabs squeezed between the electrode plates. The discharge heat was removed
through the electrodes cooled by circulating liquid methanol. Plane mirrors with
the reflectivities 99.5% (nontransparent) and 95% (partly transparent) served as
an optical resonator. Maximum radiation power of 10.1 W and 6.5% efficiency
were obtained at the cooling liquid temperature $t = -23°$C and $p = 70$ Torr with
a mixture of CO:N$_2$:He:Xe:O$_2$ =1:1:10:1:0.5. At $t = -1°$C, the power decreased

down to 4.4 W and the efficiency to 4%.

The authors of [5.122] attribute these high values to effective cooling (owing to heat conductive materials used) to elimination of impurities from the lasing medium and to a lower value of E/p as compared to a dc discharge. The smaller E/p value is thought to be associated with lower electron losses due to their oscillations in the oscillating field. However, the latter interpretation does not seem quite credible, because there are no reasons why plasma losses in an RF positive column should be smaller than in a dc column at the same geometry, gas composition and a fairly high pressure. Rather, the reason lies in the distribution pattern of ionization sources in an RF discharge with a narrow gap: the major ionization sources are localized near the gap boundaries, in the region between the plasma and the sheaths. Owing to the small gap length, 2 mm only, the plasma in its middle is sustained not only by the local field but partly by these sources. For this reason, the field in the gap middle is lower, causing a weaker glow. This happens in both α- and short γ-discharges (Sections 2.3, 2.5).

It is most important for CO lasers that the gas temperature be as low as possible, so the problem of effective heat removal is even more acute here than with CO_2 lasers. Therefore, slab systems permitting a gain in volume and power without decreasing the cooling efficiency hold considerable promise. This was confirmed by direct experiments [5.17, 5.123a]. A discharge was excited in a slab filled with a mixture of $CO:N_2:He:O_2:Xe=1:1:8:0.5:1$ at $p = 30$ Torr, wall temperature $t = -21°C$ and $f = 81$ MHz. The radiation power of 85 W was produced even without optimizing the mixture composition and the resonator parameters. Similar results were obtained by other workers [5.124, 5.125].

Promising results may also be expected from transverse RF discharges with a slab geometry to excite He-Ne lasers. Conventionally, such lasers are excited by a longitudinal dc discharge in a tube, the tube having an optimal radius R_{opt} for this laser. Small tube radii correspond to small volumes and powers. But one cannot increase the radius considerably. First, the plasma electron temperature drops in this case [5.3], decreasing the excitation efficiency of the upper laser level. Second, at larger R the diffusion of excited Ne(1s) atoms toward the walls where they are deactivated becomes slower, preventing the depletion of the lower laser level [5.126, 5.127]. The volume and power can be increased by making the cross section elliptical, with the small half-axis of the ellipse being close to the optimal radius R_{opt} of the circular cross section. On the other hand, at a large axis ratio, the discharge cannot fill up the whole elliptical tube. An optimal result can be achieved with the 4:1 ratio [5.126]. A slab design with the slab size h corresponding to optimal T_e, excitation rate and depletion of the lower laser level can provide a power gain if the slab channel area is made larger, as in CO_2 lasers. Application of transverse RF discharges for Ar-Xe laser excitation at moderate pressures also gives promising results [5.128].

In the above discussion of CO_2 lasers, we gave examples of the use of RF γ-discharges, with positive column plasma serving as an active medium. The near-electrode regions (sheaths and negative glow regions) sustained the current in the positive column. In some situations, however, it is the negative glow γ-plasma that can be used profitably as a lasing medium, while the positive column can be entirely eliminated by reducing the gap size, to save on the voltage. These measures are necessary when laser radiation sources are ion transitions [5.129], for example, in metal vapor lasers [5.130–5.132] requiring high ion densities. This is exactly what happens in the negative glow and at the beginning of the Faraday space.

A transverse RF discharge was also used to excite electrogasdynamic CO lasers operating on combustion products [5.133–5.135]. Fuel combustion products containing CO are cooled by passing them through a system of nozzles, like in a gasdynamic laser, and then the cooled mixture is excited by a transverse RF discharge with insulated electrodes ($f = 1.76$ MHz). The gas flow, the RF current, and the resonator optical axis are all perpendicular to one another. The reader can also find information on RF-excited CO lasers in [5.136–5.138].

5.8   Magnetic stabilization of slab discharges

Many attempts have been made to improve gas laser characteristics by applying a magnetic field to a discharge to affect its properties [5.139–5.152]. They all were aimed at increasing plasma homogeneity and stability, trying to bring the stability limits up to the largest possible energy inputs into the plasma. Experiments were done on dc discharges in a shaped magnetic field [5.140], on a discharge in a coaxial gap in a transverse constant magnetic field [5.142] and with a magnetic field periodically distributed in space and changing its direction at certain length intervals [5.146]. Discharges in a magnetic field with a time-rotating vector [5.148] and a stabilizing effect of a time-variable magnetic field [5.145] were studied.

Naturally, magnetic fields have also been used in the slab systems we discussed above, which hold much promise for power diffusion-cooled gas lasers without fast flow. The effect of a magnetic field was studied in both coaxial [5.147, 5.148] and plane [5.149–5.152] slab-type gaps. Especially good results were obtained in [5.151] with a three-kilowatt diffusion-cooled CO_2 laser operating in a new discharge mode (Macken discharge) produced in [5.149, 5.150].

The operation principles of the Macken discharge and of the laser are illustrated in Figure 5.22. The laser consists of two optically coupled plane slab sections, one above the other, each of the size $270 \times 19 \times 0.8$ cm^3. The resonator is designed such that a beam first goes through one and then through the other section. For simplicity and in order to focus on the discharge arrangement, we have shown in Figure 5.22 only one section. Along the largest dimension, there are a lot of small sectioned electrodes, to which constant voltage is applied through individual ballast resistances. Constant magnets are used to apply to the volume a rather uniform constant magnetic field $H \approx 600-800$ Oe normal to the slab

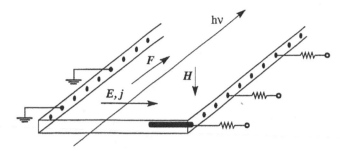

FIGURE 5.22
Macken discharge [5.151]. See text.

plane. Protruding into the volume at one end is a nonstandard electrode, whose function is to localize the discharge ignition just to this site. Ponderomotive force $F = (1/c)(j \times H)$ directed along the largest slab dimension promotes, at a sufficiently high field, a uniform filling of the space with current. At zero magnetic field, the current flows through individual filaments extending between the small electrode sections facing one another. The filaments are unstable and mobile, and two neighboring ones sometimes fuse together. The magnetic field of a certain value that varies with the system parameters rectifies this situation, contributing to the slab filling with the plasma.

Although good results have been obtained using this principle, and a fairly efficient commercial CO_2 laser has been constructed, this system possesses a serious disadvantage. One has to use in it a very large number of electrode sections, each cathode section having to be connected to a supply via an individual ballast resistance. For the explanation, see, for example, [5.3].

A possible way of circumventing this unpleasant necessity and of producing a uniform discharge was suggested and studied experimentally in [5.152], developing the ideas of [5.149–5.151]. These experiments dealt with a magnetically stabilized RF-dc combined discharge. In contrast with [5.91, 5.4], where the powers released by both currents were comparable, the RF power in this system was relatively small. The RF discharge served to initiate a dc discharge and to create the necessary initial ionization, which was exactly what the authors of [5.91, 5.4] aimed at and which was in a sense implemented in [5.101, 5.102], using repetitive pulses instead of RF field. These experiments were carried out in a gas discharge chamber of $20 \times 60 \times 1$ cm^3 formed by two metallic plates 1 (see Figure 5.23) covered from the inside with a thin dielectric layer 2. The plates strengthened the construction during evacuation of the chamber and simultaneously served as magnetic conductors providing a uniform transverse magnetic field of several hundreds (up to 1,000) of oersteds. Constant magnets located on the chamber sides along its largest dimension induced a magnetic field. Constant voltage was applied to solid electrodes 3 and 4, which could be separated by as much as a

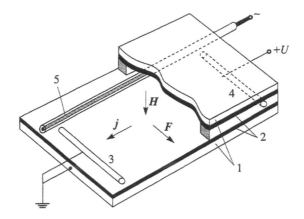

FIGURE 5.23
Experimental design [5.152] for a combined magnetically stabilized discharge.

maximum distance $L = 40$ cm. An auxiliary RF discharge was used to provide the initial localization of a dc discharge at one end of the chamber, from which the plasma 'spread' throughout the slab volume under the action of the pondero-motive force $F = (1/c)(j \times H)$. A capacitive RF discharge was operated at the frequency 13.56 MHz between dielectric-coated active electrode 5 and grounded metallic plates 1. Experiments were done with discharges in air and in the laser mixture $CO_2:N_2:He=1:1:3$ at $p = 20$ Torr.

As the dc strength increased, the plane plasma column expanded along elec-trodes 3 and 4 along the ponderomotive force F, while the constant electrode voltage remained practically unchanged—as in the case with a normal glow dis-charge. The luminosity intensity also remained more or less constant, and the glow and the current just filled the space. At a sufficiently high magnetic field, all the slab volume could be filled with a uniform plasma. Over 3 W cm^{-3} power could be put into the plasma, which is quite acceptable for creating a power slab CO_2 laser by scaling the design (for details, see [5.145]).

5.9 Plasmachemical technology

The increasing integration and decreasing response time of microschemes, as well as the development of large and superlarge integrated microschemes, require new technologies that would permit size reduction of the structure elements with a simultaneous increase in the precision, reliability, and automation of their produc-tion. Plasma technology meets these requirements and still offers much promise. There have been many books and review articles published on application of plasma technology to microelectronics (see, for example, [5.153–5.160]). Here we will briefly discuss processes that involve the use of plasmachemical reactors with an RF capacitive discharge.

The basic processes in ion-plasma treatments are ion-plasma etching and deposition of thin films and coatings.

5.9.1 Etching

Etching is aimed at total or local removal of a surface layer of material in order to clean it from contaminants or to make a pattern. Out-of-date etching technologies using, in particular, chemically active liquids, yield commercial scheme elements of 3–5 μm in size, while laboratory elements may be made as small as 1 μm— which was estimated by specialists to be the resolution limit of this technique. This is primarily due to the fact that liquid etching removes material isotropically, equally in all directions, so liquid solvents cannot penetrate into narrow grooves because of the surface tension. Besides, storage of the liquids used is inconvenient and costly. New etching technologies, often termed as elionic (meaning 'electron' plus 'ion') technologies, provide electronics elements with a resolution less than 0.1 μm, the limit achieved so far being 0.01 μm. Part of the elionic technology is lithography based on the use of large-molecular compounds, or resists, possessing the ability to change their properties under the action of various radiations. The most common lithographic technique is contact and projection photolithography, which is used to make precisely the prescribed patterns in films directly on sub-strates for semiconductor devices and integrated microschemes. Patterns can be created in dielectric (SiO_2, Si_3N_4), metallic (Al, Cu, Au, Ni) or semiconductor (Si, Ge) films. Figure 5.24 illustrates the basic etching stages of a lithographic process. For example, silicon substrate 1 with silicon dioxide film 2 and predeposited light-sensitive material 3 called as photoresist is exposed to ultraviolet radiation through a special glass mask—photomask 4—having transparent and opaque parts. The radiation changes the photoresist properties; and during the development in special chemical reagents, some parts of the photoresist layer are removed. In this way the mask pattern is transferred to the layer to be etched. The next stage is the critical one to this technology. This is plasmachemical or ion etching of a solid-state film or substrate using the exposed mask. The last stage is the removal of the photoresist residues from the treated sample. The photoresist can also be removed plasmachemically.

Among elionic technologies, there are impurity ion incorporation and plasmachemical deposition of dielectric and metallic films.

Quantitatively, etching processes are characterized by etch depth h and etch line width L, known as resolution, as well as by etch rate equal to the depth of the surface layer removed per unit time, $v_{et} = h/t$.

The major qualitative characteristics of an etching process are uniformity, anisotropy and selectivity. Uniformity is measured as the ratio of maximum-to-minimum etch rates on a treated surface area (up to 10 cm in diameter). The

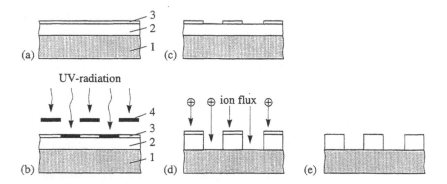

FIGURE 5.24
Basic lithographic stages in elionic technology. (a) Sample to be treated: 1, substrate, for
example a Si; 2, solid state film, for example a SiO$_2$ film, in which a pattern is made;
3, layer of polymer photoresist. (b) Sample UV-irradiated through a mask 4 with a
prescribed pattern. (c) Exposed photoresist parts removed in a solvent or in a reactive gas
in a plasmachemical reactor. (d) Etching process—transfer onto a solid state film of the
pattern by intense ion bombardment or by more selective ion-chemical and
plasmachemical processes. (e) Final stage—photoresist removal (chemically or
plasmachemically) leaving the desired pattern on the film.

highest quality of etching is achieved only if it is made mostly in the direction
normal to the treated surface. This is so-called anisotropical etching. The smaller
the pattern details (or the larger their height-to-width ratio), the higher are the
requirements on etching anisotropy, which is characterized by the etch ratio in the
normal and lateral directions.

In etching, ions bombard simultaneously the mask and the substrate, on which
the desired pattern must be obtained. The material of the substrate and of the
mask have different etch rates. Etching selectivity is characterized by the etch
rate ratio of the substrate and the mask, the etch rates being measured normal to
their surfaces. In order to obtain a desired result, the type of etching, the mask
material, the working gas and the etching regime must be chosen very carefully.
Consider the commonly used types of etching in terms of these requirements.

In *ion etching*, the surface layers of material are removed by bombarding it
with energetic inert gas ions, usually argon ions; and this process is, in a sense,
similar to pneumatic sand spraying used to clean dirty surfaces. Ion bombardment
is carried out at low pressure. The atom to be removed must gain an energy
exceeding that binding it to the surface, and this is possible only if the ion energy
$\mathcal{E}_i$ is larger than a certain threshold sputtering energy $\mathcal{E}_s$. The value of $\mathcal{E}_s$ is
one of the most important parameters of the process. At $\mathcal{E}_i < \mathcal{E}_s$, ions recoil
from the surface without knocking out the atoms or become adsorbed and then
desorbed. In this case, the ions are neutralized by electrons extracted from target
material by their electrical field. The threshold energies $\mathcal{E}_i$ for most metals and

semiconductors exceed the binding energies of surface atoms $U \approx 1{-}10$ eV (sublimation energy), because a bombarding ion gives off only some of its kinetic energy to an atom in an elastic collision. It is determined by the accommodation coefficient $K_a \approx 4M_i M_a / (M_i + M_a)^2$, where M_i and M_a are the masses of the ion and of the surface atom. For example, the experimentally measured characteristic values of $\mathcal{E}_s$ for Al, Cu and Ge bombarded by argon ions are 9.7, 12.5 and 18.5 eV, respectively.

A quantitative parameter of the sputtering process is the sputtering coefficient K_s equal to the number of atoms sputtered off by one ion hitting a surface. For example, when an Al surface is bombarded by Ar^+ ions with $\mathcal{E}_i = 1$ keV at normal incidence, $K_s \approx 1.94$ atom per ion. For Cu, Ge and Si, these coefficients are 3.64, 1.5 and 1 atom per ion, respectively.

The energy range used for ion etching of microstructures is usually $\mathcal{E}_i = 0.5{-}5$ keV. At $\mathcal{E}_i < 0.5$ keV, the rates of the process are too small for practical application. The upper limit for $\mathcal{E}_i$ is determined by the admissible damage of the treated surface and by the mask material durability.

As for ion energy, only 5% of it is usually spent for sputtering, the other 95% being spent for heating the target material, for the damage and destruction of the surface, for ion incorporation, secondary electron emission and electromagnetic radiation. So, this process has a low efficiency.

Much more efficient is *ion-chemical etching*. The principle of this method is that conventional ion sputtering is accompanied by reactions with chemically active ions of the plasma that fall onto the treated surface. In the simplest case, chemically active oxygen is added to the inert gas (argon) to change the etch rate and provide a high selectivity of the etching. For example, metallic films of Ti, Cr, Mn, Mo, Ta and Al are often used as masks, and the etch rate of the films is decreased by adding oxygen to the inert gas. Such masks are used for deep etching of materials, whose etch rate depends very little on oxygen addition (Si, SiO_2, Si_3N_4, Cu, Ag and others).

Microelectronics technology widely uses *plasmachemical etching* to remove a surface layer of material in a chemically active plasma. In this process, active gas molecules decompose in a discharge (an RF plasmachemical reactor can be used to advantage for this purpose) into reactive particles—electrons, ions and free radicals—which interact chemically with substrate surfaces or with deposited solid-state films of various materials. For instance, RF plasma electrons can decompose CF_4 into F^- ions and CF_3^+ radicals, which can etch various materials (metals, polymers) at different rates. Usually, plasmachemical etching is carried out at a pressure of $0.1{-}10$ Torr. This process provides etching resolution approximately equal to the thickness of the layer being etched. The relatively low value of the resolution is due to diffusion of plasma particles to the surface being treated. Plasma etching provides a high treatment rate (up to $2{-}10$ nm s^{-1}) but, we repeat, with a low resolution. For this reason, this method is usually used for operations not requiring high etching precision, for instance, for surface cleaning and removal of a photoresist following photolithography, which is an integral part of production technologies for semiconductor devices and integrated microschemes.

5.9.2   Deposition of thin films and coatings

A process close to etching is deposition of film coatings, because in both cases a target material is sputtered in a low-pressure discharge.

Deposition of films and coatings is another extensive area of application of plasmachemical technology. Depending on the layer thickness (from one to several micrometers) and the deposition technique used, the structural, mechanical, electrical, dielectrical, piesoelectrical, ferromagnetic and superconductive properties of the film may, or may not, coincide with those of the respective single crystal. This feature is used in making thin film insulators, capacitors, resistors, piesoelements for magnetic storage, thermocouples, semiconductor devices and microschemes. Thin films are employed not only in microelectronic devices, but also to protect materials from corrosion, for decorative coatings, to increase the durability and to improve the performance quality of cutting instruments.

Ion-plasma technologies are used to produce practically any kind of film coating, from elemental to multicomponent ones. Ion-plasma treatment can be performed in dc glow discharges and in RF discharges, including magnetron discharges.

Ion-plasma deposition is a process of making film coatings by sputtering target material in an inert gas plasma when a negative electrical potential is applied to the target. The available modifications of this method differ in the technical facilities used for plasma creation and for target bombardment. These are cathode sputtering and magnetron sputtering. Gas pressure in an ion-plasma deposition process varies from 10^{-3} to 10^{-1} Torr. The high energy of sputtered particles provides high density of films and their good adhesion to the substrates. It is possible to produce films of multicomponent materials, alloys and compounds without changing their stoichiometric composition. Ion-plasma deposition is used as part of technology for producing films of practically any materials—metals, semiconductors and dielectrics.

Reactive ion-plasma deposition is used for making multicomponent film coatings by sputtering a target in a plasma containing a reactive gas. Films are formed through chemical interaction of the sputtered material and the reactive gas, usually on the target surface. Different reactive gases are used: methane for producing metal carbide films, oxygen for oxide films, nitrogen for nitrite films, selenium vapors for selenite films, and so on. By varying the reactive gas and by regulating its partial pressure, one can change the film composition. Reactive ion-plasma deposition is performed by all kinds of ion-plasma sputtering to produce oxide, metal nitrite and semiconductor films.

To conclude, we give in Figure 5.25 a general schematic diagram of a planar RF plasmachemical reactor. We discussed the details of this type of system in Chapter 3 when analyzing discharge processes. As a rule, the electrodes used in ion etching reactors have different areas. The samples to be treated are placed on

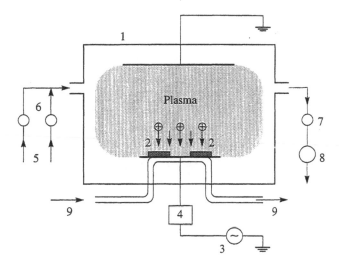

FIGURE 5.25
General scheme of a planar plasmachemical RF reactor: 1, discharge chamber; 2, samples
to be treated; 3, RF generator; 4, matching element; 5, working gases; 6, mass flowrate
control; 7, pressure control in the discharge chamber; 8, vacuum pump; 9, cooling water.

the smaller water-cooled electrodes which serve as targets, the ion bombardment
of which is more intense. Typical parameters of technological regimes for ion
etching are: pressure $5 \times 10^{-3} - 10^{-1}$ Torr, interelectrode distance 4–8 cm, RF
electrode voltage 0.8–5 kV, frequency 1.76–13 MHz, target electrode radius 1–
5 cm, etch rate $v_{et} = 0.1 - 0.8$ nm s^{-1} and argon as working gas. For a sufficiently
good transfer of the power to the load, the RF generator must be matched with the
sputtering equipment (Section 4); otherwise, the etch rate will be lower.

References

Chapter 1

1.1 Raizer, Yu. P., *Gas Discharge Physics*, Springer-Verlag, Berlin, New York, 1991.

1.2 Raizer, Yu. P., *Fiz. Plasmy*, **5**, 408, 1979.

1.3 Levitskii, S. M., *Zh. Tekh. Fiz.*, **27**, 970, 1001, 1957.

1.4 Godyak, V. A., *Fiz. Plasmy*, **2**, 141, 1976.

1.5 Velikhov, E. P., Kovalyov, A. S., and Rakhimov, A. T., *Physical Phenomena in the Gas Discharge Plasma*, Nauka, Moscow, 1987 (in Russian).

1.6 Godyak, V. A. and Kuzovnikov, A. A., *Fiz. Plasmy*, **1**, 496, 1975.

1.7 Godyak, V. A., *Zh. Tekh. Fiz.*, **41**, 1364, 1971.

1.8 Lieberman, M. A., *IEEE Trans. Plasma Sci.*, **16**, 638, 1988.

1.9 Yatsenko, N. A., *Zh. Tekh. Fiz.*, **51**, 1195, 1981.

1.10 Vidaud, P., Durrani, S. M. A., and Hall, D. R., *J. Phys. D.*, **21**, 57, 1988.

Chapter 2

2.1 Barkalov, A. D., Gavrilyuk, V. D., Gladush, G. G., Glova, A. F., Golubev, V. S., and Lebedev, F. V., *Teplofiz. Vys. Temp.*, **16**, 265, 1968.

2.2 Kovalyov, A. S., Rakhimov, A. T., and Feoktistov, V. A., *Fiz. Plasmy*, **7**, 1411, 1981.

2.3 Raizer, Yu. P. and Shneider, M. N., *Fiz. Plasmy*, **13**, 471, 1987; **14**, 226, 1988.

2.4 Yatsenko, N. A., *Zh. Tekh. Fiz.*, **51**, 1195, 1981.

2.5 Yatsenko, N. A., *Teplofiz. Vys. Temp.*, **20**, 1044, 1982.

2.6 Yatsenko, N. A., *Zh. Tekh. Fiz.*, **50**, 2480, 1980.

2.7   Gladush, G. G. and Samokhin, A. A., *Zh. Prikl. Mekh., Tekh. Fiz.*, **6**, 16, 1986.

2.8   Raizer, Yu. P. and Surzhikov, S. T., *Teplofiz. Vys. Temp.*, **26**, 1044, 1982.

2.9   Raizer, Yu. P. and Shneider, M. N., *Fiz. Plasmy*, **17**, 1362, 1991.

2.10 Smirnov, A. S., *Zh. Tekh. Fiz.*, **54**, 61, 1984.

2.11 Kovalyov, A. S., Nazarov, A. I., Rakhimov, A. T., Suetin, N. V., and Feoktistov, V. A., *Fiz. Plasmy*, **12**, 1264, 1986.

2.12 Godyak, V. A. and Khanneh, A. S., *IEEE Trans. Plasma Sci.*, **PS-14**, 112, 1986.

2.13 Belenguer, Ph. and Boeuf, J. P., *Phys. Rev. A*, **41**, 4447, 1990.

2.14 Godyak, V. A. and Ganna, A. Kh., *Fiz. Plasmy*, **6**, 676, 1980.

2.15 Lieberman, M. A., *IEEE Trans. Plasma Sci.*, **16**, 638, 1988; *ibid*, **17**, 338, 1989.

2.16 Kovalyov, A. S., Muratov, E. A., Ozerenko, A. A., Rakhimov, A. T., and Suetin, N. V., *Fiz. Plasmy*, **11**, 882, 1985.

      Velikhov, E. P., Kovalyov, A. S., and Rakhimov, A. T., *Physical Phenomena in the Gas Discharge Plasma*, Nauka, Moscow, 1987 (in Russian).

2.17 Smirnov, A. S. and Tsendin, L. D., *Zh. Tekh. Fiz.*, **60**, 56, 1990; **61**, 64, 1991.

2.18 Raizer, Yu. P., *Gas Discharge Physics*, Springer Verlag, Berlin, New York, 1991.

2.19 Yatsenko, N. A., *Zh. Tekh. Fiz.*, **52**, 1220, 1982.

2.20 Vidaud, P., Durrani, S. M. A., and Hall, D. R., *J. Phys. D.*, **21**, 57, 1988.

2.21 Raizer, Yu. P. and Shneider, M. N., *Fiz. Plasmy*, **16**, 878, 1990.

2.22 Yatsenko, N. A., *Zh. Tekh. Fiz.*, **58**, 294, 1988.

2.23 Rusanov, V. D. and Fridman, A. A., *Physics of Chemically Active Plasma*, Nauka, Moscow, 1984 (in Russian).

2.24 Mnatsakanyan, A. Kh. and Naidis, G. V., *Teplofiz. Vys. Temp.*, **23**, 640, 1985.

2.25 Borisov, N. D., Gurevich, A. B., and Milih, G. M., *An Artificial Area in the Atmosphere*, IZMIR RAN, Moscow, 1986 (in Russian).

2.26 Biberman, L. M., Vorobyev, V. S., and Yakubov, I. T., *Kinetics of Nonequilibrium Low-Temperature Plasma*, Plenum, New York, 1987, Chap. 9.

2.27 Raizer, Yu. P. and Shneider, M. N., *Teplofiz. Vys. Temp.*, **25**, 1008, 1987.

2.28 Raizer, Yu. P. and Shneider, M. N., *Fiz. Plasmy*, **18**, 1476, 1992.

2.29 Mewe, R., *Physica*, **47**, 373, 1970.

2.30 Raizer, Yu. P. and Shneider, M. N., *Fiz. Plasmy*, **19**, 994, 1993.

2.31 Raizer, Yu. P. and Shneider, M. N., *Teplofiz. Vys. Temp.*, **29**, 1041, 1991.

2.32 Kaneda, T., Kubota, T., Ohuchi, M., and Chang, J. S., *J. Phys. D*, **23**, 1642, 1990.

2.33 Myshenkov, V. I. and Yatsenko, N. A., *Zh. Tekh. Fiz.*, **51**, 2055, 1981.

2.34 Kolesnichenko, Yu. A., Matyukhin, V. D., Muravyov, V. F., and Smazkov, S. I., *Dokl. Akad. Nauk SSSR*, **246**, 1091, 1979.

2.35 Yatsenko, N. A., *Proc. XX Int. Conf. on Phenom. in Ionized Gases*, Pisa, Italy, Vol. 5, p. 1159, 1991.

2.36 Yatsenko, N. A., *Ph.D. Dissertation*, Moscow Physical Technical Institute, 1978.

2.37 Godyak, V. A. and Khanneh, A. S., *J. Phys. (France)*, **40**, C7-147, 1979.

2.38 Zvyagintsev, A. V., Mitin, R. V., and Pryadkin, N. I., *Zh. Tekh. Fiz.*, **45**, 278, 1976.

2.39 Rykalin, N. M., Kulagin, I. D., Sorokin, L. M., and Gugnyak, A. V., *Zh. Tekh. Fiz.*, **46**, 730, 1976.

2.40 Mitin, R. V., Shamrayev, V. T., and Yaremenko, V. I., *Teplofiz. Vys. Temp.*, **21**, 595, 1983.

2.41 Schwab, H. A. and Manka, C. K., *J. Appl. Phys.*, **40**, 696, 1969.

2.42 Schwab, H. A. and Hotz, R. F., *J. Appl. Phys.*, **41**, 1503, 1970.

2.43 Schwab, H. A., *Proc. IEEE*, **59**, 613, 1971.

2.44 Grigorovich, R. and Kristesku, J., *Opt. Spektrosk.*, **6**, 129, 1959.

2.45 Kachanov, A. V., Trekhov, E. S., and Fetisov, E. P., *Zh. Tekh. Fiz.*, **40**, 340, 1256, 1976.

2.46 Yatsenko, N. A., *III All-Union Conf. on Gas Discharge Phys., Abstr.*, Kiev, Vol. 1, p. 143, 1986 (in Russian).

2.47 Yatsenko, N. A., *Preprint No. 338*, Institute for Problems in Mechanics, Russian Academy of Sciences, 1988 (in Russian).

2.48 Yatsenko, N. A., *Inzh.-Fiz. Zh.*, **62**, 739, 1992.

Chapter 3

3.1 Coburn, J. W. and Kay, E., *J. Appl. Phys.*, **43**, 4965, 1972.

3.2 Koenig, H. R. and Maissel, L. I., *IBM J. Res. Dev.*, **14**, 168, 1970.

3.3 Lieberman, M. A., *J. Appl. Phys.*, **65**, 4186, 1989.

3.4 Lieberman, M. A. and Savas, S. E., *J. Vac. Sci. Technol. A*, **8**, 1632, 1990.

3.5 Köhler, K., Coburn, J. W., Horne, D. E., Kay, E., and Keller, J. H., *J. Appl.*

Phys., **57**, 59, 1985.

3.6   Raizer, Yu. P. and Shneider, M. N., *Fiz. Plasmy*, **14**, 226, 1988.

3.7   Gould, R. W., *Phys. Lett.*, **11**, 236, 1964.

3.8   Christensen, O. and Brunot, M., *Levide, Les Couches Minces*, **165**, 37, 1973.

3.9   Chapman, B., *Glow Discharge Processes—Sputtering and Plasma Etching*, Wiley-Interscience, New York, 1980.

3.10 Vossen, J. L., *J. Electrochem. Soc.*, **126**, 2345, 1979.

3.11 Horwitz, C. M., *J. Vac. Sci. Technol. A*, **1**, 60, 1983.

3.12 Cox, T. I., Deshmukh, V. G., Hope, P. A., Hydes, A. J., Baithwaite, N. St., and Benjamin, N. M., *J. Phys. D*, **20**, 820, 1987.

3.13 Yeom, G. Y., Thornton, J. A., and Kushner, M. J., *J. Appl. Phys.*, **65**, 3825, 1989.

3.14 Godyak, V. A., *Fiz. Plazmy*, **2**, 141, 1976.

3.15 Godyak, V. A., *Soviet Radio Frequency Discharge Research*, Delphic Associates Inc., Falls Church, VA, 1986.

3.16 Godyak, V. A. and Khanneh, A. S., *J. Phys. (France)*, **40**, C7-147, 1979.

3.17 Raizer, Yu. P. and Shneider, M. N., *Plasma Sources Sci. Technol.*, **1**, 102, 1992.

3.18 Godyak, V. A., Piejak, R. B., and Alexandrovich, B. M., *IEEE Trans. Plasma Sci.*, **19**, 660, 1991.

3.19 Raizer, Yu. P., *Gas Discharge Physics*, Springer-Verlag, Berlin, New York, 1991.

3.20 Biehler, S., *Appl. Phys. Lett.*, **54**, 317, 1989.

3.21 Kaganovich, I. D. and Tsendin, L. D., *Pis'ma Zh. Tekh. Fiz.*, **16**, 4, 1990.

3.22 Smirnov, A. S. and Tsendin, L. D., *Proc. XIX Int. Conf. on Phenom. in Ionized Gases*, Belgrad, Vol. 3, p. 456, 1989.

3.23 Lieberman, M. A., *IEEE Trans. Plasma Sci.*, **16**, 638, 1988.

3.24 Barnes, M. S., Colter, T. J., and Elta, M. E., *J. Appl. Phys.*, **61**, 81, 1987.

3.25 Meijer, P. M. and Goedheer, W. J., *IEEE Trans. Plasma Sci.*, **19**, 170, 1991.

3.26 Anaratone, B. M., Ku, V. P., and Allen, J. E., *Proc. XXI Int. Conf. on Phenom. in Ionized Gases*, Bochum, Vol. 1, p. 31, 1993.

3.27 Aleksandrov, A. F., Godyak, Y. A., Kusovnikov, A. A., and Sammani, A. Y., *Proc. VIII Int. Conf. on Phenom. in Ionized Gases*, Viena, p. 165, 1967.

3.28 Yatsenko, N. A., *Inzh.-Fiz. Zh.*, **62**, 739, 1992.

3.29 Levitskii, S. M., *Zh. Tekh. Fiz.*, **27**, 970, 1001, 1957.

3.30 Bohm, D., in *Characteristics of Electrical Discharges in Magnetic Fields*, Eds. A. Cutrie and R. K. Wakerling, McGraw-Hill, New York, Ch. 2, 1949.

3.31 Rieman, K.-U., *J. Appl. Phys.*, **65**, 999, 1989.

3.32 Garscadden, A. and Emeleus, K. G., *Proc. Phys. Soc.*, **79**, 535, 1962.

3.33 Butler, H. S. and Kino, G. S., *Phys. Fluids*, **6**, 1346, 1963.

3.34 Keller, J. H. and Pennebaker, W. B., *IBM J. Res. Dev.*, **23**, 3, 1979.

3.35 Mantei, T. D., *J. Electrochem. Soc.*, **130**, 1958, 1983.

3.36 Schneider, F., *Z. Angew. Phys.*, **1**, 456, 1956.

3.37 Benoit-Cattin, P. and Bernard, L. C., *J. Appl. Phys.*, **39**, 5723, 1968.

3.38 Tsui, R. T. C., *Phys. Rev.*, **168**, 107, 1968.

3.39 Pointu, A. M., *J. Appl. Phys.*, **60**, 4113, 1986.

3.40 Metze, A., Ernie, D. W., and Oskam, H. J., *J. Appl. Phys.*, **60**, 3081, 1986.

3.40a Metze, A., Ernie, D. W., and Oskam, H. J., *J. Appl. Phys.*, **65**, 993, 1989.

3.41 Rieman, K.-U., *Phys. Fluids*, **24**, 2163, 1981.

3.42 Vallinga, P. M. and de Hoog, F. J., *J. Phys. D*, **22**, 925, 1989.

3.43 Suzuki, K., Ninomiya, K., Nishimatsu, S., Thoman, J. W., and Steinfeld, J. I., *Jpn. J. Appl. Phys.*, **25**, 1569, 1986.

3.44 Meijer, P. M. and Goedheer, W. J., *Proc. XIX Int. Conf. on Phenom. in Ionized Gases*, Belgrad, Vol. 2, p. 386, 1989.

3.45 Godyak, V. A. and Oks, S. N., *J. Phys. (France)*, **40**, C7-809, 1979.

3.46 Godyak, V. A. and Popov, O. A., *Sov. J. Plasma Phys.*, **5**, 227, 1979.

3.47 Stekolnikov, A. F., *Fiz. Plasmy*, **17**, 1516, 1991.

3.48 Vallinga, P. M., Meijer, P. M., and de Hoog, F. J., *J. Phys. D*, **22**, 1650, 1989.

3.49 Pointu, A. M., *Appl. Phys. Lett.*, **50**, 316, 1987.

3.50 Lieberman, M. A., *IEEE Trans. Plasma Sci.*, **17**, 338, 1989.

3.51 Godyak, V. A. and Ganna, A. Kh., *Fiz. Plasmy*, **6**, 676, 1980.

3.52 Godyak, V. A. and Sternberg, N., *Phys. Rev. A*, **42**, 2290, 1990.

3.53 Stekolnikov, A. F., Brainthwaite, N. St. J., and Allen, J. E., *Proc. IX Int. Conf. on Gas Discharges and Their Applic.*, Venice, p. 391, 1988.

3.54 Pennebaker, W. B., *IBM J. Res. Dev.*, **23**, 16, 1979.

3.55 Kushner, M., *J. Appl. Phys.*, **58**, 4024, 1985.

3.56 Goedheer, W. J. and Meijer, P. M., *Proc. XIX Int. Conf. on Phenom. in Ionized Gases*, Belgrad, Vol. 2, p. 388, 1989.

3.57 Kovalyov, A. S., Muratov, E. A., Ozerenko, A. A., Rakhimov, A. T., and Suetin, N. V., *Fiz. Plasmy*, **11**, 882, 1985.

3.58 Velikhov, E. P., Kovalyov, A. S., and Rakhimov, A. T., *Physical Phenomena in the Gas Discharge Plasma*, Nauka, Moscow, 1987 (in Russian).

3.59 Godyak, V. A., Popov, O. A., and Hanna, A. H., *Proc. XIII Int. Conf. on Phenom. in Ionized Gases*, Berlin, p. 347, 1977.

3.60 Godyak, V. A., *Phys. Lett. A*, **89**, 80, 1982.

3.61 Godyak, V. A. and Sternberg, N., *IEEE Trans. Plasma Sci.*, **18**, 159, 1990.

3.62 Goedheer, W. J. and Meijer, P. M., *IEEE Trans. Plasma Sci.*, **19**, 245, 1991.

3.63 Riemann, K.-U., *Phys. Fluids B*, **4**, 2693, 1992.

3.64 Banerji, D. and Bhattacherya, *Philos. Mag.*, **17**, 313, 1934.

3.65 Hay, J., *Can. J. Res. A*, **16**, 191, 1938.

3.66 Erö, J., *Acta Phys. Hung.*, **5**, 391, 1956; *Nucl. Instrum.*, **3**, 303, 1958.

3.67 Cook, C. J., Heinz, O., Lorents, D. C., and Peterson, J. R., *Rev. Sci. Instrum.*, **33**, 649, 1962.

3.68 Vasile, M. J., *J. Appl. Phys.*, **51**, 2503, 1980.

3.69 Bruce, R. M., *J. Appl. Phys.*, **52**, 7064, 1981.

3.70 Köhler, K., Horne, D. E., and Coburn, J. W., *J. Appl. Phys.*, **58**, 3350, 1985.

3.71 Thompson, B. E., Allen, K. D., Richards, A. D., and Saurn, H. H., *J. Appl. Phys.*, **59**, 1890, 1986.

3.72 Wild, C. and Koidl, P., *J. Appl. Phys.*, **69**, 2909, 1991.

3.73 Otsubo, T. and Ohara, K., *Jpn. J. Appl. Phys.*, **30**, 1882, 1991.

3.73*a* Bragin, V. E., Bykanov, A. N., Matyuhin, V. D., and Muravjev, V. F., *Proc. XX Int. Conf. on Phenom. in Ionized Gases*, Pisa, Italy, Vol. 1, p. 297, 1991.

3.73*b* Annaratone, B. M., Ku, V. P. T., and Allen, J. E., *Proc. XXI Int. Conf. on Phenom. in Ionized Gases*, Bochum, Germany, Vol. 1, p. 29, 1993.

3.74 Knypers, A. D. and Hopman, H. J., *J. Appl. Phys.*, **67**, 1229, 1990.

3.75 Vender, D. and Boswell, R. W., *IEEE Trans. Plasma Sci.*, **18**, 725, 1990.

3.76 Surendra, M. and Graves, D. B., *Appl. Phys. Lett.*, **59**, 2091, 1991.

3.77 Boswell, R. W. and Porteous, R. K., *J. Appl. Phys.*, **62**, 3123, 1987.

3.78 Graves, D., *AIChE J.*, **35**, 1, 1989.

3.79 Emeleus, K. G. and Wolsey, G. A., *Discharges in Electronegative Gases*, Taylor and Francis, London, 1970.

3.80 Flamm, D. L. and Donelly, V. M., *J. Appl. Phys.*, **59**, 1052, 1986.

3.81 Gottsho, R. A., Burton, R. H., Flamm, D. L., Donelly, V. M., and Davis, G. P., *J. Appl. Phys.*, **55**, 2707, 1984.

3.82 Gottsho, R. A. and Gaebe, C. E., *IEEE Trans. Plasma Sci.*, **14**, 92, 1986.

3.83 Gottsho, R. A., *Phys. Rev. A*, **36**, 2233, 1987.

3.84 Gogolides, E., Nicolai, J. P., and Sawin, H. H., *J. Vac. Sci. Technol. A*, **7**, 1001, 1989.

3.85 Jourbort, O., Jelleier, J., and Arnal, V., *J. Appl. Phys.*, **65**, 5096, 1989.

3.86 Bletzinger, B., *J. Appl. Phys.*, **67**, 130, 1990.

3.87 Tsu, D. V., Kim, S. S., and Theil, J. A., *J. Vac. Sci. Technol.*, **39**, 33, 1990.

3.88 Eliabaly-Reisman, A., *J. Electrochem. Soc.*, **138**, 1061, 1991.

3.89 Klopovskii, K. S., Kovalyov, A. S., and Lopayev, D. V., *Fiz. Plasmy*, **18**, 1606, 1992.

3.90 Kushner, M. J., *J. Appl. Phys.*, **54**, 4958, 1983.

3.91 Kushner, M. J., *IEEE Trans. Plasma Sci.*, **14**, 188, 1986.

3.92 Boeuf, J. P., *Phys. Rev. A*, **36**, 2782, 1987.

3.93 Graves, D. B., Jensen, K. F., *IEEE Trans. Plasma Sci.*, **14**, 78, 1986.

3.94 Boiko, V. V., Mankelevich, Yu. A., Rakhimov, A. T., Suetin, N. V., Feoktistov, V. A., and Filippov, S. S., *Fiz. Plasmy*, **15**, 218, 1989.

3.95 Boiko, V. V., Mankelevich, Yu. A., Rakhimov, A. T., Suetin, N. V., and Filippov, S. S., *Fiz. Plasmy*, **15**, 867, 1989.

3.96 Paranjpe, A. P., McVittie, J. P., and Self, S. A., *Phys. Rev. A*, **41**, 6949, 1990.

3.97 Sato, N. and Tagashira, H., *IEEE Trans. Plasma Sci.*, **19**, 102, 1991.

3.98 Oh, Y.-H., Choi, N.-H., and Choi, D.-I., *J. Appl. Phys.*, **67**, 3264, 1990.

3.99 Park, S.-K. and Economou, D. J., *J. Appl. Phys.*, **68**, 3904, 1990.

3.100 Shveigert, V. A., *Fiz. Plasmy*, **17**, 844, 1991.

3.101 Meyappan, M. and Govindan, T. R., *IEEE Trans. Plasma Sci.*, **19**, 122, 1991.

3.102 Sommerer, T. J. and Kushner, M. J., *J. Appl. Phys.*, **71**, 1654, 1992.

3.103 Gogolides, E. and Sawin, H. H., *J. Appl. Phys.*, **72**, 3971, 3988, 1992.

3.104 Klopovskii, K. S., Popov, A. M., Rakhimov, A. T., Rakhimova, T. V., and Feoktistov, V. A., *Fiz. Plasmy*, **19**, 910, 1993.

3.105 Kaganovich, I. D. and Tsendin, L. D., (private communication).

3.106 *Special Issue on Modelling Collisional Low-Temperature Plasma*, Ed. S. J. Gitomer, *IEEE Trans. Plasma Sci.*, **2**, 19, 1991.

3.107 Godyak, V. A. and Khanneh, A. S., *J. Phys. (France)*, **40**, C7-125, 1979; *IEEE Trans. Plasma Sci.*, **14**, 112, 1986.

3.108 Yatsenko, N. A., *Proc. XX Int. Conf. on Phenom. in Ionized Gases*, Pisa, Italy, Vol. 5, p. 1159, 1991.

3.109 Raizer, Yu. P. and Shneider, M. N., *Fiz. Plasmy*, **18**, 1476, 1992.

3.110 Belenguer, Ph. and Boeuf, J. P., *Phys. Rev. A*, **41**, 4447, 1990.

3.111 Gill, M. D., *Vak.-Tech.*, **34**, 357, 1984.

3.112 Kuzovnikov, A. A. and Savinov, V. P., *Padiotekh., Elektron.*, **28**, 816, 1973.

3.113 Godyak, V. A. and Piejak, R. B., *Phys. Rev. Lett.*, **65**, 996, 1990.

3.114 Godyak, V. A., Piejak, R. B., and Alexandrovich, B. M., *Phys. Rev. Lett.*, **68**, 40, 1992.

3.115 Dilecce, G., Capitelli, M., and De Benedictis, S., *J. Appl. Phys.*, **69**, 121, 1991.

3.116 Kazantsev, S. A., Svelokuzov, A. E., and Subbotenko, A. V., *Zh. Tekh. Fiz.*, **56**, 1091, 1986.

3.117 Kazantsev, S. A., Kovalevskii, V. A., Kuzovnikov, A. A., Motorygina, M. B., Rys, A. G., and Savinov, V. P., *Vestn. Leningr. Univ. (USSR)*, **4**, 26, 1990.

3.118 Khasilev, V. Ya., Mikhalevskii, V. S., and Tolmachev, G. M., *Fiz. Plasmy*, **6**, 430, 1980.

3.119 Vidaud, P., Durrani, S. M. A., and Hall, D. R., *J. Phys. D*, **21**, 57, 1988.

3.120 Boeuf, J. P. and Belenguer, P. Jn., *Nonequilibrium Processes Partially Ionized Gases*, Ed. M. Capitelli and N. J. Bardsley, New York, p. 155, 1990.

3.121 Godyak, V. A., Piejak, R. B., and Alexandrovich, B. M., *Plasma Sources Sci. Technol.*, **1**, 36, 1992.

3.122 Vahedi, V., Di Peso, G., Birsdall, C. K., Lieberman, M. A., and Rognlien, T. P., *Plasma Sources Sci. Technol.*, **2**, 261, 273, 1993.

3.123 Somerrer, T. J., Hitchon, W. N. G., and Lawler, J. E., *Phys. Rev. Lett.*, **63**, 2361, 1989.

3.124 Hebner, G. A., Verdeyen, J. T., and Kushner, M. J., *J. Appl. Phys.*, **63**, 2226, 1988.

3.125 Godyak, V. A., *Zh. Tekh. Fiz.*, **41**, 1364, 1971.

3.126 Akhiezer, A. I. and Bakai, A. S., *Dokl. Akad. Nauk (SSSR)*, **201**, 1074, 1971; *Fiz. Plasmy*, **2**, 654, 1976.

3.127 Godyak, V. A., *Fiz. Plasmy*, **2**, 141, 1976.

3.128 Popov, O. A. and Godyak, V. A., *J. Appl. Phys.*, **57**, 53, 1985.

3.129 Misium, G. R., Lichtenberg, A. J., and Lieberman, M. A., *J. Vac. Sci. Technol. A*, **7**, 1007, 1989.

3.130 Goedde, C. G., Lichtenberg, A. J., and Lieberman, M. A., *J. Appl. Phys*, **64**, 4375, 1988.

3.131 Wendt, A. E. and Hitchon, W. N. G., *J. Appl. Phys.*, **71**, 4718, 1992.

3.132 Zaslavskii, G. M. and Sagdeev, R. Z., *Introduction to Nonlinear Physics: From Pendulum to Turbulence and Chaos*, Nauka, Moscow, 1988 (in Russian).

3.133 Lichtenberg, A. J. and Lieberman, M. A., *Regular and Stochastic Motion*, Springer-Verlag, New York, 1983.

3.134 Kaganovich, I. D. and Tsendin, L. D., *IEEE Trans. Plasma Sci.*, **20**, 66, 86,

1992.

3.135 Lister, G. G., *J. Phys. D*, **25**, 1649, 1992.

3.136 Popov, A. M., Rakhimov, A. T., and Rakhimova, T. V., *Fiz. Plasmy*, **19**, 1241, 1993.

3.137 Richards, A. D., Thompson, B. E., and Sawin, H. H., *Appl. Phys. Lett.*, **50**, 492, 1987.

3.138 Graves, D. B., *J. Appl. Phys.*, **62**, 88, 1987.

3.139 Volkova, E. A., Popov, A. M., Popovicheva, O. V., Rakhimova, T. V., and Feoktistov, V. A., *Fiz. Plasmy*, **17**, 481, 1991; **18**, 1452, 1992.

3.140 Wester, R. and Seiwert, S., *J. Phys. D*, **24**, 1371, 1991.

3.141 Kline, L. E., *IEEE Trans. Plasma Sci.*, **14**, 145, 1986.

3.142 Kline, L. E., Partlow, W. D., and Bies, W. E., *J. Appl. Phys.*, **65**, 70, 1989.

3.143 Okazaki, K., Makabe, T., and Yamaguchi, Y., *Appl. Phys. Lett.*, **54**, 1742, 1989.

3.144 Surendra, M., Graves, D. B., and Jellum, G. M., *Phys. Rev. A*, **41**, 1112, 1990.

3.145 Shveigert, V. A. and Shveigert, I. V., *Fiz. Plasmy*, **14**, 347, 1988.

3.146 Raizer, Yu. P. and Shneider, M. N., *Teplofiz. Vys. Temp.*, **27**, 431, 1989.

3.147 Hammersley, J. M. and Handscomb, D. C., *Monte-Carlo Methods*, Wiley, New York, 1964.

3.148 Skullerud, H. R., *J. Phys. D*, **1**, 1567, 1968.

3.149 Lin, S. L. and Bardsley, J. N., *J. Chem. Phys.*, **66**, 435, 1977.

3.150 Boeuf, J. P. and Marode, J., *J. Phys. D*, **15**, 2169, 1982.

3.151 Kaufman, V., *J. Phys. D*, **21**, 442, 1988.

3.152 Kushner, M. J., *IEEE Trans. Plasma Sci.*, **14**, 188, 1986.

3.153 Tsai, J.-H. and Wu, Ch.-H., *J. Phys. D*, **26**, 496, 1993.

3.154 Hockney, R. W. and Eastwood, J. W., *Computer Simulation Using Particles*, McGraw-Hill, New York, 1985.

3.155 Berezin, Yu. A. and Vshivkov, V. A., *Method of Particles in the Dynamic of Rarefied Plasma*, Nauka, Novosibirsk, 1980 (in Russian).

3.156 Birsdall, Ch. K. and Langdon, A. B., *Plasma Physics via Computer Simulation*, McGraw-Hill, New York, 1985.

3.157 Boswell, R. W. and Morey, I. J., *Appl. Phys. Lett.*, **52**, 21, 1988.

3.158 Birsdall, C. K., *IEEE Trans. Plasma Sci.*, **19**, 65, 1991.

3.159 Date, A., Kitamori, K., Sakai, Y., and Tagashira, H., *J. Phys. D*, **24**, 442, 1992.

3.160 Vahedi, V., Lieberman, M. A., Alves, M. V., Verboncoeur, J. P., and Birsdall, C. K., *J. Appl. Phys.*, **69**, 1008, 1991.

3.161 Alves, M. A., Lieberman, M. A., Vahedi, V., and Birsdall, C. K., *J. Appl. Phys.*, **69**, 3823, 1991.

3.162 Trombley, H. W., Terry, F. L., and Ebba M. E., *IEEE Trans. Plasma Sci.*, **19**, 158, 1991.

3.163 Turner, M. M. and Hopkins, H. B., *Phys. Rev. Lett.*, **69**, 3511, 1992.

3.164 Sommerer, T. J., Hitchon, W. N. G., and Lawler, J. E., *Phys. Rev. A*, **39**, 6356, 1989.

3.165 Sommerer, T. J., Hitchon, W. N. G., and Lawler, J. E., *Phys. Rev. A*, **43**, 4452, 1991.

3.166 Mankelevich, Yu. A., Rakhimov, A. T., and Suetin, N. V., *Fiz. Plasmy*, **17**, 1017, 1991.

3.167 Okano, H., Yamazaki, T., and Horiike, Y., *Solid State Technol.*, **25**, 166, 1982.

3.168 Naumovich, V. G., Yagola, V. V., *Preprint 87–25*, Institute of Nuclear Research, Kiev, 1987 (in Russian).

3.169 Yeom, G. Y., Thornton, J. A., and Kushner, M. J., *J. Appl. Phys.*, **65**, 3816, 1989.

3.170 Knypers, A. D., Granneman, E. H. A., and Hopman, H. J., *J. Appl. Phys.*, **63**, 1899, 1988.

3.171 Lin, I., *J. Appl. Phys.*, **58**, 2981, 1985.

3.172 Gurin, A. A. and Chernova, N. I., *Fiz. Plasmy*, **11**, 244, 1985.

3.173 Lukyanova, A. V., Rakhimov, A. T., and Suetin, N. V., *Fiz. Plasmy*, **16**, 1367, 1990.

3.174 Porteous, R. K. and Graves, D. B., *IEEE Plasma Sci.*, **19**, 204, 1991.

3.175 Lieberman, M. A., Lichtenberg, A. J., and Savas, S. E., *IEEE Trans. Plasma Sci.*, **19**, 189, 1991.

3.176 Glushko, V. A., Naumovets, V. G., Pasechnik, L. L., et al, *Proc. XV Int. Conf. on Phenom. in Ionized Gases*, Vol. 2, p. 705, 1981.

3.177 Lukyanova, A. V., Rakhimov, A. T., and Suetin, N. V., *Fiz. Plasmy*, **17**, 1012, 1991.

Chapter 4

4.1 Yatsenko, N. A., *Thesis, D. Sci. (Phys. and Math)*, Institute of High Temperatures, Russian Academy of Science, Russia, 1991 (in Russian).

4.2 Yatsenko, N. A., *Zh. Tekh. Fiz.*, **58**, 296, 1988.

4.3 Beck, H. Z., *Physik*, **97**, 355, 1935.

4.4 Levitskii, S. M., *Zh. Tekh. Fiz.*, **27**, 970, 1001, 1957.

4.5 Andreev, A. D., *Zh. Prikl. Spektrosk.*, **5**, 145, 1966.

4.6 Yatsenko, N. A., *Inzh.-Fiz. Zh.*, **62**, 739, 1992.

4.7 Godyak, V. A. and Popov, O. N., *Zh. Tekh. Fiz.*, **47**, 766, 1977.

4.8 Gagne, R. R. J. and Cantin, A., *J. Appl. Phys.*, **43**, 2639, 1972.

4.8*a* Bragin, V. E. and Bykanov, A. N., *Proc. XXI Int. Conf. on Phenom in Ionized Gases*, Bochum, Germany, Vol. 2, p. 171, 1993.

4.8*b* Annaratone, B. M., Allen, M. W., and Allen, J. E., *J. Phys. D*, **25**, 417, 1992.

4.8*c* Annaratone, B. M., Counsell, G. F., Kawano, H., and Allen, J. E., *Plasma Sources Sci. Technol.*, **1**, 232, 1992.

4.9 Yatsenko, N. A., *Zh. Tekh. Fiz.*, **51**, 1195, 1981.

4.10 Yatsenko, N. A., *Teplofiz. Vys. Temp.*, **20**, 1044, 1982.

4.11 Kuzovnikov, A. A. and Savinov, V. P., *Vestn. Mosk. Univ., Fiz.*, **2**, 215, 1973.

4.12 Andreev, A. D., Il'in, V. D., and Lobanov, Yu. N., *Zh. Tekh. Fiz.*, **36**, 1636, 1966.

4.13 Kalmykov, A. V., Nezhentsev, B. Yu., and Smirnov, A. S., *Zh. Tekh. Fiz.*, **59**, 93, 1989.

4.14 Yatsenko, N. A., *II All-Union Conf. on Gas Discharge Physics, Abstr.*, Kiev, Vol. 1, p. 143, 1986 (in Russian).

4.15 Graves, D. B., *AIChE J.*, **35**, 1, 1989.

4.16 Moore, C. A., Davis, G. P., and Gottsho, R. A., *Phys. Rev. Lett.*, **52**, 538, 1984.

4.17 Mandich, M., Gaebe, C. E., and Gottsho, R. A., *J. Chem. Phys.*, **83**, 3349, 1985.

4.18 Gottsho, R. A. and Gaebe, C. E., *IEEE Trans. Plasma Sci.*, **14**, 92, 1986.

4.19 Gangnly, B. N. and Garscadden, A., *Appl. Phys. Lett.*, **46**, 540, 1985.

4.20 Doughty, D. K. and Lawler, J. E., *Appl. Phys. Lett.*, **45**, 611, 1984.

4.21 Walkup, R., Deyfus, R. W., and Avouris, Ph., *Phys. Rev. Lett.*, **50**, 1846, 1983.

4.22 Taillet, J. C. R., *Acad. Sci. Paris*, **269**, 52, 1969.

4.23 Kramer, J., *J. Appl. Phys.*, **60**, 3072, 1986.

4.24 Den Hartog, E. A., Doughty, D. A., and Lawler, J. E., *Phys. Rev. A*, **38**, 2471, 1988.

4.25 Lawler, J. E., Den Hartog, E. A., and Hitchon W. N. G., *Phys. Rev. A*, **43**,

4427, 1991.

4.26 Gottsho, R. A., Mitchell, A., Scheller, G. R., and Chan, Y. -Y., *Phys. Rev. A*, **40**, 6407, 1989.

4.27 He, D., Baker, C. J., and Hall, D. R., *J. Appl. Phys.*, **55**, 4120, 1984.

4.28 Lipatov, N. I., Pashinin, P. P., Prokhorov, A. M., and Yurov, V. Yu., *Proc. of the General Physics Institute, Russian Academy of Sciences*, Nauka, Moscow, Vol. 17, 53, 1989 (in Russian).

4.29 Hatch, A. J. and Henckroth, L. E., *J. Appl. Phys.*, **41**, 1701, 1970.

Chapter 5

5.1 Javan, A., Bennett, W. R., and Herriot, D. R., *Phys. Rev. Lett.*, **6**, 106, 1961.

5.2 Patel, C. K. N., *Phys. Rev. Lett.*, **13**, 617, 1964.

5.3 Raizer, Yu. P., *Gas Discharge Physics*, Springer-Verlag, Berlin, New York, 1991.

5.4 Brown, C. O. and Davis, J. W., *Appl. Phys. Lett.*, **21**, 480, 1972.

5.5 Crocker, A. and Wills, M., *Electron. Lett.*, **5**, 63, 1969.

5.6 Eletskii, A. V., Mishenko, L. G., and Tychinskii, V. V., *Zh. Prikl. Spektrosk.*, **8**, 425, 1968.

5.7 Goikhman, V. Kh. and Goldfarb, V. M., *Zh. Prikl. Spektrosk.*, **21**, 465, 1974.

5.8 Yatsenko, N. A., *Ph.D. Thesis*, MFTI, 1978 (in Russian).

5.9 Kalmykov, A. V., Moiseev, V. T., Smirnov, A. S., and Tomashevich, S. V., *All-Union Conf. on Laser Optics, Abstr.*, Leningrad, p. 60, 1986 (in Russian).

5.10 Karapusikov, A. I. and Troshin, V. I., *V All-Union Conf. on Laser Optics, Abstr.*, p. 72, 1986 (in Russian).

5.11 He, D. and Hall, D. R., *Appl. Phys. Lett.*, **43**, 726, 1983.

5.12 Yatsenko, N. A., *Preprint No. 338*, Institute for Problems in Mecanics, Russian Academy of Sciences, 1988 (in Russian).

5.13 Yatsenko, N. A., *Zh. Tekh. Fiz.*, **51**, 1195, 1981.

5.14 Yatsenko, N. A., *Bull. Acad. Sci. USSR, Phys. Ser.*, **56**, 1901, 1992.

5.15 Akimov, A. G., Koba, A. V., Lipatov, N. I., Mineev, A. P., Pashinin, P. P., and Prokhorov, A. M., *Kvant. Elektron. Mosk.*, **16**, 938, 1989.

5.16 Svich, V. A., Tkachenko, V. M., and Topkov, A. N., *Kvant. Elektron. Mosk.*, **17**, 690, 1990.

5.17 Yatsenko, N. A., *Ph.D. Thesis*, Institute of High Temperatures, Russian Academy of Sciences, Moscow, 1991.

5.18 Zybin, D. N., Protsenko, D. E., and Tikhomirova, T. A., *Preprint No. 6*, Institute of General Physics, Russian Academy of Sciences, Moscow, 1993.

5.19 He, D. and Hall, D. R., *IEEE J. Quantum Electron.*, **20**, 509, 1984.

5.20 Hugel, H. E., *Proc. SPIE*, Vol. 650, p. 2, 1986.

5.21 Yatsiv, S., *Proc. VI Int. Symp. Flow and Chemical Lasers*, Rosenwarhs S. Ed., Springer-Verlag, Berlin, p. 252, 1987.

5.22 Nowack, R., Opower, H., Wessel, K., Krüger, H., Haas, W., and Wenzel, N., *Laser Optoelektron.*, **23**, 68, 1991.

5.23 Yelden, E. F., Seguin, H. J. J., Capjack, C. E., and Nikumb, S. K., *Opt. Commun.*, **82**, 503, 1991.

5.24 Wester, R., Seiwert, S., and Wagner, R., *J. Phys. D*, **24**, 1796, 1991.

5.25 Kozgunov, S. V., Novgorodov, M. Z., and Smirnova, E. P., *Laser Physics*, **3**, 84, 1993.

5.26 Ehrlichmann, D., Habich, V., and Plum H. D., *J. Phys. D*, **26**, 183, 1993.

5.27 France Patents, No. 2092912, No. 2108912, HOIS 3/00, 1972.

5.28 Kozlov, G. I., Kuznetsov, V. A., and Masyukov, V. A., *Pis'ma Zh. Tekh. Fiz.*, **4**, 129, 1978.

5.29 Antyukhov, V. V., Bondarenko, A. I., Globa, A. F., Golubev, V. S., Kachurin, O. R., Kolesov, L. L., Lebedev, E. A., Lebedev, F. V., Suslov, Yu. F., and Timofeev, V. A., *Kvant. Elektron. Mosk.*, **8**, 2234, 1981.

5.30 Hoffmann, P. *Proc. SPIE*, Vol. 650, p. 23, 1986.

5.31 Lipatov, N. I., Pashinin, P. P., Prokhorov, A. M., and Yurov, V. Yu., *Proc. of the General Physics Institute, Russian Academy of Sciences*, Nauka, Moscow, Vol. 17, p. 53, 1989 (in Russian).

5.32 Yatsenko, N. A., *Preprint No. 465*, Institute for Problems in Mechanics, Russian Academy of Sciences, 1990 (in Russian).

5.33 Yatsenko, N. A., *Preprint No. 338*, Institute for Problems in Mechanics, Russian Academy of Sciences, Moscow, 1988 (in Russian); *Izv. Ros. Akad. Nauk, ser. Fiz.*, **57**, No. 12, 110, 1993.

5.34 Kim, Y.-M., Yaun, C. E., and Ra, J. W., *J. Appl. Phys.*, **67**, 1127, 1990.

5.35 Colley, A. D., Baker, H. J., and Hall, D. R., *Appl. Phys. Lett.*, **61**, 136, 1992.

5.36 Yatsenko, N. A., *Proc. Int. Conf. 'RF Discharge in Wave Fields and RF Pumping of Gas Lasers,'* Tashkent, p. 3, 1992 (in Russian).

5.37 Yatsenko, N. A., *Proc. VI All-Union Conf. 'Physics of Low Temperature Plasma,'* Leningrad, Vol. 2, p. 51, 1983 (in Russian).

5.38 Yatsenko, N. A., *Proc. II All-Union Conf. 'Physics of Electrical Breakdown of Gases,'* Tartu, Vol. 2, p. 342, 1984 (in Russian).

5.39 Colley, A. D., Abramskii, K. M., Baker, H. J., and Hall, D. R., *CLEO-93*, Baltimor, Vol. 2, p. 288, 1993.

5.40 Lapucci, A., Gangioli, G., *Appl. Phys. Lett.*, **62**, 7, 1993.

5.41 Yelden, E. F., Seguin, H. J. J., Capjack, C. E., and Nikumb, S. K., *Appl. Phys. Lett.*, **58**, 693, 1991.

5.42 Fusayama, T., Sekiguohi, T., *Jpn. J. Appl. Phys.*, **14**, 735, 1975.

5.43 Laakman, K. D., U.S. Patent No. 4.169.251, 1979.

5.44 Christensen, C. P., Powell, F. X., and Djeu, N., *IEEE J. Quantum Electron.*, **16**, 949, 1980.

5.45 Alloock, G. and Hall, D. R., *Opt. Commun.*, **37**, 49, 1981.

5.46 Lovold, S. and Wang, G., *Appl. Phys. Lett.*, **40**, 13, 1982.

5.47 Bakarev, A. E., Vasilenko, L. S., and Skhimnikov, O. M., *Kvant. Elektron. Mosk.*, **9**, 1729, 1982.

5.48 Mirzayev, A. T. and Sharakhimov, M. Sh., *Kvant. Elektron. Mosk.*, **11**, 1236, 1984.

5.49 He, D. and Hall, D. R., *J. Appl. Phys.*, **56**, 856, 1984.

5.50 Sinclair, R. L. and Tulip, J., *Rev. Sci. Instrum.*, **55**, 1539, 1984.

5.51 Vidaud, P., He, D., and Hall, D. R., *Opt. Commun.*, **56**, 185, 1985.

5.52 Newman, L. A., Hart, R. A., Kennedy, J. T., and Dellaria, A. J., *Conf. on Laser and Electro-Optics*, San Francisco, p. 162, 1986.

5.53 Hochuli, V. E. and Halemann, P. R., *Rev. Sci. Instrum.*, **57**, 2238, 1986.

5.54 Abramski, K. M. and Besztak, H., *Opt. Appl.*, **18**, 109, 1988.

5.55 McArthur, B. A. and Tulip, J., *Rev. Sci. Instrum.*, **59**, 23, 1987.

5.56 Bruno, W., *Proc. SPIE*, Vol. 1020, p. 57, 1988.

5.57 Plinski, E. F., *Opt. Appl.*, **19**, 63, 1989.

5.58 Macken, J. A., Yagnik, S. K., and Samis, M. A., *IEEE J. Quantum Electron.*, **25**, 1695, 1989.

5.59 Xin, J. G., Yan, P., and Wei, G. H., *Appl. Phys. Lett.*, **59**, 3363, 1991.

5.60 Balakin, S. V., Leontyev, V. G., Rakhvalov, V. V., Stepanov, V. A., Shishkanov, E. F., and Yukhimchuk, A. A., *Zh. Prikl. Spektrosk.*, **54**, 939, 1991.

5.61 Akimov, A. G., Zybin, D. N., Lipatov, N. J., Nevedov, S. M., Protsenko, D. E., and Tikhomirova, T. E., *Laser Physics*, **3**, 980, 1993.

5.62 Yatsenko, N. A., *Pis'ma Zh. Tekh. Fiz.*, **17**, 72, 1993.

5.63 Vitteman, V. Ya., Iliyeva, M., Ilyukhin, B. I., Kolesnikov, V. N., Ochkin, V. N., and Udalov, Yu. B., *Kvant. Elektron. Mosk.*, **19**, 945, 1992.

5.63*a* Heeman-Ilieva, M. B., Udalov, Yu. B., Witteman, W. J., Peters, P. J. M., Hoen, K., and Ochkin, V. N., *J. Appl. Phys.*, **74**, 4786, 1993.

5.64 Myshenkov, V. I. and Yatsenko, N. A., *Kvant. Elektron. Mosk.*, **8**, 2121, 1981.

5.65 Gabai, A., Hertzberg, R., and Yatsiv, S., *CLEO-84*, Vol. 84, p. 28, 1984.

5.66 Xin, J. G. and Hall, D. R., *Opt. Commun.*, **58**, 420, 1986.

5.67 Yatsiv, S., *Proc. VI Int. Symp. Gas Flow and Chem. Lasers*, Jerusalem, p. 252, 1986.

5.68 Xin, J. G., and Hall, D. R., *Appl. Phys. Lett.*, **51**, 469, 1987.

5.69 Vitruk, P. P. and Yatsenko, N. A., USSR Patents No. 1644269 and No. 16442270, 1988.

5.70 Vitruk, P. P. and Yatsenko, N. A., *Pis'ma Zh. Tekh. Fiz.*, **15**, 1, 1989.

5.71 Jackson, P. E., Baker, H. J., and Hall, D. R., *Appl. Phys. Lett.*, **54**, 1950, 1989.

5.72 Abramski, K. M., Colley, A. D., Baker, H. J., and Hall, D. R., *Appl. Phys. Lett.*, **54**, 1833, 1989.

5.73 Hall, D. R., Baker, H. J., *Laser Focus World.*, 77, October, 1989.

5.74 Yatsiv, S., Gabay, A., Sterman, B., and Sintov, Y., *Proc. SPIE*, Vol. 1397, 1991; *Proc. VIII Int. Symp. on Gas Flow and Chem. Lasers*, Madrid, p. 319, 1990.

5.75 Zhang, X. S., Baker, H. J., and Hall, D. R., *J. Phys. D*, **26**, 359, 1993.

5.76 Newman, L. A. and Hart, R. A., *Laser Focus: Electro-Optics*, **23**, 80, 1987.

5.77 Xin, J. G. and Hall, D. R., *Opt. Commun.*, **58**, 420, 1986.

5.78 Jackson, P. E., Baker, H. J., and Hall, D. R., *Appl. Phys. Lett.*, **54**, 1950, 1989.

5.79 Likhanskii, V. V. and Napartovich, A. P., *Usp. Fiz. Nauk*, **160**, 101, 1990.

5.80 Vasil'tsov, V. V., *Izv. Ros. Akad. Nauk Ser. Fiz.*, **57**, No. 12, 150, 1993.

5.81 Glova, A. F., Dreizin, Yu. A., Kachurin, O. R., Lebedev, F. V., and Pis'menny, V. D., *Pis'ma Zh. Tekh. Fiz.*, **11**, 249, 1985.

5.82 Wilcox, J. Z., Jansen, M., Yang, J., Peterson, G., Silver, A., Simmons, W., Ou, S. S., and Sergant, M., *Appl. Phys. Lett.*, **51**, 631, 1987.

5.83 Ackley, D. E., Engelmann, R. W. H., *Appl. Phys. Lett.*, **39**, 27, 1981.

5.84 Yomans, D. G., *Appl. Phys. Lett.*, **44**, 365, 1984.

5.85 Newman, L. A., Hart, R. A., Kennedy, J. T., Cantor, A. T., De Maria, A. J., and Bridges, W. B., *Appl. Phys. Lett.*, **48**, 1701, 1986.

5.86 Hart, R. A., Newman, L. A., Cantor, A. J., and Kennedy, J. T., *Appl. Phys. Lett.*, **51**, 1057, 1987.

5.87 Abdullina, G. R., Mirinoyatov, M. M., Niyazov, B. A., Solovyov, I. A., and Stepanov, V. A., *Proc. Int. Conf. 'RF Discharge in the Wave Fields and RF Pumping of Gas Lasers,'* Tashkent, p. 7, 1992 (in Russian).

5.88 Shackleton, C. J., Abramski, K. M., Baker, H. J., and Hall, D. R., *Opt. Commun.*, **89**, 423, 1992.

5.89 Xin, J. G., Duncan, A., and Hall, D. R., *Appl. Opt.*, **28**, 45, 76, 1989.

5.90 Abramski, K. M., Colley, A. D., Baker, H. J., and Hall, D. R., *IEEE Quantum Electron.*, **26**, 711, 1990.

5.91 Eckbreath, A. C. and Davis, J. W., *Appl. Phys. Lett.*, **21**, 25, 1972.

5.92 Eckbreath, A. C. and Blaszuk, P. R., AIAA Paper No. 72–723, 1972.

5.93 Allis, W. P., Brown, S. C., and Everhart, E., *Phys. Rev.*, **84**, 519, 1951.

5.94 Brown, S. C., *Basic Data of Plasma Physics*, MTI Press, Cambridge, MA, 1959.

5.95 Yatsenko, N. A., *Teplofiz. Vys. Temp.*, **20**, 1044, 1982.

5.96 Rakhimova, T. V. and Rakhimov, A. T., *Fiz. Plazmy*, **1**, 854, 1975; *Teplofiz. Vys. Temp.*, **14**, 1313, 1976.

5.97 Raizer, Yu. P. and Shapiro, G. I., *Fiz. Plazmy*, **4**, 850, 1978.

5.98 Shapiro, G. I., *Pis'ma Zh. Tekh. Fiz.*, **2**, 451, 1976.

5.99 Kuteev, V. V. and Smirnov, A. S., *Pis'ma Zh. Tekh. Fiz.*, **4**, 111, 1976.

5.100 Zhilinskii, A. P., Kuteev, V. V., Smirnov, A. S., and Shevchenko, Yu. I., *Zh. Tekh. Fiz.*, **48**, 2260, 1978.

5.101 Generalov, N. A., Zimakov, V. P., Kosynkin, V. D., Raizer, Yu. P., and Roitenburg, D. I., *Pis'ma Zh. Tekh. Fiz.*, **1**, 431, 1975; *Fiz. Plazmy*, **3**, 626, 634, 1977; *ibid*, **6**, 1152, 1980.

5.102 Generalov, N. A., Zimakov, V. P., Kosynkin, V. D., Raizer, Yu. P., and Solov'yov, N. G., *Kvant. Elektron. Mosk.*, **9**, 1549, 1982.

5.103 Kozlov, G. I. and Yatsenko, N. A., *Pis'ma Zh. Tekh. Fiz.*, **4**, 424, 1978.

5.104 Hügel, H., *Proc. VI Int. Symp. Gas Flow and Chem. Lasers*, Jerusalem, p. 258, 1986.

5.104a Schock, W., Wittwer, W., Giesen, A., Hall, T., and Hügel, H., *Proc. III Int. Conf. on Laser Manufact.*, Paris, Kempston, p. 271, 1986.

5.105 Myshenkov, V. I. and Yatsenko, N. A., *Zh. Tekh. Fiz.*, **52**, 2055, 1981; *Fiz. Plasmy*, **4**, 704, 1982.

5.106 Sasnett, H. W., *Laser Focus: Electro-Optics*, **24**, 48, 1988.

5.107 Nichols, D. E. and Brandenberg, W. M., *IEEE J. Quantum Electron.*, **8**, 718, 1972.

5.108 Akirtava, O. S., Dzhikiya, V. L., Kvitiya, Z. A., and Shengeliya, N. A., *Pis'ma Zh. Tekh. Fiz.*, **7**, 1231, 1981.

5.109 Schock, W., Hügel, H., and Hoffmann, P., *Laser Electro-Optic.*, **2**, 78, 1981.

5.110 Kovalyov, A. S., Rakhimov, A. T., and Feoktistov, V. A., *Fiz. Plazmy*, **7**, 1411, 1981.

5.111 Kovalyov, A. S., Muratov, E. N., Ozerenko, A. A., Rakhimov, A. T., and Suetin, N. V., *Pis'ma Zh. Tekh. Fiz.*, **10**, 1139, 1984.

5.112 Schock, W. and Hügel, H., *Gas Flow and Chem. Lasers*, Ed. M. Okorato, Plenum Press, New York, London, p. 435, 1985.

5.113 Gladush, G. G. and Samokhin, A. A., *Zh. Prikl. Mekh. Tekh. Fiz.*, **6**, 16, 1986.

5.114 Smirnov, A. S., Frolov, K. S., and Shevchenko, Yu. I., *Zh. Tekh. Fiz.*, **57**, 1310, 1987.

5.115 Akirtaya, O. S., Artamonov, A. V., Artemov, V. M., Dzhikiya, V. L., Ky-itiya, Z. A., and Rogozhina, G. P., *Kvant. Electron. Mosk.*, **14**, 2454, 1987.

5.116 Beck, R., *Appl. Phys. B*, **42**, 233, 1987.

5.117 Wegmann, H. G. and Sckeyvaerts, *Laser Magazin*, **3**, 26, 1989.

5.118 Schock, W., Walz, B., Wessel, K., and Wildermuth, E., *Proc. SPIE*, Vol. 1276, p. 41, 1989.

5.119 Wester, R. and Seiwert, S., *J. Phys. D*, **24**, 1371, 1991.

5.120 Wester, R., Seiwert, S., and Wagner, R., *J. Phys. D*, **24**, 1796, 1991.

5.121 Wester, R., *J. Appl. Phys.*, **70**, 3449, 1991.

5.122 Pearson, G. N. and Hall, D. R., *Appl. Phys. Lett.*, **50**, 1222, 1987.

5.123 Aleinikov, V. S. and Masychev, V. I., *CO-Lasers*, Radio i Svyaz, Moscow, 1990 (in Russian).

5.123a Yatsenko, N. A., *Proc. XIV Int. Conf. on Coherent and Nonlinear Optics*, Leningrad, Vol. 2, p. 52, 1991.

5.124 Yatsiv, S., Gabay, A., Sterman, B., and Sintov, Y., *Proc. SPIE*, Vol. 1397, p. 319, 1991.

5.125 Zhao, H., Baker, H. J., and Hall, D. R., *Appl. Phys. Lett.*, **59**, 1281, 1991.

5.126 Kondilenko, I. I., Korotkov, P. A., and Khizhnyak, A. I., *Lasers Physics*, Vyshya Shkola, Kiev, 1984 (in Russian).

5.127 Karlov, N. V., *Lectures on Quantum Electronic*, CRC Press, Boca Raton, 1993.

5.128 Udalov, Yu. B., Peters, P. J. M., Heeman-Jlieva, M. B., Ernst, F. H. J., Ochkin, V. N., and Witteman, W. J., *Appl. Phys. Lett.*, **63**, 721, 1993.

5.129 Goldsborough, J. P., Hodges, E. B., and Bell, W. E., *Appl. Phys. Lett.*, **8**, 218, 1966.

5.130 Latush, E. L., Mikhalevskii, V. S., and Sam, M. F., *Pis'ma Zh. Eksp. Teor.*

Fiz., **24**, 81, 1976.

5.131   Strokan', G. G. and Tolmachev, G. N., *Avtometriya*, **1**, 62, 1984.

5.132   Mikhalevskii, V. S., Tolmachev, G. N., and Khasilev, V. Ya., *Kvant. Electron. Mosk.*, **7**, 1537, 1980.

5.133   Baranov, G. A., Baranov, I. Ya., Boreisho, A. S., and Timoshyk, T. V., *XI ESCAMPIG*, St. Petersburg, p. 188, 1992.

5.134   Baranov, G. A., Baranov, I. Ya., Boreisho, A. S., and Timoshyk, I. V., *Kvant. Electron. Mosk.*, **20**, 222, 1993.

5.135   Baranov, I. Ya., *Fiz. Plazmy*, **19**, 1495, 1993.

5.136   Schok, W., Schall, W., Hügel, H., and Hoffmann, P., *Appl. Phys. Lett.*, **36**, 793, 1980.

5.137   Terunuma, K., Noguchi, A., Sato, S., Saito, H., and Fujioka, T., *Proc. VI Int. Symp. Gas Flow and Chem. Lasers*, Jerusalem, p. 232, 1986.

5.138   Baranov, G. A., Efremov, Yu. V., Smirnov, A. S., Frolov, K. S., and Shevchenko, Yu. I., *Kvant. Electron. Mosk.*, **16**, 261, 1989.

5.139   Buczek, C. J., Wayne, R. J., Chenausky, P., and Freiberg, R. J., *Appl. Phys. Lett.*, **16**, 321, 1970.

5.140   Seguin, H. J. J., Capjack, C. E., Antoniuk, D., and Nam, K. A., *Apll. Phys. Lett.*, **37**, 130, 1980.

5.141   Seguin, H. J. J., Capjack, C. E., Antoniuk, D. M., and Seguin, V. A., *Appl. Phys. Lett.*, **39**, 203, 1981.

5.142   Capjack, C. E., Seguin, H. J. J., Antonuik, D. M., and Seguin, V. A., *Appl. Phys. Lett.*, **26**, 161, 1981.

5.143   Yessik, M. and Macken, J. A., *J. Appl. Phys.*, **54**, 1693, 1983.

5.144   Seguin, V. A., Seguin, H. J. J., Capjack, C. E., and Nikumb, S. K., *Appl. Phys. Lett.*, **43**, 127, 1987.

5.145   Harry, J. E. and Evans, D. R., *J. Appl. Phys.*, **62**, 4708, 1987.

5.146   Ilyukhin, A. A., Lipatov, N. I., Mineev, A. G., Myshenkov, V. I., Pashinin, P. P., and Prokhorov, A. M., *Pis'ma Zh. Tekh. Fiz.*, **11**, 25, 1985.

5.147   Golubev, V. S., Krivenko, Yu. N., Leonov, P. G., and Flerov, V. B., *Pis'ma Zh. Tekh. Fiz.*, **14**, 1522, 1988.

5.148   Nath, A. K., Chaubey, R. S., Sree RamKumar, U. V., Chowdhary, P., Kumar, M., and Abhinandan, L., *IEEE J. Quantum Electron.*, **27**, 476, 1991.

5.149   Macken, J., U.S. Patent No. 4.755.999, 1988.

5.150   Macken, J., *Lasers Optronics*, **7**(5), 19, 1988.

5.151   Macken, J., *Proc. Laser Advanced Materials Processing (LAMP '92)*, Vol. 1, p. 67, 1992.

5.152   Yatsenko, N. A. and Masyukov, I. V., *Pis'ma Zh. Tekh. Fiz.*, **19**, 17, 1993.

5.153 Chapman, B., *Glow Discharge Processes—Sputtering and Plasma Etching*, Wiley-Interscience, New York, 1980.

5.154 Flamm, D. L. and Donelly, V. M., *Plasma Chem. Plasma Process.*, **1**, 319, 1981.

5.155 Flamm, D. L., Donelly, V. M., and Ibbotson, D. E., *J. Vac. Sci. Technol. B*, **1**, 23, 1983.

5.156 Kireev, V. Yu., Danilyan, B. S., and Kuznetsov, V. I., *Plasma-Chemical and Ion-Chemical Microstructure Etching*, Radio i Svyaz, Moscow, 1983 (in Russian).

5.157 *VLSI Electronics Microstructure Science*, Vol. 8, *Plasma Processing for VLSI*, Einspruch, N. G. and Brown, D. M., Eds., Academic Press, 1984.

5.158 Ivanovskii, G. F. and Petrov, V. I., *Ion-Plasma Materials Processing*, Radio i Svyaz, Moscow, 1986 (in Russian).

5.159 Graves, D., *AIChE J.*, **35**, 1, 1989.

5.160 Parkhomenko, V. D., Tsibulov, P. M., and Krasnokutskii, Yu. I., *Technology of Plasma-Chemical Processing*, Vyshya Shkola, Kiev, 1991 (in Russian).

Index

Amplification coefficient, 227, 231, 241
Amplitude of electron oscillation,
    drift, 5, 29
    free, 4
    magnetized, 186
Anode stage, 49, 80, 96, 168
Attachment of electrons to
    atoms, 20, 163, 167
    dielectrics, 30, 52

Ballast capacitive resistance, 104
Beam of electrons, 172, 174, 181

Capacity of sheaths,
    equivalent, 126
    stray, 193, 196, 199
Cathodic stage, 49, 81, 96
Child–Langmuir law, 120, 161
Cross section for momentum trans-
    fer, 4
Current,
    conduction, 8, 11
    displacement in vacuum, 8, 26
    of polarization, 8, 11
    total of charges, 8, 9
Current-voltage characteristic of
    cathode sheath, 60
    γ-discharge, 81, 82
    positive column, 21, 61, 65, 66,
        68, 70, 76, 234, 237
    sheath, 101, 150, 154
    RF discharges, 37, 39, 56, 57,
        60, 65, 66, 68, 70, 72, 76,

123, 170, 236

Debye radius, 10, 146, 150, 151,
    152, 153, 155, 156, 163
Diffusion,
    characteristic length, 19
    computational, 52
    extrapolational length, 119

Effective electrical field, 17
Electron-sheath collision regime,
    174, 175, 183
Energy of electrons, 6, 7
Equation of
    diffusion-drift approxi-
        mation, 45
    electrostatics, 23, 147, 164
    hydrodynamics of
        electrons, 139
        ions, 155
    Maxwell, 7

Faraday dark space 40, 82, 84, 86,
    87, 171, 172, 207, 244,
    245
Frequency of
    electron attachment, 20
    energy relaxation, 17
    ionization, 17
      in RF field, 18
    Larmour, 9, 185
    plasma, 139, 140

Hysteresis, 57